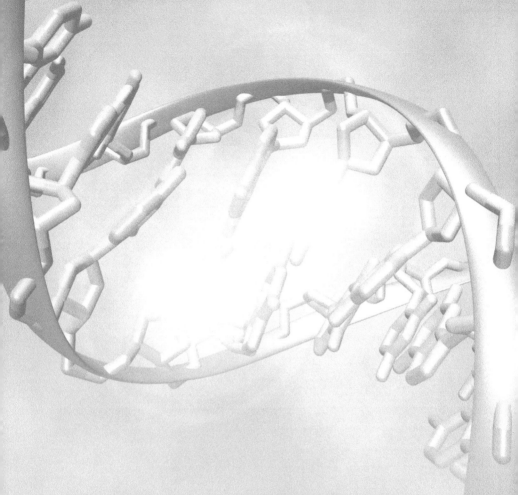

NANOMEDICINE

NANOMEDICINE

Science, Business, and Impact

Michael Hehenberger

PAN STANFORD PUBLISHING

Published by

Pan Stanford Publishing Pte. Ltd.
Penthouse Level, Suntec Tower 3
8 Temasek Boulevard
Singapore 038988

Email: editorial@panstanford.com
Web: www.panstanford.com

British Library Cataloguing-in-Publication Data
A catalogue record for this book is available from the British Library.

Nanomedicine: Science, Business, and Impact

Copyright © 2015 Pan Stanford Publishing Pte. Ltd.

The cover image of epigenetics/DNA methylation (highlighted with bright dots for two bases) was generated by Christoph Bock, Max Planck Institute for Informatics, Saarbrücken, and was submitted to Wikipedia by Dr. Bock. It was merged with the author's photograph of a forest near Sandviken, Sweden.

ISBN 978-981-4613-76-7 (Hardcover)
ISBN 978-981-4613-77-4 (eBook)

Printed in the USA

This book is dedicated to the memory of

Ernst Hehenberger, my Austrian father, who believed in education and supported the aspirations of his son to become a scientist.

Professor Per-Olov Loewdin, my Swedish teacher, who inspired numerous students across both developed and developing countries to embrace quantum science, the splendor of Norway's mountains and the charm of Florida's Sanibel beaches and wildlife.

Contents

Purpose   xi
Acknowledgments   xiii

1 Introduction   **1**
   1.1 What Is Nanomedicine?   1
   1.2 What Is Translational Medicine?   3
   1.3 What Is Innovation?   6
   1.4 What Is Intellectual Property?   7
   1.5 From IP to IPO   11
   1.6 What Is Impact?   16

2 From Atoms to Proteins   **17**
   2.1 Atoms and Molecules   17
   2.2 Atoms of Life   23
   2.3 Molecules of Life   33
   2.4 Amino Acids   55
   2.5 Proteins   63

3 Genetics and DNA Sequencing   **77**
   3.1 DNA and the Genetic Code   77
   3.2 From DNA to Proteins and Cells   81
   3.3 History of Genetics   83
   3.4 Molecular Basis of Genetics   86
   3.5 DNA Sequencing and the Human Genome Project   106
   3.6 Sequencing Technologies   109
      3.6.1 Chemical Sequencing   109
      3.6.2 Sanger Sequencing, NGS, and the $1000
            Genome Challenge 110
      3.6.3 Second-Generation DNA Sequencing 114

3.6.4 Third-Generation DNA Sequencing	121
3.6.5 Fourth-Generation DNA Sequencing	124
3.7 Sequencing Data Analysis	130
3.7.1 Data Reduction and Management	133
3.7.2 Downstream Analysis	134
3.8 Bioethics	135
3.9 Business Aspects of Sequencing Technology and the Human Genome Project	143
3.10 Post–Human Genome Project "-Omics"	151
4 Biopharmaceutical R&D	**157**
4.1 Stages of Biopharmaceutical Drug Discovery and Clinical Development	159
4.2 Stage 1: Drug Discovery	160
4.2.1 Target Identification and Validation	160
4.2.2 Lead Identification and Optimization	167
4.3 Stage 2: Preclinical Research	171
4.4 Stages 3–4: Clinical Trials and FDA Review	174
4.5 The Emerging Importance of "Biologics"	176
4.6 Biomarkers and Stratified/Personalized Medicine	183
4.7 The Past and Future of Biopharmaceutical R&D	192
4.8 Semiconductors: Progress through Collaboration	199
4.9 Life Sciences Industry: Exclusive IP Deals and Limited Collaboration	201
4.10 IMI and AMP	202
4.10.1 Innovative Medicines Initiative	202
4.10.2 Accelerating Medicines Partnership	203
5 Nanomedicine	**205**
5.1 Computers in Nanomedicine	205
5.2 Biocompatible Nanoparticles and Targeted Drug Delivery	217
5.2.1 Dendrimers	217
5.2.2 Liposomes	223
5.2.3 Targeted Drug Delivery	225
5.2.4 The Business of Nanotherapeutics	232
5.3 Biomedical and Molecular Imaging	234
5.4 Nanodiagnostics	243

5.5 Regenerative Medicine: Stem Cells, Gene Therapy, and
Immunotherapy   251

6 Impact of Nanomedicine   285
6.1 The Gut Microbiome   285
6.2 Central Nervous System: Brain and Spinal Cord   290
    6.2.1 The Human Brain   291
          6.2.1.1 Brain function   292
          6.2.1.2 Left and right brain   293
          6.2.1.3 The cerebral cortex   293
          6.2.1.4 Language   295
          6.2.1.5 Thalamus   295
          6.2.1.6 Hippocampus   296
          6.2.1.7 Brain damage and disease   298
          6.2.1.8 Brain metabolism   298
          6.2.1.9 Dementia and Alzheimer's disease: a
                   global health priority   298
          6.2.1.10 Parkinson's disease   299
    6.2.2 The Spinal Cord   300
6.3 Cancer and Immunology   304
6.4 Cardiovascular Disease   310
6.5 Diabetes   312
6.6 Infectious Disease   315
    6.6.1 Pathogenic Bacteria and Antibiotics   317
          6.6.1.1 Diagnosis   319
          6.6.1.2 Antibiotic resistance   319
    6.6.2 Viral Pathogens and Therapies   322
          6.6.2.1 Diagnosis   322
          6.6.2.2 Classification   323
    6.6.3 Pandemic Risk   328
6.7 Tissue and Organ Transplantation   330

7 The Healthcare Ecosystem and Biomedical Research Funding   333
7.1 It's All about the Patient/Consumer   334
7.2 Providers of Medical Care   334
    7.2.1 Biomedical Research: Academic Medical
          Research Centers and Government Research   335

| | | 7.2.2 | Drug Developers and Manufacturers and Clinical Research Organizations | 339 |

7.2.2 Drug Developers and Manufacturers and
Clinical Research Organizations   339

7.2.3 Charitable Organizations: Advocacy Groups
and Foundations   339

7.2.4 Public and Private Payers of Health Insurance   339

7.2.5 IT and Technology Solution Providers   340

7.2.6 Government Regulators   340

7.2.7 Medical Devices and Diagnostics Companies   341

7.3 The Future of Funding for Basic Biomedical and
Translational Research   342

8 Public Health and Global Health Economics   **357**

8.1 Global View of Healthcare Costs, Infant Mortality,
and Life Expectancy   359

8.2 US Health Statistics   362

9 Conclusions   **371**

References   375

Index   397

Purpose

I feel that there is a need for a book that covers the emerging importance of nanomedicine and associated concepts of molecular science, biology and nanotechnology.

My second reason for writing this book is to inspire young readers to appreciate science and perhaps consider a deeper dive into any of the many exciting topics that are bound to bring about the transformation of medicine to a more science and information based field. Although occasionally rather technical, I have tried to avoid jargon and am not assuming knowledge of field-specific acronyms and terms. Wherever such terms are introduced, my intent is to explain them in plain English.

My third reason is to point to the huge challenge associated with the term "translation," and to emphasize the roles to be played by politicians, investors, drug and medical device developers, regulators, bioethicists, scientists and medical professionals in bringing about the changes in clinical care of patients enabled by nanomedicine. The impact of nanomedicine on patient care will largely depend on research funding, successful commercialization of research results and the willingness of healthcare stakeholders to embrace change. In other words, a prerequisite for "translational nanomedicine" is a thriving and impactful nanomedicine business environment.

Acknowledgments

In writing this book I have relied heavily on Wikipedia; the excellent web site of the Nobel Foundation, www.nobelprize.org; and on US Government and WHO websites. There may be quality issues with some Wikipedia entries but the built-in self-correcting mechanism is a great concept and it seems to work. I have seen immediate reactions whenever I found a detail that needed correction.

Most figures in this book have been copied from Wikipedia and I'd like to thank the Wikimedia Foundation for making them freely available. As to the Swedish way of successfully managing the selection of Nobel laureates for more than 100 years, I'd like to express my respect for the job done by the various Swedish academies. I have had the privilege of knowing several members of Swedish Nobel Committees, such as the late Profs. Loewdin and Roos at Uppsala University and Lund University, respectively, and was always impressed by their dedication to this important task and their personal and scientific integrity.

One strong reason for writing this book on nanomedicine was the first invitation by the European Foundation for Clinical Nanomedicine (CLINAM), 2012, to give a keynote presentation on DNA sequencing. I have since attended the annual CLINAM events, held in Basel, Switzerland, and greatly enjoyed the scientific programs as well as the social events organized by CLINAM CEO Beat Loeffler. I'd like to thank him, his partner Prof. Patrick Hunziker, and his scientific advisors for putting together such excellent conferences.

During my over 28 years at IBM, I was able to interact with numerous exceptional scientists, both inside IBM Research and among IBM's Life Sciences academic and industrial clients, in particular leaders of global bio-pharmaceutical R&D organizations.

Among IBM scientists I'd like to single out Dr. Ajay Royyuru and his Computational Biology team. Ajay and friends have taught me a lot during many stimulating discussions, including numerous research seminars.

I'd like to thank my daughters, Karin Hehenberger, MD, PhD; Lisa Hehenberger, PhD; Anna Hehenberger, JD, for reading parts of the manuscript and for making suggestions for improvements. In addition, BGI cofounder and chairman Prof. Huanming (Henry) Yang has read several sections of the book and provided valuable input. I thank him for his encouragement and wish him a lot of success with BGI's ambitious "trans-omics" projects.

I'd also like to thank the publisher, Pan Stanford, and its editorial staff, for help and advice.

Finally, I want to thank my family, in particular my wife, Ulla, and several friends and colleagues for putting up with my occasional grumpiness during the course of my writing of this book: it's not always easy to overcome innate procrastination and develop the necessary discipline to finish a book.

Chapter 1

Introduction

1.1 What Is Nanomedicine?

The word "nano" is of Greek origin. In Greek, "nanos" means *dwarf* or *gnome*. As a prefix, nano- denotes things that are extremely small, or minute. As a metric prefix, it has a precise meaning:

The word "nano" (n) denotes "one billionth," or 10^{-9},

in a similar way as

"deci" (d) denotes "one tenth," or 10^{-1},
"centi" (c) denotes "one hundredth," or 10^{-2},
"milli" (m) denotes "one thousandth," or 10^{-3},
"micro" (μ) denotes "one millionth," or 10^{-6},
"pico" (p) denotes "one trillionth," or 10^{-12},
"femto" (f) denotes "one quadrillionth," or 10^{-15},
"atto" (a) denotes "one quintillionth," or 10^{-18},
"zepto" (z) denotes "one sextillionth," or 10^{-21},
"yocto" (y) denotes "one septillionth," or 10^{-24}, and
"googol" is the very large number 10^{100}, that is, the digit 1 followed by 100 zeroes!

The units deci, centi, and milli were introduced as part of the metric system in 1795 by the Conférence générale des poids et mesures (CGPM). In 1960, the CGPM added micro, nano, and pico. The

Nanomedicine: Science, Business, and Impact
Michael Hehenberger
Copyright © 2015 Pan Stanford Publishing Pte. Ltd.
ISBN 978-981-4613-76-7 (Hardcover), 978-981-4613-77-4 (eBook)
www.panstanford.com

remaining four units femto, atto, zepto, and yocto were introduced as late as 1991 by the CGPM which still meets in Sèvres (southwest of Paris) every four to six years.

So what's so special with nano?

A good answer would be that a nanometer is a convenient unit of length or size in molecular biology, one (and perhaps the most important) of the modern "life sciences." For instance, the diameter of a DNA molecule is 2–3 nm.

When we talk about "nanoparticles," we are usually referring to particles measuring between 1 and 100 nm.

Although considered a discovery of modern science, nanoparticles actually have a very long history. Nanoparticles were used by artisans as far back as the ninth century in Mesopotamia, the area of the Tigris–Euphrates river system corresponding to modern-day Iraq, Kuwait, northeast Syria, southeast Turkey, and smaller parts of southwestern Iran. Those ancient silver and copper nanoparticles generated a glittering effect, or "luster," on the surface of pots, dispersed homogeneously in the glassy matrix of the ceramic glaze.

In 1857, Michael Faraday provided the first scientific description of the optical properties of nanometer-scale metals, thereby explaining what caused "luster."

Nanoparticles are of great scientific and technological interest as they often possess unexpected optical and other properties caused by quantum effects: as we approach molecular and atomic dimensions, the laws of Newtonian classical physics are no longer fully adequate.

Nano-medicine (spelled simply as "nanomedicine" from now on) is the application of nanotechnology to medical diagnostics and therapy. Being based on nanotechnology, nanomedicine is a new field at the interface of the physical and life sciences. The European Science Foundation[1] defined nanomedicine as the "science and technology of diagnosis, treatment and prevention of disease, injury, pain and the improvement of human health." The aim of nanomedicine is further defined as "use of engineered devices and nanostructure at a molecular level for the control, repair, defense and improvement of all human biological systems." There are ongoing debates about "nanoscale" and the emerging consensus is that nanomaterials should fall into a size range of 1 and (several) 100 nm.

In this book the definition of nanomedicine will be quite broad and will include topics such as DNA and RNA sequencing, stem cells, regenerative medicine, and even computer technology. To create new nanomedical breakthroughs, advances in physics, chemistry, and technologies such as structural mechanics and materials science—as, for example, used in the development and fabrication of semiconductors—have to be combined with advances in biology, genetics, physiology, anatomy, and other life sciences.

The impact of nanomedicine depends on how well nanomedical breakthroughs can be translated into clinical care of patients.

1.2 What Is Translational Medicine?

In 1967, the Nobel Prize for physiology or medicine was awarded jointly to Ragnar Granit, Haldan Keffer Hartline, and George Wald "for their discoveries concerning the primary physiological and chemical visual processes in the eye."[2]

The underlying story of visual perception illustrates a key theme of this book: to advance the art of medicine, it is necessary to understand physiology on a molecular basis. By "translating" molecular science into biological pathways and connecting them to clinical observations in patients, we create knowledge that enables new medical treatments.

When Alfred Nobel[3] decided in his will to award prizes in physics, chemistry, physiology and medicine, literature, and peace, he restricted the number of recipients per prize to three and he would not allow posthumous awards. Whereas he entrusted decisions about the first four Nobel Prize categories to various Royal Swedish Academies, he handed the responsibility to select Nobel Peace Prize winners to the Norwegian Parliament. The first Nobel Prizes were awarded in 1901.[4] In 1968, the Swedish Central Bank then made the decision to add a Nobel Prize for economics.

The 1967 Nobel Prize for physiology or medicine is of particular interest for several reasons because it illustrates the diverse geographic and cultural heritage of the laureates, showcases the interdisciplinary character of "life sciences" research fields leading to big breakthroughs, and points to the importance of stimulating environments during the formative stages of research careers:

Ragnar Granit was born in Helsinki, Finland, in 1900, to Swedish-speaking parents. After initial studies of experimental psychology he decided to add a full medical degree to his education. He did his research in Oxford, England, at the University of Pennsylvania, Philadelphia, at the University of Helsinki, and eventually, at the Karolinska Institute in Stockholm. From 1920 to around 1947 Granit's main research was in the field of vision, beginning with psychophysics in the twenties and ending up with electrophysiological work from the early thirties onward. He then turned his interest to muscle spindles and their motor control, passing over to the spinal cord.

Haldan Keffer Hartline was born in Pennsylvania to parents who were teachers. His father was a biology professor. After spending summers doing research at the Marine Biology Lab in Woods Hole, Massachusetts, Hartline studied physics at Johns Hopkins University in Baltimore, Maryland. Supported by a traveling E.R. Johnson Foundation fellowship from the University of Pennsylvania (U Penn), he then spent three semesters working with two renowned German physicists, namely, Werner Heisenberg (Leipzig)[5] and Arnold Sommerfeld (Munich).[6] After his return to the United States in the spring of 1931, he engaged in his research on the optical nerve at U Penn and at the Cornell Medical College in New York. He started out with horseshoe crabs and frogs, turning to human vision in the early 1940s. After many productive years at Johns Hopkins, he finally accepted the position of professor of visual physiology at Rockefeller University in New York.

George Wald was born in New York to German-speaking immigrant parents with roots in the Austro–Hungarian empire. After studying biology and zoology at Columbia University, NY, he spent two stimulating and highly productive years in Europe at the laboratories of Otto Warburg (Berlin–Dahlem)[7] and Paul Karrer (Zurich).[8] It was in Berlin where he first identified vitamin A in the retina. He confirmed the experiments in Zurich and then spent time with Otto Meyerhof[9] in Heidelberg. After a brief stay at the

University of Chicago, he then spent the rest of his career at Harvard University. Perhaps more than anybody else, Wald emphasized the role played by large biological molecules such as vitamin A and rhodopsin in visual perception.

After having dissected the human organism into smaller and smaller building blocks, he frequently concluded his lectures at Harvard by referring to man as "just a collection of molecules."[10]

In the award ceremony speech, the Nobel Committee for Physiology and Medicine of the Royal Karolinska Institute (KI) emphasized the importance of an increased molecular understanding of visual processes for medicine and, ultimately, for patients. The KI presenter started out by explaining the (quantum) **physics** of visual perception: "The light is composed of packets of energy, which combine the properties of waves and particles. When these particles—the quanta—strike the retina of the eye they are caught by the specialized sense cells—rods and cones. It is known that one quantum, which represents the least possible amount of light, is sufficient to initiate a reaction in a single rod." To explain this "reaction in a single rod," we have to turn to **chemistry**. We know now that molecules such as rhodopsin,[11] a G-protein-coupled receptor,[12] enable this signal. Finally, we have to move to **biology** because we need to describe the "excitation of the sensory cells resulting in messages directed toward the brain," leading to visual perception. "As there are no direct connections from the eye to the brain the messages must be transmitted through several relays that combine signals from several sensory cells and translate the message into a language that can be understood by the brain. The primary relay is in the retina itself, represented by an intricate nerve net, the structural beauty of which was revealed by the neurohistologist Ramón y Cajal, Nobel laureate of 1906. In this complex structure messages from a great number of sensory cells converge on a far smaller number of optic nerve fibers and this results in a transformation of the pattern of signals."[13]

When today's ophthalmologists check the eyes of their patients, they rely on breakthroughs rewarded by Nobel Prizes, and the subsequent detailed studies of various aspects of vision, spanning scientific disciplines such as physics, chemistry, biology, zoology, experimental psychology, physiology, and neuroscience.

Deep scientific knowledge and understanding have been and are still translated into medical and clinical practice.

Today's definition of "translational research" is an extension and generalization of the example given above.

It is based on the integration of scientific facts that are then used to move from "evidence-based medicine" to clinical practice and, in extension, to sustainable solutions for public health problems.

1.3 What Is Innovation?

Translation from the science of nanomedicine to the business of nanomedicine is a necessary requirement for innovation. For the purpose of this book we define "innovation" as the process of translating an idea or invention into a product or service that creates value. "Value" can be defined as commercial value, as something customers will be willing to pay for, but also as something that benefits society and contributes to the common good.

In the medical field, value must be associated with improving health, preventing and curing disease, relieving pain and suffering, and prolonging human life.

New ideas or inventions are not always leading to innovation. To be called an innovation, an idea must be replicable at an economical cost and must satisfy a specific need. In business, innovation often results when ideas are applied by the company in order to further satisfy the needs and expectations of the customers. In a social context, innovation helps create new methods for improved quality of life of a majority of citizens. In finance, innovation should create value both for the innovating financial institution and for society. For instance, the "mortgage" idea was an important financial innovation: It improved the living standards of house owners, and it made it possible for families to provide better living spaces by making a long-term commitment to the mortgage lender to pay back the principal with interest. It lifted all participants in a complex system of stakeholders and led to stable economic growth.

Innovations are sometimes divided into:

- evolutionary innovations caused by incremental advances in technology or processes

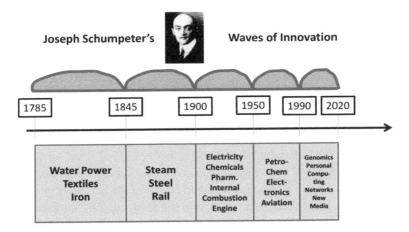

Figure 1.1 Schumpeter's waves of innovation, 1785–2020.

- revolutionary innovations that are disruptive, discontinuous, and new

One of the 20th century's great thinkers, the Austrian economist Joseph Schumpeter is well known for his analysis of activities that lead to economic growth in capitalist economies. He considered entrepreneurial innovations as the key driver of economic growth. Schumpeter argued that competition among market participants leads to a desire to seek out new ways to improve technology, new ways to do business, and other types of advantages that would increase profit margins and directly impact the entrepreneur's standard of living. Figure 1.1 shows an illustration of his way of thinking.[14]

1.4 What Is Intellectual Property?

A necessary requirement for "innovation" is a legal framework that protects intellectual property (IP), defined as legally recognized exclusive rights to creations of the mind. Under IP law, owners are granted certain exclusive rights to a variety of intangible assets, such as discoveries and inventions; musical, literary, and artistic works; and even words, phrases, symbols, and designs. Common types of IP rights include patents, copyright, trademarks, and industrial design rights.

The first known patent law that granted inventors exclusive rights to their inventions was passed in 1474 in Venice, as a result of an economic policy. The Venice statute had all the basic elements of a modern patent system—a requirement of novelty, a requirement of proof of usefulness, and a requirement that the patentee describe and explain the invention.

The Statute of Monopolies, passed on May 25, 1624, was an act of the Parliament of England notable as the first expression of English *patent law*. It has also been described as one of the landmarks in the transition of England's economy from the feudal to the capitalist.

Similarly, the Statute of Anne, an act of the Parliament of Great Britain, passed in April 1710 during the reign of Queen Anne, was the first statute to provide for *copyright* regulated by the government and courts, rather than by private parties. Under the statute, copyright was for the first time vested in authors rather than publishers.

In America's colonial period, the earliest grant of an exclusive commercial right to utilize a new process was given in 1641 by the Massachusetts General Court: this first American "patent" was granted to Samuel Winslow for a new process of making salt and gave him an exclusive right for 10 years.

The US Constitution (1787) included a provision "To promote the Progress of Science and useful Arts, by securing for limited Times to Authors and Inventors the exclusive Right to their respective Writings and Discoveries." After two preliminary Patent Acts (1790, 1793), the United States created its United States Patent and Trademark Office in 1836 and transferred it to the Department of Commerce in 1925.

Modern usage of the term "intellectual property" goes back as far as 1867 with the founding of the North German Confederation, whose constitution granted legislative power over the protection of IP (Schutz des geistigen Eigentums) to the confederation. After establishments of the Paris Convention (1883) and the Berne Convention (1886), merged in 1893 and located in Berne, Switzerland, the term "intellectual property" was officially adopted. The final step was taken when this international organization relocated to

Geneva in 1960 and, in 1967, became an agency of the United Nations, called the World Intellectual Property Organization (WIPO). It was only at this point that the term did enter popular usage. In the United States, the Bayh–Dole Act of 1980 contributed strongly to the emphasis on IP in all discussions of science, technology, and innovation. The key change made by the Bayh–Dole Act was in ownership of inventions made with federal funding. Before the Bayh–Dole Act, federal research funding contracts and grants obligated inventors (where ever they worked) to assign inventions they made using federal funding to the federal government. After 1980, any university, small business, or nonprofit institution may elect to pursue ownership of an invention, thus unleashing a wave of IP licensing agreements and the creation of start-up companies in areas such as the life sciences.

According to WIPO, a *patent* grants an inventor the right to exclude others from making, using, selling, offering to sell, and importing an invention for a limited period of time (usually 20 years) in exchange for the public disclosure of the invention. An invention is a solution to a specific technological problem, which may be a product or a process. The procedure for granting patents varies between countries, but typically, a granted patent application must include one or more *claims* that define the invention. A patent may include many claims, each of which defines a specific property right. These claims must meet relevant patentability requirements, such as novelty and nonobviousness. The exclusive right granted to a patentee in most countries is the right to prevent others from commercially making, using, selling, importing, or distributing a patented invention without permission.

In most countries, both individuals and corporate entities may apply for a patent. In the United States, only the inventor(s) may apply for a patent, although it may be *assigned* to a corporate entity. Frequently, inventors may be required to assign inventions to their employers under an employment contract. In most European countries, ownership of an invention may pass from the inventor to their employer by rule of law if the invention was made in the course of the inventor's normal or specifically assigned employment duties. The inventors, their successors, or their assignees become the proprietors of the patent when and if it is granted. The ability to assign ownership rights increases the liquidity of a patent as property. Inventors can obtain patents and then sell them to third parties who then own the patents and have the same rights to prevent others from exploiting the claimed inventions.

When a patent application is published, the invention disclosed in the application becomes *prior art* and enters the public domain.

In 2011, the Leahy–Smith America Invents Act (AIA) changed the US patent system from "first to invent" to "first inventor to file," thereby aligning the United States with the rest of the world.

The applicability of patents to substances and processes wholly or partially natural in origin has been and continues to be subject of much debate. An emerging consensus, still subject to interpretation, seems to be that natural biological substances themselves can be patented (apart from any associated process or usage) only if they are sufficiently "isolated" from their naturally occurring states.

A *copyright* gives the creator of an original work exclusive rights to it, usually for a limited time, currently (in the United States) lasting for 70 years after the death of the author.[15] Copyright may apply to a wide range of creative, intellectual, or artistic forms, or "works." Copyright does not cover ideas and information themselves, only the form or manner in which they are expressed. Copyright infringements (e.g., recorded music, movies) are often called "piracy."

An *industrial design* right protects the visual design of objects that are not purely utilitarian. An industrial design consists of the creation of a shape, configuration or composition of pattern or color, or a particular combination of pattern and color in 3D form representing aesthetic value. A design patent is generally granted protection for 14 years measured from the date the design patent is granted.

A *trademark* is a recognizable sign, design, or expression that distinguishes products or services of a particular trader from the similar products or services of other traders. A US trademark[16] generally lasts as long as the trademark is used in commerce and defended against infringement. The term of a US federal trademark registration is ten years, with 10-year renewal terms. However, between the fifth and the sixth year after the date of initial trademark registration, an "affidavit of use" must be filed and an additional fee paid to keep the registration alive.

A *trade secret* is a formula, practice, process, design, instrument, pattern, or compilation of information that is not generally known or reasonably ascertainable, by which a business can obtain an economic advantage over competitors or customers. A famous example of a trade secret is the secret formula used by Coca-Cola to produce the taste of its soft drink. A trade secret can be protected indefinitely as long as the secret is commercially valuable, its value derives from the fact that it is secret, and the owner takes reasonable precautions to maintain its secrecy.

The stated objective of IP law (with the exception of trademarks) is to "promote progress." By exchanging limited exclusive rights for disclosure of inventions and creative works, society and the respective patentee/copyright owner mutually benefit, and an incentive is created for inventors and authors to create and disclose their work.

In the Western world, it is the generally accepted view that a society that protects private property is more effective and prosperous than societies that do not. For instance, the development of the patent system has had a significant impact on innovation since the 19th century. By providing innovators with return on their investment of time, labor, and other resources, IP rights seek to maximize social utility. Public welfare is encouraged by the creation, production, and distribution of IP.

There is no doubt that this is particularly true for life sciences and nanomedicine.

Finally, it should be pointed out that the field of nanomedicine benefits from IP across all traditional fields of science and technology, from chemistry, biology, and medicine to electrical engineering and physics. There is significant "innovation crossover" between information technology (IT) and biotechnology (BT).

1.5 From IP to IPO

Having protected a given invention in the area of nanomedicine by means of a patent application, scientists have to decide whether to pursue commercialization. If so, the first step will be to identify a nanomedical application that satisfies an unmet medical need and therefore represents "added value."

The second step will be to create a project plan that leads to a "proof-of-concept" (PoC) designed to validate the commercialization idea.

Part of the project plan should be a realistic timeline and budget estimate that includes all the labor and materials needed to finish the PoC.

Step three is about initial (or "seed") funding: Who will make the first (Series A round) investment, and who will take the initial risk?

Scientists associated with a university may want to approach their business development function; others may have to ask family and friends for help or approach their local bank or an "angel investor."

An *angel investor* is an affluent individual who provides capital for a business start-up, usually in exchange for convertible debt or ownership equity. The word "angel" originally came from Broadway, where it was used to describe wealthy individuals who provided money for theatrical productions. Angel capital can fill the gap in start-up financing between "friends and family," who provide initial seed funding and formal venture capital. It is usually difficult to raise more than a few hundred thousand dollars from friends and family. On the other hand, most traditional venture capital funds are unable to make or evaluate small investments, under $1–2 M. Thus, angel investment is an occasional first and/or a common second round of financing for high-growth start-ups.

An alternative funding option comes from "venture philanthropy". Venture philanthropy is a methodology that provides a combination of funding and management support to help build stronger organizations,[17] similar to venture capital but normally without a financial return expectation. Applied to the funding of research, venture philanthropy can bridge the "valley of death" between funding of basic research from public and purely philanthropic sources and more commercial funders such as venture capitalists. The term "valley of death" refers to the high probability that a startup firm will die off before a steady stream of revenues is established. After a startup company receives its first round of

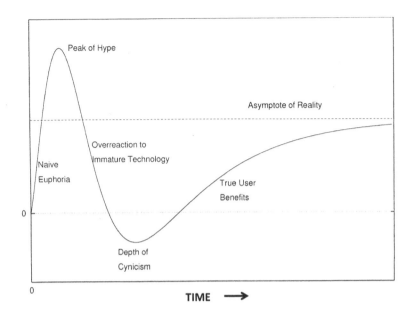

Figure 1.2   Peaks and valleys of a startup company's public expectation.[18]

financing, it usually incurs new operating costs caused by office space requirements, hiring of staff, etc., while not yet earning significant income. Unless a company can effectively manage itself through the death valley curve, it will fall victim to negative cash flows.

In addition, public expectation with regard to IP that is subject to commercialization, often has its peaks and valleys, as illustrated in Fig. 1.2.

Charitable organizations such as the Michael J Fox Foundation[19] advocate a venture philanthropy approach as researchers often lack the necessary management skills apart from funding to set up successful businesses.

Having secured seed funding, it's time to formally create and register a company, a step that nowadays presents few problems or bureaucratic hurdles, but should be discussed with a certified public accountant (CPA) to avoid tax issues. After months of hard work, execution of the project plan will hopefully generate the desired answers to the key questions regarding the PoC. If the results are positive and the PoC has turned out to be a success, generating the desired validation of the inventor's hypothesis, it's time to approach either angel investors for a second round or approach venture capitalists who are looking for attractive new investments with high potential return.

Venture capital (VC) is financial capital provided to early-stage, high-potential, growth start-up companies such as BT start-ups focused on nanomedicine. The venture capital fund earns money by owning equity in the companies it invests in. The typical venture capital investment occurs after the seed (Series A) funding round, in the interest of generating a return through an eventual realization (or "Exit") event, such as an *initial public offering* (IPO) or sale of the company. VC is a type of private equity, a financial asset class consisting of equity securities and debt in operating companies that are not publicly traded on a stock exchange. VC is attractive for new companies with limited operating history that are too small to raise capital in the public markets and have not reached the point where they are able to secure a bank loan or complete a debt offering. In exchange for the high risk that venture capitalists assume by investing in smaller and less mature companies, venture capitalists usually get significant control (e.g., via preferred shares) over company decisions, in addition to a significant portion of the company's ownership (and consequently value).

In the United States, both angel investors and venture capitalists are valid options for start-ups, with the number of angels exceeding the number of venture capitalists by about a factor of 50. According to the Center for Venture Research, there were 258,000 active angel investors investing a total of \$26 B (across all technology areas) in the United States in 2007.[21] VC investments in the United States added up to a similar total amount of funding.

An IPO or stock market launch is a type of public offering in which shares of stock in a company usually are sold to institutional investors that in turn sell to the general public, on a securities exchange, for the first time. Through this process, a private company transforms into a public company. IPOs are used by companies to raise expansion capital, to possibly monetize the investments of early private investors, and to become publicly traded enterprises. After the IPO, when shares trade freely in the open market, money passes between public investors. Although an IPO offers many advantages, there are also significant disadvantages, among them higher costs and the requirement to disclose more information about both the company and the "product." Details of the proposed offering are disclosed to potential purchasers in the form of a lengthy document known as a *prospectus*. Most companies undertake an IPO with the assistance of an investment banking firm acting in the capacity of an underwriter.

An IPO must be planned carefully, considering the following checklist:

- √ Put together an impressive management and professional team.
- √ Present the company's business with an eye to the public marketplace.
- √ Obtain audited financial statements using IPO-accepted accounting principles.
- √ Develop good corporate governance and attract reputable board members and scientific advisors.
- √ Take advantage of IPO windows created by market conditions.

By going public, the previously private company gains the following advantages:

- Enlarging and diversifying the equity base
- Enabling cheaper access to capital
- Increasing exposure, prestige, and public image
- Attracting and retaining better management and employees through liquid equity participation
- Facilitating acquisitions (potentially in return for shares of stock)
- Creating multiple financing opportunities: equity, convertible debt, cheaper bank loans, etc.

On the other hand, there will be some disadvantages compared to keeping the company private:

- Significant and ongoing legal, accounting, and marketing costs

- Requirement to disclose financial and business information on a quarterly basis, leading to short-term focus and stress on management to deliver toward plans
- Significant time, effort, and attention required to manage investor relations
- Required funding probably not being raised due to market conditions or mistakes made during IPO preparation
- Public dissemination of information probably useful to competitors and other stakeholders
- Loss of control due to new shareholders who may be active

Bottom line: Don't rush the IPO! Be sure to have a clear and complete understanding of the company's product positioning and core business opportunities before embarking on the very time-consuming and complex IPO process.

1.6 What Is Impact?

In mechanics, IMPACT is defined is a high force or shock applied over a short time period when two or more bodies collide. For our purpose, we rather prefer the definition to read like the force exerted by a new scientific idea, concept, or technology. Having taken a scientific breakthrough through the path of Intellectual Property and Commercialization (including an IPO or the acquisition of the associated startup company by an established industry leader), there is still no guarantee of "Impact". What is required for the new nanomedical idea to have real impact is the successful translation into clinical care of patients, including adoption by the medical community and willingness of health insurance payers to recognize a benefit for patients and for society. In this book we will trace Nanomedicine from Science to Business and Impact.

Chapter 2

From Atoms to Proteins

Wer immer strebend sich bemüht,
Den können wir erlösen.

(Whoever strives with all his might,
That man we can redeem.)

—say the Angels in *Faust II/5* by Johann Wolfgang von Goethe

2.1 Atoms and Molecules

For the purpose of this book, and generally to discuss concepts of chemistry and biology, it is sufficient to look at atoms as consisting of protons, neutrons and electrons.[22] An atom's mass is concentrated in the *nucleus*, made of protons and neutrons. Protons carry a positive charge, neutrons don't carry any charge, but electrons are negatively charged. There are about ninety naturally occurring different arrangements of protons, neutrons and electrons and they are called the *elements*. All elements are electrically neutral, their nucleic (proton) charges are balanced by electronic charges. An element's atomic number is therefore equal to the number of its protons/electrons.

Nanomedicine: Science, Business, and Impact
Michael Hehenberger
Copyright © 2015 Pan Stanford Publishing Pte. Ltd.
ISBN 978-981-4613-76-7 (Hardcover), 978-981-4613-77-4 (eBook)
www.panstanford.com

The simplest atom—hydrogen (H)—consists of only one proton and one electron. In early atomic models, the electron of H was depicted as a "planet" circling the proton "sun." Quantum theory resolved the ad hoc nature of previous atomic models and provided an elegant mathematical foundation for the electronic structure of atoms. In mathematical terms, electronic motion can be formulated as an *eigenvalue problem*, as first shown by physicist and Nobel laureate Erwin Schrödinger.[23] Assuming a fixed nucleus, the negatively charged electron is interacting with the positively charged proton and trying to reach a stable state of lowest energy. It turns out that the 3D problem can be reduced to a simple ordinary (1D) differential equation with the following boundary conditions:

The electron cannot crash into the proton (zero probability for the electron to move to distance $x = 0$) and it cannot escape from the H atom (zero probability for the electron to exist at $x = \infty$) as long as the element is neutral and stable.

Stated as a mathematical equation, this statement reads

$$H \Psi(x) = E \Psi(x)$$

where H is referred to as the "Hamiltonian"[24] of the atomic system, and where $\Psi(x)$ is Schrödinger's famous wave function.

For certain values of E, namely, $E_0, E_1, E_2, \ldots,$ $\Psi(x)$ is satisfying the boundary conditions described above. Those E values E_i, $i = 0,1,2$, etc., are called eigenvalues and E_0 is referred to as the ground state of the atomic system. The corresponding solution $\Psi_i(x)$ is a wave with a discrete number "i" of nodes, starting with zero and increasing to 1, 2, 3, etc. In its physical interpretation, the eigenvalues are the energy levels of the electron and the (complex) squares of the values Ψ of the wave function for distances x from zero to infinity can be related to probabilities of the electron to occupy an orbit of distance x from the nucleus.

What we just discussed is the essence of quantum theory: on the atomic level, particles have a dual nature; they can also appear as waves and their nature (particle versus wave) is only revealed by experimental observation. In classical (Newtonian) mechanics, the observer does not influence the result of his/her observation. In quantum mechanics, which applies to very small atomic dimensions, the observer is no longer detached from his/her measurement.

It turns out that Schrödinger's concept of "quantization as an eigenvalue problem" can be extended to all atoms and even as atoms are combined into molecules and the solid state, giving rise to the new field of *quantum chemistry*. For a many-electron system, all we have to do is to extend the concept of a wave function to as many variables as needed to describe the system. Assuming a fixed N-proton nucleus, element N has N electrons and the wave function would read $\Psi(\mathbf{r}_1, \mathbf{r}_2, \ldots \mathbf{r}_N)$, where $\mathbf{r}_i$ would denote the distance (in three dimensions) of electron "i" from the nucleus. In practice, the process of solving the many-particle Schrödinger equation can be quite tricky and requires a lot of compute power. The concept, however, is a straightforward extension of the trivial hydrogen atom example introduced above.

That the naturally occurring $\sim$90 elements can be arranged in a *periodic table* of elements was suggested by Dmitri Mendeleev in 1869. The underlying rules were found empirically and later fully explained by quantum theory. Useful concepts in building the periodic table of elements are the definitions of orbitals and electron shells: An orbital is the quantum equivalent of a classic orbit as it defines the space where the electron—described by a wave—resides. For the H atom, the orbitals are fuzzy spheres around the nucleus. For higher quantum numbers, the orbitals are farther away from the center and can assume complex shapes. If a higher-level electron in an excited state returns to a lower-level state or the lowest-level ground state, it loses energy that will be emitted as a quantum of light (or a photon) with a wavelength uniquely defined by the electronic structure of the atom. Atomic spectra are like fingerprints that can be used to prove the existence of atoms, even light years away in space.

The lowest orbital of the H atom is called the 1s orbital. Because of another quantum property of electrons, namely, spin, the 1s orbital can contain up to two electrons—one with spin "up" and the other with spin "down." As we are building heavier elements with increasing number of protons, we have to add orbitals at larger distances from the nucleus. Stable elements are built from the ground up, with lower orbitals first filled up until outer orbitals are added. The first shell (1s) can only contain two electrons. The second shell with quantum number 2 can contain up to eight electrons, two of type $2s$ and then another six of type $2p_x$, $2p_y$, or $2p_z$. As opposed to the spherical s-orbitals, p-orbitals are shaped like dumbbells and exist in three different versions due to the three spatial dimensions characterized by x, y, and z. The third shell contains again up to two s-electrons ($3s$) and six $3p$ electrons. Elements with fully filled shells are expected to be particularly stable, and indeed, the rare gases helium (He), element #2; neon (Ne), element #10; and argon (Ar), element #18 are often referred to as inert, or unwilling to interact with other elements. Elements with only partially filled shells, on the other hand, are eager to interact, as discussed further below.

The fourth shell exhibits a new type of orbital, named d. There are five types of d-orbitals with complex shapes, sometimes referred to as d_{zz}, $d_{x^2-y^2}$, d_{xy}, d_{zx}, and d_{yx}. Elements containing d-orbitals are all metals, a concept explained further below. They also mix well with each other. The fourth shell includes a total of $2 + 6 + 10 = 18$ elements. Element #36 is krypton (Kr), another rare gas.

The fifth shell contains another 18 elements characterized by s-, p-, and d-orbitals. Element #54 is xenon (Xe), a rare gas.

The sixth shell contains another new type of orbital, named f, with 7 types of different orbitals and hence 14 slots to be filled up. The sixth shell contains s-, p-, d-, and f-electrons and accommodates $2 + 6 + 10 + 14 = 32$ new elements, among them the rare earths, or lanthanides. Element $54 + 32 = 86$ is radon (Rn), still a rare (noble) gas but so heavy that it's also highly radioactive (a concept to be explained below).

A periodic table of elements (status 2013) is shown below (Fig. 2.1). Note that only about 90 elements are occurring naturally.[25] The others are synthesized in laboratories and have short lifetimes due to radioactive decay.

At and above atomic number 86 (radon with a nucleus of 86 protons and an even higher number of neutrons), we are approaching the limit of stability and elements like #88 radium (Ra) and #92 uranium (U) are known to be radioactive, that is, prone to nuclear instability and decay. The most common decay mechanisms are alpha decay and beta decay. In alpha decay, a large and unstable nucleus expels an alpha particle, which is composed of two protons and two electrons. By losing an alpha particle, the number of protons

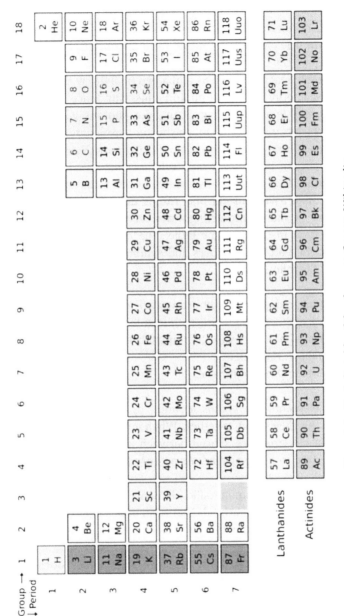

Figure 2.1 Periodic table of the elements. Source: Wikipedia.

is reduced by two, leading to transmutation of an element into another.

In beta decay, a neutron spontaneously changes into a proton and an electron. The electron is emitted from the nucleus at high speed, as a beta particle.

In nuclear reactions such as alpha and beta decay the nucleus is also losing energy in form of photon emissions. Since such nuclear energy losses tend to be significant, the photons will have high energy, which translates into much shorter wavelengths than visible light, often shorter than ultraviolet (UV) light and X-rays. Such rays of very energetic photons are also called gamma rays.

The concept of elements as isolated atoms is an oversimplification, although a very useful one. Atoms do not exist in isolation; they *bond* in many interesting ways, very much depending on their position in the periodic table. The color coding of squares with inscribed atomic numbers and symbols in the periodic table indicates the following:

- Pink, red, purple, light brown/beige: Metals
- Olive green: Metalloids (semiconductors)
- Light blue: Rare gases (completely filled shells)
- Yellow: Halogens
- Green: Other nonmetals

Most elements are *metals*, defined by their free, delocalized electrons as they bond together. When metal atoms bond, their outermost electrons are shared between all the metal nuclei and therefore free to move around. As a consequence, metals are good conductors of electricity.

The electrons in nonmetallic elements are not free, they are typically held in shared orbitals that form bonds between atoms. As a result, the green nonmetals such as element #16, sulfur (S), are typically *insulators*.

The olive green metalloids are insulators under normal circumstances (in room temperature), but their electrons can be promoted into a delocalized state (conduction band) by heat or by electromagnetic radiation. The best example of a *semiconductor* is silicon (Si).

Nonmetals such as hydrogen (H), nitrogen (N), and oxygen (O) exist as gases with small molecules (atoms binding with themselves) as basic components. A good example is the hydrogen molecule, consisting of two H atoms.

As two H atoms are combined into a diatomic hydrogen molecule, they form a strong bond: the spherical s-orbitals of the atom are combined into a σ-orbital of the molecule, shared by the two nuclei. The bond involving shared electrons between two or more nonmetallic atoms is called *covalent*. Other good examples of a covalent bond are the water molecule (H_2O), ammonia (NH_3), carbon dioxide (CO_2), and benzene (C_6H_6).

Even stronger bonds are formed between atoms that are situated at opposite sides of the periodic table, in particular between alkali metals and halogens: Alkali metals are situated at the far left, in group 1. They all have a single s-electron in their outermost shell. On the other hand, halogens are situated just to the left of the rare gases, in group 17 of our periodic table. They have an incomplete outer shell; they are just missing one electron compared to their neighbors, the rare gases. Such *ionic* bonds are formed as atoms lose/gain electrons to form closed (complete) shells. When sodium (Na) meets chlorine (Cl), the resulting molecule NaCl will be held together by an ionic bond between positively charged Na^+ and negatively charged Cl^- ions that typically will form a repeating (in all three dimensions) crystal structure, well known as table salt.

2.2 Atoms of Life

Life on earth involves molecules developed most certainly via synthesis of a small number of original atmospheric constituents, namely, carbon dioxide, carbon monoxide, the nitrogen and oxygen molecules, water and perhaps ammonia, methane, and the hydrogen molecule. Catalyzed by UV light from the sun and the electrical discharges of lightning, more complicated organic molecules such as the amino acids and nucleotides (the basic building blocks of DNA, to be discussed later) could have been created as a "primordial organic soup."

There is a very small number of "atoms of life" involved in what we call organic chemistry and they are all in the green category of the periodic table, as shown above. Let's therefore discuss hydrogen, carbon, nitrogen, phosphorus, and oxygen in more detail. In addition we will also take a peek at silicon, sulfur, the metals sodium, potassium, magnesium, calcium, iron, copper, technetium, and gold, and the halogens fluorine, chlorine, and iodine.

Hydrogen (H) is the most abundant of all the elements, constituting more than 75% of all ordinary (as opposed to "dark") matter in the universe by mass, and accounting for ~90% of all atoms.[26] On earth, most of hydrogen is in water molecules but also in all molecules involved in life processes. Hydrogen is unique among atoms because it will readily donate its single electron (turning into a positive H^+ ion, much like the alkali metals of group 1, as well as accept an electron to fill its $1s$ shell and become a negative H^- ion, like the halogens of group 17. The hydrogen molecule H_2 is a gas at room temperature but can behave like a metal under extreme pressure. Hydrogen was first discovered by Cavendish but later named by Lavoisier, who picked the Greek word for *generating water*. The water molecule H_2O can easily separate into H^+ and OH^- ions. Acidic solutions have greater concentration of hydrogen ions [H^+] than pure water. The commonly used measure for acidity in a liquid solution, pH, is really a measure of the concentration of H^+ ions. When put in contact with a metal, an acid solution will bind with metal and release molecules of hydrogen gas. In large organic molecules, the presence of hydrogen is adding stability through a special type of chemical bond, the *hydrogen bond*. A hydrogen bond is the electromagnetic attractive interaction between polar molecules in which hydrogen (H) is bound to a highly electronegative[27] atom, such as nitrogen (N), oxygen (O), or fluorine (F). In water, hydrogen bonding strongly affects the crystal structure of ice, helping to create an open hexagonal lattice (Fig. 2.2).

The density of ice is less than the density of water at the same temperature; thus, the solid phase of water floats on the liquid, unlike most other substances, thereby preserving life in winter. Liquid water's high boiling point is due to the high number of hydrogen bonds each molecule can form, relative to its low

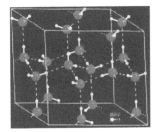

Figure 2.2 Crystal structure of hexagonal ice. Gray dashed lines indicate hydrogen bonds. Source: Wikipedia.

molecular mass. The high boiling pint is caused by the high energy required to breaking these bonds.

We will return to the importance of the hydrogen bond when discussing the structure of molecules of life, such as DNA.

Carbon (C) is perhaps the most versatile of all elements. It is also a very abundant element on earth, where it occurs as limestone $CaCO_3$ and in fossil fuels such as coal and oil. Its versatility can be attributed to its electron configuration: As all elements in group 14—like silicon (Si), germanium (Ge), tin (Sn), and lead (Pb)—its outer electrons are configured as $s^2 p^2$. In other words, there is still space for another four electrons in the outer p-shell. Therefore, carbon's coordination number is 4, that is, it can bind with up to four other elements in a molecule. But carbon can do much more—it can appear as diamond, as graphite, and as graphene. Those various forms of carbon are also referred to as *allotropes*.

Diamonds are formed under high pressure and high temperature and are incredibly hard and strong due to a special 3D structure (sp^3 hybridization,[28]) where each carbon atom is bonded to four others by covalent bonds, forming repeating tetrahedrons. All the electrons are involved in these bonds. Therefore, no electrons are available to absorb light, rendering diamonds fully transparent. As there are no electrons available to conduct electricity, diamond is an isolator.

In *graphite*, carbon's electron configuration is turned into a planar sp^2 configuration with a spare electron that becomes delocalized and shared by the planar arrangement of carbon atoms. Many strong covalent bonds are holding the graphite structure together, but only in two dimensions. Bond angles are 120 degrees.

The layers are free to slide easily over one another. Graphite is opaque and is a conductor. Graphite powder is therefore used as a lubricant and in making pencils. Graphite can also be mixed with polymers to fabricate tough carbon fiber–reinforced plastics used in consumer goods and industrial automotive and aerospace products.

Less well known but very interesting allotropes of carbon are *fullerenes*. The first fullerene to be discovered, in 1985, was *buckminsterfullerene* (or *buckyball*), a spherical molecule consisting of 60 carbon atoms joined in hexagonal and pentagonal rings. Research into these spherical molecules led to the discovery of carbon nanotubes—sheets of graphite rolled into tiny cylinders with widths of a few nanometers (10^{-9} m). Nanotubes are remarkably strong and have interesting electrical properties. Another carbon allotrope is *graphene*, carbon atoms arranged in hexagonal rings across a flat plane.

Finally, there is a*morphous carbon* as found in charcoal and soot. Activated charcoal is used in water filters and gas masks.

More proof to the versatility of carbon is the fact that there are more than 10 million known carbon compounds. It can connect with other atoms in single, double, and triple bonds and it can easily form rings. Life as we know it is based on carbon chemistry or *organic chemistry*.

Nitrogen (N) and phosphorus (P) are also extremely important to all life. Nitrogen is a defining ingredient in all proteins and DNA. Phosphorus is also a crucial part of DNA and also of life's system of energy based on the molecules' adenosine-5′-triphosphate (ATP) and adenosine diphosphate (ADP).

Their outer shell electron configuration is $s^2 p^3$, that is, a filled s-shell and a half-filled p-shell. Since s is spherical and the three p-orbitals combine to form a spherical shape as well, all elements of the nitrogen group (group 15) are spheres when occurring singly. However, they usually exist as two-atom (diatomic) molecules. N_2 is an important component of the air we breathe. N_2 is a colorless odorless gas.

Nitrogen is the most abundant element in the earth's atmosphere. It has also earned a reputation as a key ingredient in explosives. The Swedish chemist Alfred Nobel invented *dynamite* by incorporating nitroglycerine into clay, thereby making it safer to transport and easier to use. Nitroglycerine is also used in medicine to widen arteries and increase blood flow (via release of NO). Another important use of nitrogen is in fertilizers, mostly as ammonia (NH_3). Breaking down the triple bond in N_2 is difficult, but nitrogen-fixing bacteria (diazotrophs) can do the job and let nitrogen react with hydrogen to form ammonia. In 1908, the industrial process for producing ammonia—enabling mass production of fertilizers—was invented by Haber and Bosch. It was a breakthrough for the world's food production. Although nitrogen is nontoxic, cyanides are among the most actively acting poisons. Cyanide molecules include a carbon atom joined to a nitrogen atom via a triple bond. Cyanide ions $(CN)^-$ interfere negatively with cellular respiration.

Nitrogen oxide (N_2O), sometimes called *laughing gas*, is inhaled as an anesthetic and pain killer or used as an additive to fuels where extra oxygen makes fuels burn more rapidly.

Phosphorus is used in matches as its nontoxic red allotrope and in weapons (smoke bombs) in its toxic white form. In nature, phosphorus is nearly always bound to four oxygen atoms, making a phosphate ion $(PO_4)^{3-}$ found in phosphate minerals and in all living things.

Phosphate ions form part of the sugar–phosphate "backbone" of every molecule of DNA. An adult human contains about 750 g of phosphorus, mostly in bones and teeth, as hydroxylapatite $(Ca_5(PO_4))_3(OH)$.

Oxygen (O) is the third most abundant element in the universe, after hydrogen and helium, and the second most abundant element on earth. Oxygen accounts for 90% of the mass of pure water and accounts for 21% of the volume of pure air. With its electron configuration of $(1s^2)(2s^2)(2p^4)$ it lacks two p-electrons for a full shell and is therefore highly reactive. It easily binds with other elements. Oxygen atoms account for 65% of the mass of the human body. To constantly replenish oxygen in the atmosphere and the oceans, plants, and certain types of bacteria photosynthesize: During the first stage of photosynthesis, energy from sunlight is used to split water molecules into hydrogen ions, oxygen gas (O_2), and free electrons. During the second stage, the hydrogen ions and free electrons are reacting with carbon dioxide (CO_2) to synthesize energy-storing carbohydrates such as glucose ($C_6H_{12}O_6$) and subsequently larger organic molecules such as sucrose and starch. Larger molecules made from units of glucose (including cellulose) are used by plants to build cell walls.

The reverse of photosynthesis is aerobic respiration where oxygen reacts with carbohydrates inside cells to generate carbon dioxide and water and to produce energy. Respiration is controlled combustion or burning, it releases the energy stored during photosynthesis. Oxygen plays hence a key role in respiration and combustion and is essential for life, for industrial processes such as steelmaking, and for heating and transportation.

Silicon (Si) is the second element in the carbon group, the second (after oxygen) most abundant element (by mass) of the earth's crust. Si has been known for a long time for forming silicon dioxide (SiO_2) or silica, which is found in quartz and is the main ingredient in glassmaking. Today silicon is best known, however, for its application in electronics due to its versatility as semiconductor. Silicon chips are made from ultrapure crystalline silicon, often "doped" with other elements to create transistors,[29] the fundamental building blocks of modern electronic devices. Another important application is solar cells.

Sulfur (S) is the second element in the oxygen group and is an essential element for all living things. In particular, it is present in many proteins. It can be found in its native (elemental) state near volcanoes and hot springs. Sulfur-containing compounds often carry the prefix *thio-*, rooted in the Greek word *theion* for sulfur.

Lithium (Li), sodium (Na), and potassium (K) are three important representatives of the alkali metals, to be found in group 1 of the periodic system and having only one *s*-electron in their outer shell. They easily give up this electron and are therefore extremely reactive, forming ionic bonds. Their preferred bonding partners are the halogens (F, Cl, etc.) in group 17 that lack an electron to completely fill their outer shell. Compounds of Li are found in batteries and in the glass and ceramics industry. Lithium carbonate (Li_2CO_3) is the active ingredient in medicines for various psychological disorders such as bipolar (or mano-depressive) disorder where it could be linked to a decline in suicide rates.[30]

Sodium is the sixth most abundant element in the earth's crust. As salt or sodium chloride (NaCl) it is dissolved in seawater. Na is essential for proper functioning of nerve cells and for the regulation (as electrolyte) of the body's level of hydration. Too much salt can cause the body to retain too much fluid, giving rise to elevated blood pressure.

Potassium is the seventh most abundant element in the earth's crust, mostly found as potassium chloride (KCl), potassium nitride (KNO_3), and potassium hydroxide (KOH), the first two used a lot in fertilizers. In humans and other animals, K is key to the transmission of nerve impulses, and plays a role as electrolytesimilar to Na. Potassium ions, K^+, can be found dissolved in red blood cells and as regulators of heart function. In the field of cell biology, potassium channels are the most widely distributed type of ion channel and are found in virtually all living organisms. They form potassium-selective pores that span cell membranes. In cardiac muscle, malfunction of potassium channels may cause life-threatening arrhythmias.

They also regulate cellular processes such as the secretion of hormones (e.g., insulin release from beta cells in the pancreas).

Magnesium (Mg) and calcium (Ca) are two important representatives of the alkaline earth metals, to be found in group 2 of the periodic system.

Mg is the eighth's most abundant element in the earth's crust and is found in rock-forming minerals. The human body contains about 25 g of Mg and hundreds of biochemical reactions depend on Mg. Atoms of magnesium lie at the heart of the green pigment chlorophyll, part of green vegetables and plants.

Ca is the fifth most abundant element in the earth's crust and is found in rocks as calcium carbonate ($CaCO_3$), calcium magnesium carbonate or dolomite ($CaMgCO_3$), calcium sulfate, and calcium fluoride. Lime is the general term for any mineral rich in Ca compounds. Marble is metamorphic rock, limestone altered by pressure and heat. Calcium is an essential element in nearly all living things and is the most abundant metallic element in humans, making up about 1.2 kg, mostly tied up in calcium phosphate, the main constituent of bones and teeth. Calcium is also involved in vital functions, including cell division, nerve and muscle function, the release of hormones, and the control of blood pH. Food sources for Ca are milk and dairy products and green vegetables. However, Ca uptake often requires the presence of vitamin D. Vitamin D deficiency can therefore inhibit the growth and maintenance of bones. Calcium ions play an important regulating the role in the calcium channels of cell membranes. Certain signals cause the channels to open, letting calcium rush into the cell. The resulting increase in intracellular calcium has different effects in different types of cells. Calcium channel blockers prevent or reduce the opening of these channels and thereby reduce these effects. They may reduce blood pressure by causing an increase in arterial diameter (a phenomenon called vasodilation), by acting on cardiac muscles (myocardium) and reducing the contraction of the heart, and by slowing down the conduction of electrical activity—thereby slowing down the heartbeat.

Iron (Fe) is one of the most important metals and has—in addition to its well-known industrial uses—highly relevant biological function. In ancient Greek and Roman mythology, iron was associated with Mars, the God of war, and the name of the red planet, which owes its color to the dust of red iron (III) oxide (Fe_2O_3) across the planet's surface. On Earth, Fe accounts for about one-third of the total mass. In the earth's crust, iron is only the fourth most abundant element, after oxygen, silicon, and aluminum. The Earth has an iron-rich core, consisting of 90% iron. The inner core is solid and spins within the liquid outer core, causing the Earth's magnetic field. Together with nickel (element #27) and cobalt (element #28), iron (element #26) is one of only three ferromagnetic elements: it can be magnetized and made to retain this property as help in navigation as magnetized needles (of a compass) point in the direction of the Earth's North and South Poles.

In living things, iron is also extremely vital. Many important reactions, including the synthesis of DNA and photosynthesis, rely on iron. More than half of the iron in a typical human body is bound in hemoglobin, the protein in red blood cells that is responsible for transporting oxygen in the bloodstream. Iron deficiency will limit the amount of oxygen delivered to the tissues and subsequently lead to fatigue and lowered immunity, a condition often referred to as anemia. On the other hand, iron excess is toxic and iron overload can be fatal.

Copper (Cu) has already been known in ancient times with the oldest artefacts made of copper about 10,000 years old. Combined with tin (Sn), it has been made into bronze around 5000 years ago. Combined with zinc (Zn), the resulting alloy is brass, which has been popular since the Romans. Along with gold (Au), copper is the only metal that's not silver-gray in color. Due to its high conductivity, copper is used as electric wiring, including the windings of electromagnets, transformers, and motors. Cu is of great biological importance because of its presence in enzymes that are involved in building the hemoglobin molecule. Some organisms even use copper (instead of iron) in their blood: Octopuses and many other molluscs are "blue-blooded"—they use the Cu-based protein hemocyanin for transporting oxygen. An adult human requires $\sim$1 mg Cu per day, provided by liver, egg yolk, cashew nuts, avocados, etc. Copper has antimicrobial action, making copper alloys a popular choice in hospitals and schools.

Technetium (Tc) is a radioactive transition metal with atomic number 43. It's not included here because it is an "atom of life" but because of its nanomedical importance. Its name means *artificial* because it was the first element to be discovered only after being produced in a particle accelerator by bombarding molybdenum with deuterons (particles composed of one proton and one neutron). It is located in group 7 along with manganese (Mn) and rhenium (Re). Only tiny amounts of Tc occur naturally. Although two of the Tc isotopes have half-lives of more than a million years, there is no completely stable Tc isotope. It exists on earth as the result of disintegration of uranium nuclei in uranium ore. The excited state of technetium-99, called Tc-99m, is a short-lived gamma ray–emitting nuclear isomer. It is used in nuclear medicine for a wide variety of diagnostic tests after injection into a patient's bloodstream. It is well suited to the role because it emits readily detectable 140 keV gamma rays,[20] and its half-life is 6.01 hours (meaning that about 94% of it decays to Tc-99 in 24 hours). Tc-99m can be bound to a variety of nonradioactive compounds and can therefore be used for a multitude of diagnostic tests. There are more than 50 commonly used radiopharmaceuticals based on Tc-99m for imaging and functional studies of the brain, myocardium, thyroid, lungs, liver, gallbladder, kidneys, skeleton, blood, and tumors.

Iodine (I) is a halogen, along with fluorine, chlorine, bromine, and the radioactive element astatine (At). It exists in its elemental form as a solid composed of diatomic molecules, I_2. It has a deep blue-black color with an almost metallic sheen. Seawater contains a small amount of I and it occurs in seaweed and other algae. Iodine is essential in humans, although the total amount of the element present in the body is only ~20 mg. It plays an important role in the synthesis of two hormones produced in the thyroid. The hormones are involved in regulation of metabolism, body temperature, heart rate, and growth. Iodine deficiency has serious consequences, including fatigue, weight gain, stunted physical growth in children, and even impaired mental development (cretinism). Much of the world's table salt has iodide (compounds containing iodine in its oxidation state, namely, I^{-1}) such as KI (potassium iodide) added to avoid the iodine deficiency problem. On the other hand, too much intake of iodine can lead to a condition called hyperthyroidism, leading to an increased metabolic rate and hyperactivity.

Gold (Au) is often found native, similar to its group 11 partners copper and silver (Ag). Due to its filled electron shell, Au is very unreactive (i.e., noble). Even when hot, it is not attacked by water or oxygen. It survives the acid test—it will not dissolve in a drop of nitric acid. Although the concentration is very low, there is gold in seawater. Gold nanoparticles have many uses. Attached to specific proteins or tiny structures within cells, Au nanoparticles are used to enhance contrast in electron microscopy. They can also be made to enter cancerous cells and then be a target for light amplification by stimulated emission of radiation (laser) light irradiation. Alternatively, gold nanoparticles can carry cancer drugs directly into the cancerous cells' nuclei. Gold is nontoxic.

2.3 Molecules of Life

Science knows no country, because knowledge belongs to humanity, and is the torch which illuminates the world. Science is the highest personification of the nation because that nation will remain the first which carries the furthest the works of thought and intelligence.

—Louis Pasteur (1822–1895)

As discussed previously, the characteristics of atomic systems change discontinuously, they are quantized. In particular, there are discrete energy levels. Transitions from a given level to another can be referred to as quantum jumps and define the spectroscopic properties: If the second state of the system has greater energy than the first, at least the difference of the two energies has to be supplied from the outside to make the transition possible. To a lower level the system can change spontaneously, releasing the energy difference as radiation.

Moving from a single atom to a selection of atoms, there may be a lowest energy level associated with a stable spatial configuration of the atomic nuclei and their respective electron clouds. Atoms in such a state form a molecule. The electronic configuration can be calculated by solving Schrödinger's wave equation for the many-electron system. As a first approximation, the nuclei are assumed to be fixed point charges and the electrons of all participating atoms are allowed to move freely, interacting with each other and with the atomic nuclei. That's what quantum chemistry, a field that has been enabled by and has flourished with the advancement of computer technology, is all about.

As atoms form molecules, their outer atomic orbitals are merging and are forming molecular orbitals. The inner shell orbitals stay localized around the atomic nuclei, but the outer valence orbitals are becoming delocalized. In quantum chemistry, the ground state of a molecule is the state of lowest electron energy for a given geometric arrangement of atomic nuclei. The ground state is calculated by solving the Schrödinger equation of the N-electron system, N being the total number of electrons contributed to the molecule by the participating atoms. For instance, the water molecules has 10

Figure 2.3 Formation of molecular σ-bonds and π-bonds from atomic s- and p-orbitals. Source: Wikipedia.

electrons: oxygen contributes eight electrons (of type $1s^2$, $2s^2$, $2p^4$) and the two hydrogen atoms add one ($1s$) each.

Figure 2.3 shows how two atomic s-orbitals merge to form a molecular orbital of σ type (left picture) and how p-orbitals merge to form a molecular orbital of π type (right picture).

σ-bonds are the strongest type of covalent chemical bond. σ-bonding is most clearly defined for diatomic molecules (like H_2 or O_2) but is also keeping carbon atoms attached to each other in cyclic compounds such as benzene, C_6H_6. A σ-bond is symmetrical with respect to rotation about the bond axis.

π-bonds are also covalent chemical bonds where two lobes of one involved atomic orbital overlap two lobes of the other involved atomic orbital. Each of these atomic orbitals is zero at a shared nodal plane, passing through the two bonded nuclei. The same plane is also a nodal plane for the molecular orbital of the π-bond. π-bonds are usually weaker than σ-bonds, for example, the C–C double bond (consisting of one σ- and one π-bond) has a bond energy less than twice the C–C single σ-bond energy.

An interesting case is the above-mentioned benzene molecule, an important building block of organic chemistry. Carbon has the atomic configuration $(1s^2)(2s^2)(2p^2)$, a total of 4 s-electrons and 2 p-electrons. Benzene has therefore a total number of 42 electrons, contributed by 6 carbon atoms ($6 \times 6 = 36$ electrons) and 6 hydrogen atoms ($6 \times 1 = 6$ electrons). However, since the inner shell ($1s^2$) carbon electrons remain localized around the six carbon nuclei, we only need to consider 30 electrons when discussing the bonding configuration.

Figure 2.4 illustrates how the benzene molecule is formed with 24 electrons occupying σ-bonds and 6 electrons engaged in π-bonds that will be delocalized across the hexagonal geometry.

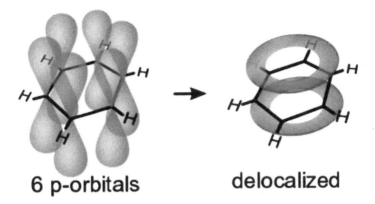

6 p-orbitals   **delocalized**

Figure 2.4   Carbon *p*-orbitals forming delocalized π-orbitals in a benzene molecule. Source: Wikipedia.

This delocalization of π-bonds in the benzene ring has puzzled chemists for a long time and could first be resolved in a satisfactory way by quantum theory: By solving the Schrödinger equation for the electronic configuration of benzene, it is clearly seen that the lowest energy ground state of the molecule is characterized by an exact hexagonal structure and by completely delocalized π-orbitals. An interesting consequence of π-electron delocalization is the associated mobility that is important for biology and is sometimes referred to as aromaticity. Aromatic molecules such as benzene typically display enhanced chemical stability but also have distinctive pleasant smells caused by reactivity due to electron mobility. Another way of detecting aromaticity is to analyze the nuclear magnetic resonance (NMR) signal, to be discussed in more detail below (see end of Section 2.4): the circulating π-electrons in an aromatic molecule produce ring currents that oppose the applied magnetic field.

In the ground state of any isolated single molecule, be it water or benzene or any other molecule, the electrons will occupy the lowest-possible molecular orbitals, where "lowest" refers to energy levels. There will also be excited energy levels where electrons will occupy higher-level molecular orbitals that may be not part of the ground state. What will typically happen with such excited states is that the system will try to reach the stable ground state by emitting a photon with an energy corresponding to the energy difference between the excited and ground state of the molecule. By comparing the result of a quantum chemistry calculation with the observed wavelength of an observed photon, it is then possible to check the accuracy of the calculation and the validity of quantum theory.

Another straightforward result of quantum chemistry calculations is the determination of ionization energies: By performing two separate calculation with N electrons (molecular ground state) and $(N-1)$ electrons, it is possible to compute the "ionization energy", defined as the energy required to cause an electron to escape from the molecule.

The delocalization of molecular orbitals can be pushed further when considering crystal structures instead of single molecules. In crystal structures, the atoms are arranged in regular geometric arrangements that extend in all three dimensions.

As more and more atoms are brought together, the molecular orbitals grow larger and extend over big distances. At the same time, the energy levels of the growing molecule will become increasingly dense. Eventually, the collection of atoms forms a giant molecule, also called solid. For this solid, the energy levels are so close that they can be considered to form a continuum or a band.

Band theory provides a good explanation of the electrical properties of crystals.

The lowest energy band will be filled by inner shell electrons.

Moving up in energy, the next band will be associated with fully occupied valence electrons.

Moving up even further, we will find either partly occupied or fully occupied valence electron levels: If this band is fully occupied, the solid crystal will be an insulator because the electrons filling up the band will have no freedom of moving around.

If the highest valence electron band is only partly filled, the band will be called a conduction band. Electrons can move around feely and conduct electricity and heat, that is, the crystal will behave like a metal.

The conduction band quantifies the range of energy required to free an electron from its bond to an atom. Once freed from this bond, the electron becomes a delocalized electron, moving freely within the atomic lattice of the material to which the atom belongs. Solid-

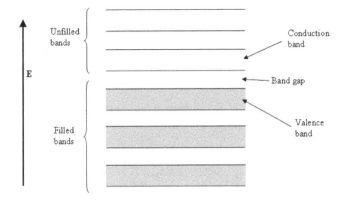

Figure 2.5 Band gap between valence and conduction bands. Source: Wikipedia.

state materials may be classified by their band gap, defined as the difference between the valence and conduction bands, as illustrated in Fig. 2.5:

- In *insulators* (or nonconductors) the conduction band is clearly separated from the valence band, requiring very high energies to excite the valence electrons. There is a significant band gap.
- In *semiconductors*, the band gap is small. It takes very little energy (in the form of heat or light) to excite semiconductor electrons and move them into the conduction band where they will readily delocalize and conduct electricity.
- In *metals*, the highest valence band is only partly occupied and therefore becomes the conduction band.

In conclusion, solid-state theory can be considered an extension of molecular theory.

The electronic properties of metals, semiconductors, and insulators have great importance in electronics and other areas of technology, including medical diagnostics and medical devices.

To examine what will happen to a molecule at body temperature, we need to heat our molecule and apply principles of statistical mechanics, pioneered in the 19th century by Gibbs and Boltzmann. It suffices here to note that most molecules will require considerable energy to be lifted from their stable state of lowest energy (or ground state) to their next higher state. We will return to this concept when discussing molecular genetics, in particular the molecular basis of mutations. Before going there, let's introduce some important "molecules of life."

Water (H$_2$O) is playing a vital role in all life processes and acts as a universal solvent in biology (Fig. 2.6).

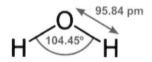

Figure 2.6   Water molecule. Bond angle 104.45° and distance 95.84 × 10^{-12} pm. Source: Wikipedia.

Water is the most abundant compound on earth, covering about 70% of its surface. It makes up 55%–78% of the human body, with newborn babies having the highest percentage.

Water is a liquid at normal temperature and pressure, but it often coexists on earth with its solid state (ice) and gaseous state (water vapor or steam). Since the water molecule is nonlinear and the oxygen atom carries a slight negative charge balanced by slightly positive charges of the two hydrogen atoms, water is a polar molecule with an electrical dipole moment. Each water molecule also can form up to four intermolecular hydrogen bonds. These factors lead to strong attractive forces between molecules of water. Water molecules stay close to each other (*cohesion*), due to the collective action of hydrogen bonds between water molecules. Water also has high *adhesion* properties because of its polar nature. In biological cells and organelles, water is in contact with membrane and protein surfaces that are *hydrophilic*, that is, surfaces that have a strong attraction to water.

Water's strong cohesion is giving rise to high surface tension and capillary forces. The capillary action refers to the tendency of water to move up a narrow tube against the force of gravity. This property is relied upon by all vascular plants, such as trees.

Pure water has a low electrical conductivity, but this increases with the dissolution of a small amount of ionic material such as sodium chloride (NaCl, salt). On earth, 96.5% of the planet's water is found in seas and oceans, 1.7% in groundwater, 1.7% in glaciers and the ice caps of Antarctica and Greenland, and a small fraction in other large water bodies, and 0.001% in the air as vapor, clouds (formed of solid and liquid water particles suspended in air), and precipitation.

Elements such as Li, Na, K, and Ca (more electropositive than H) can displace hydrogen in water and form hydroxides. Being a flammable gas, the hydrogen given off is dangerous—reactions of water with these elements may be violently explosive.

Carbohydrates ($C_m(H_2O)_n$) are defined as organic compounds comprising only carbon, hydrogen, and oxygen, usually with a hydrogen:oxygen atom ratio of 2:1. In other words, the generalized formula for carbohydrates is $C_m(H_2O)_n$, where m can be different from n. In biochemistry, carbohydrates are mostly called *saccharides*, divided into four chemical groupings: monosaccharides, disaccharides, oligosaccharides, and polysaccharides.

Carbohydrates perform numerous roles in living organisms. In particular polysaccharides are involved in the storage of energy and serve as structural components (e.g., cellulose in plants).

Monosaccharides and disaccharides are smaller and carry a lower molecular weight than oligo- and polysaccharides.

Sugars denote a class of chemically related sweet-flavored substances with chemical formulas $C_n(H_2O)_n$, where n can assume values between 3 and 7. The names of the sugars very often end in the suffix -ose. For example, grape sugar is the monosaccharide *glucose*, cane sugar is the disaccharide *sucrose*, and milk sugar is the disaccharide *lactose*.

Sugars are chemically a special case of carbohydrates with $m = n$.

Glucose ($C_6H_{12}O_6$) (Fig. 2.7) is also called dextrose or grape sugar and occurs naturally in fruits and plant juices. Most ingested carbohydrates are converted into glucose during digestion and it is the form of sugar that is transported around the bodies of animals in the bloodstream. Cells use glucose as the primary source of energy and as a metabolic intermediate.

Note that the graphical representation used in Fig. 2.7 and below does not identify carbon atoms—they are assumed to occupy all positions not labeled explicitly. Also, whenever carbon has one or

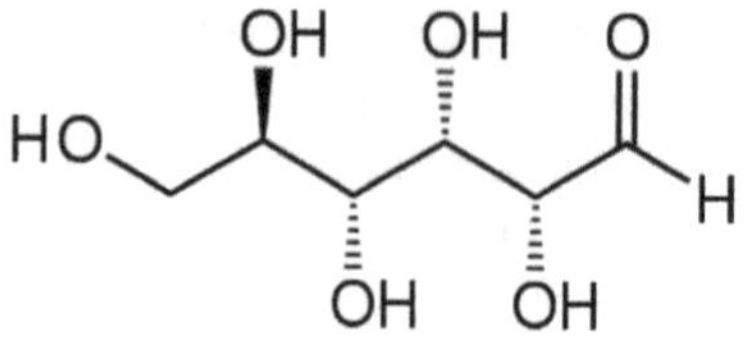

Figure 2.7   Glucose $(C_6H_{12}O_6)$. Source: Wikipedia.

Figure 2.8   Generic alcohol molecule including a hydroxyl (OH) functional group. Source: Wikipedia.

several of its four bonds available it will bind to one or several hydrogen atoms that will not be explicitly labeled.

Glucose is the ubiquitous fuel in biology. It is used as an energy source in most organisms, from bacteria to humans. Glucose is also the primary source of energy for the brain. Its availability influences psychological processes. When glucose is low, mental efforts such as self-control and decision making are impaired. On the other hand, elevated glucose levels are leading to diabetic complications in humans.

Alcohol $(C_nH_{2n+1}OH)$ (Fig. 2.8.) is the common name for a class of organic compounds in which the *hydroxyl* functional group (–OH) is bound to a carbon atom. The hydroxyl group is the defining functional group in alcohols. In addition, this carbon center should be saturated, having only single bonds to three other atoms.

The most common alcohol, ethanol (C_2H_5OH), is found in alcoholic beverages.

Propylalcohol, C_3H_7OH, is an interesting molecule because it illustrates the concept of *isomerism*. The atoms forming this molecule can unite in more than one way, as shown in Fig. 2.9 below.

In the language of quantum chemistry, all three configurations are stable molecules; they behave as though they were lowest states. Configurations I and II are both alcohols and are chemically not too different. However, configuration III is completely different:

Figure 2.9 Structural isomers of C_3H_7OH or C_3H_8O. Source: Wikipedia.

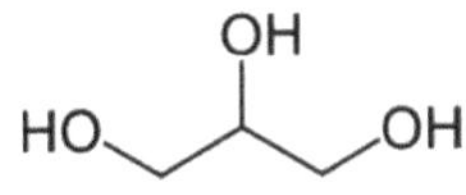

Figure 2.10 Glycerol (glycerin). Source: Wikipedia.

methoxyethane is no longer an alcohol since the oxygen is connected to two carbon atoms only (not to hydrogen in a hydroxyl, –OH, functional group).

There are no spontaneous transitions from either state toward the other. Transitions between the three configurations can only occur via intermediate configurations of significantly higher energy than the ground states. An overly simple explanation would be that the oxygen has to be first extracted from one position and then inserted into the other. That cannot be done without passing through an intermediate state (or *transition state*) of higher energy, comparable to a "mountain separating two valleys of stability."

Glycerol (or glycerine, glycerin, chemical formula $C_3H_8O_3$) (Fig. 2.10) is a simple sugar alcohol compound. It is a colorless, odorless, viscous liquid that is sweet tasting and of low toxicity. Glycerol has three hydroxyl groups that are responsible for its solubility in water and its hygroscopic nature.

A carboxylic acid (Fig. 2.11) is an organic acid characterized by the presence of at least one carboxyl group. The general formula of a carboxylic acid is R–COOH, where R is some monovalent functional group.

A *carboxyl* group (or carboxy) is a functional group that has the formula –C(=O)OH, usually written as –COOH. Carboxylic acids are proton (H^+) donors and the most common type of organic

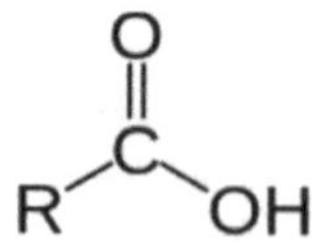

Figure 2.11 Structure of a carboxylic acid. Source: Wikipedia.

acid. Among the simplest examples are formic acid H–COOH, which occurs in ants, and acetic acid CH_3–COOH, which gives vinegar its sour taste. Carboxylic acids often have strong odors. Acids with two or more carboxyl groups are called dicarboxylic, tricarboxylic, etc. Carboxylic acids react with alcohols to give esters.

Esters (Fig. 2.12) are chemical compounds formed by condensing an acid with an alcohol. The hydrogen in the compound's carboxyl group is replaced with a hydrocarbon group.

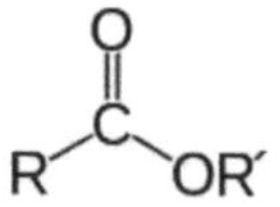

Figure 2.12 A carboxylate ester. Source: Wikipedia.

Esters are ubiquitous. Most naturally occurring fats and oils (e.g., triglycerides) are the fatty acid esters of glycerol (see above). A *fatty acid* is a carboxylic acid with a long tail (chain), usually consisting of an even number of carbon atoms, from 4 to 28. Fatty acids are important sources of fuel because, when metabolized, they yield large quantities of ATP (discussed further below). Many cell types can use either glucose or fatty acids for this purpose.

Esters with a low molecular weight are commonly used as fragrances and are found in essential oils.

Triglycerides are esters derived from glycerol and three fatty acids. They are classified as saturated compounds—all available places where hydrogen atoms could be bonded to carbon atoms are occupied—and unsaturated compounds (Fig. 2.13) that include double bonds (C=C) between carbon atoms, thereby reducing the number of places where hydrogen atoms can bond to carbon atoms. Triglycerides are a blood *lipid,* responsible for the storage of energy, for signaling, and for acting as structural components of

Figure 2.13 Example of an unsaturated fat triglyceride. Source: Wikipedia.

cell membranes. There are many triglycerides depending on the oil source and their level of saturation.

In Fig. 2.13, glycerol is occupying the left part. On the right-hand side we can see the (saturated) palmitic acid on top and the unsaturated fatty acids oleic acid and α-linolenic acid below.

In the human body, high levels of triglycerides in the bloodstream have been linked to *atherosclerosis* (thickening of blood artery walls as a result of the accumulation of fatty materials, therefore increasing the risk of heart disease and stroke). However, the relative negative impact of raised levels of triglycerides compared to that of cholesterol is as yet not fully understood.

Cholesterol is a **steroid**, a type of organic compound that contains a characteristic arrangement of four cycloalkane rings that are joined to each other (Fig. 2.14). Alkanes are organic molecules that only contain carbon and hydrogen atoms and that are fully saturated, having chemical structure formulas C_nH_{2n+2} (for $n = 1,2,3,...$). Methane, CH_4, is the simplest alkane, followed by ethane ($n = 2$), propane ($n = 3$), and butane ($n = 4$). Cycloalkanes are types of alkanes that have one or more rings of carbon atoms in the chemical structure of their molecules. A general chemical formula for cycloalkanes would be $C_nH_{2(n+1-g)}$, where $n =$ number of carbon atoms and $g =$ number of rings in the molecule.

The core of steroids (such as cholesterol) is composed of 20 carbon atoms bonded together that take the form of 4 fused rings: 3 cyclohexane rings (designated as rings A, B, and C in the figure to the right) and 1 cyclopentane ring (the D ring).

Cholesterol is an essential structural component of animal cell membranes and is required for proper membrane permeability and fluidity. Cholesterol is an important precursor molecule for the synthesis of vitamin D and the steroid hormones, including the adrenal gland hormones cortisol and aldosterone, as well

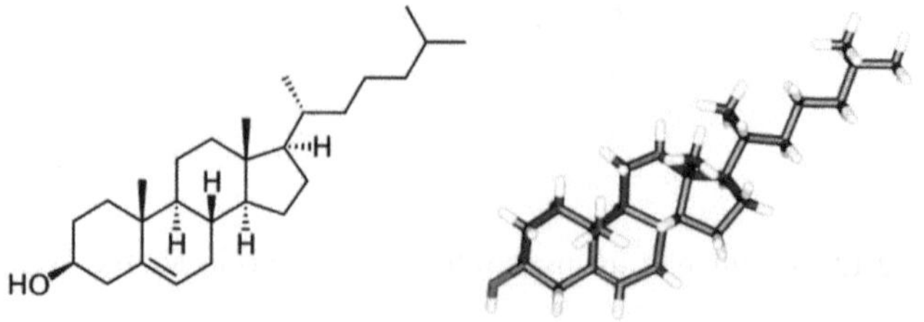

Figure 2.14 Cholesterol structure, including a 3D "ball and stick" view. Source: Wikipedia.

as the sex hormones progesterone, estrogens, and testosterone, and their derivatives. Since cholesterol is insoluble in blood, it is transported in the circulatory system within *lipoproteins*, complex particles that have an exterior composed of amphiphilic (possessing both hydrophilic [water-loving, polar] and lipophilic [fat-loving] properties) proteins and lipids whose outward-facing surfaces are water soluble and inward-facing surfaces are lipid soluble.

According to a generally accepted lipid hypothesis, abnormal cholesterol levels—actually higher concentrations of low-density lipoprotein (LDL) and very-low-density lipoprotein (VLDL) particles and lower concentrations of functional high-density lipoprotein (HDL) particles—are strongly associated with cardiovascular disease (CVD) because these promote atherosclerosis. This disease process leads to myocardial infarction (heart attack), stroke, and peripheral vascular disease. Since higher blood LDL, especially higher LDL particle concentrations and smaller LDL particle size, contribute to this process more than the cholesterol content of the HDL particles. Hence LDL and VLDL are called bad cholesterol and HDL is called good cholesterol.

Phospholipids are a class of lipids that are a major component of all cell membranes as they can form lipid bilayers. Most phospholipids contain a diglyceride, a phosphate group, and a simple organic molecule such as choline, $C_5H_{14}NO$, a water-soluble essential nutrient usually grouped within the B-complex vitamins. Choline is used in the synthesis of the constructional components in the body's cell membranes and. It provides structural integrity and signaling roles for cell membranes (to be discussed in more

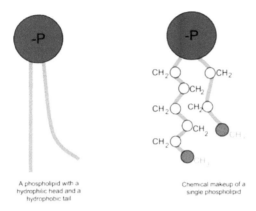

Figure 2.15   Single phospholipid. Source: Wikipedia.

detail in Section 3.4), in particular for neurotransmission. Choline is important for expecting mothers as it's needed to support the fetus's developing nervous system.

The first phospholipid identified in biological tissues was lecithin, or phosphatidylcholine, in the egg yolk, by Theodore Nicolas Gobley, a French chemist and pharmacist, in 1847. As shown in Fig. 2.15, the structure of the phospholipid molecule generally consists of hydrophobic (repelled by water and therefore forced to aggregate) tails and a hydrophilic (attracted to water) head.

Figure 2.16 shows how phospholipids can form lipid bilayers.

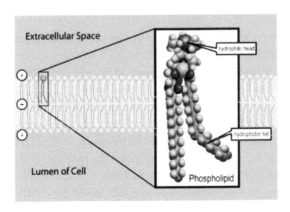

Figure 2.16   Lipid bilayer formed by phospholipids. Source: Wikipedia.

Fats are defined broadly as a group of compounds that are generally soluble in organic solvents and generally insoluble in water. Chemically, fats are triglycerides—triesters of glycerol and any of several fatty acids.

Fats may be either solid or liquid at room temperature, depending on their structure and composition. Both fats and oils are subcategories of lipids on the basis of their behavior at room temperature: oils are liquid and fats are solid. Also, the term "lipids" is generally used in a medical or biochemical context.

After having discussed water, carbohydrates/sugars, alcohol, esters, and lipids, let's jump to the *nucleobases*, nitrogen-containing biological compounds found within DNA, RNA, nucleotides, and nucleosides. Their ability to form base pairs and to stack upon one another lead directly to the helical structure of DNA and RNA.

Adenine (A) is a nucleobase and one of the two purines in DNA (the other one being guanine) (Fig. 2.17).

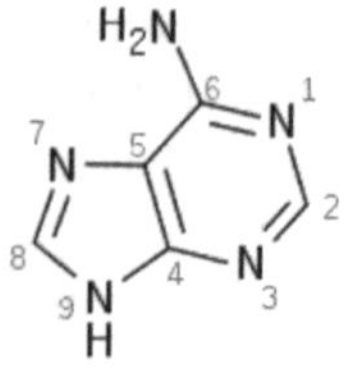

Figure 2.17 Adenine $(C_5H_5N_5)$. Source: Wikipedia.

In Fig. 2.17, the red numbers indicate the conventional labeling of atoms by position. For example, the carbon atoms in positions 2 and 8 have a free bond and therefore a hydrogen atom attached. The carbon atom bonds in positions 4, 5, and 6 are fully occupied, resulting in naked carbons in those positions.

The shape of A is complementary to either thymine (T) in DNA or uracil (U) in RNA: In DNA, A binds to T via two hydrogen bonds, thereby stabilizing the nucleic acid structures. In RNA, A binds to U.

Adenine is playing variety of roles in biochemistry, including cellular respiration, in the form of both the energy-rich ATP and the cofactors nicotinamide adenine dinucleotide (NAD) and flavin adenine dinucleotide (FAD), and protein synthesis, as a chemical component of DNA and RNA. Although called vitamin B_4 in older literature, adenine is no longer considered a true vitamin or part of the vitamin B complex. However, two B vitamins, niacin and riboflavin, bind with adenine to form NAD and FAD, respectively. The acronyms (DNA, RNA, ATP, etc.) used in this paragraph will be explained in further detail below.

Guanine (G) is a nucleobase and one of the two purines in DNA (the other one being adenine—see above) (Fig. 2.18). In DNA, guanine is paired with cytosine.

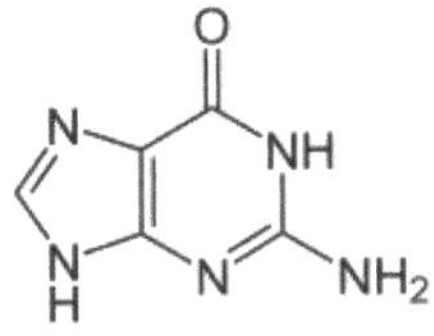

Figure 2.18   Guanine ($C_5H_5N_5O$). Source: Wikipedia.

G binds to cytosine (C) through three hydrogen bonds at the O, NH, and NH_2 positions. G got its name from *guano*, excreta of seabirds from which guanine was first isolated. Guano was used as a source of fertilizers.

Cytosine (C) is a nucleobase and one of the two pyrimidines in DNA/RNA (the other one being thymine/uracil) (Fig. 2.19).

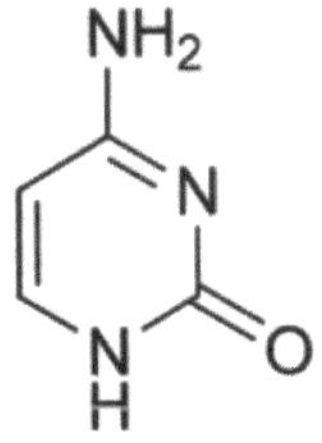

Figure 2.19   Cytosine ($C_4H_5N_3O$). Source: Wikipedia.

In DNA, cytosine is forming three hydrogen bonds (NH_2, N, and O positions) with guanine.

Cytosine is inherently unstable and can change into uracil (U) by spontaneous deamination, that is, the removal of an amine group (see Fig. 2.20).

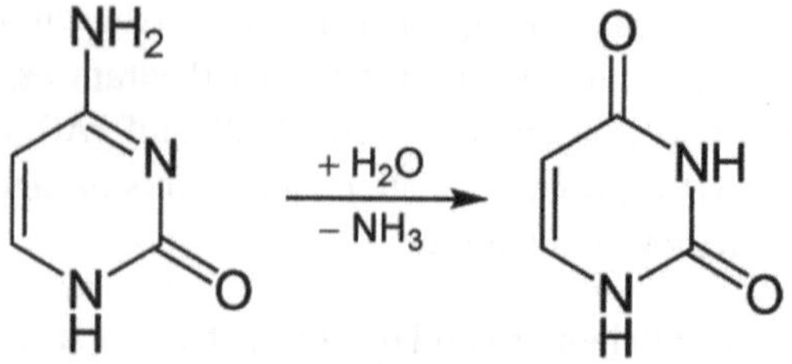

Figure 2.20   Deamination changes cytosine into uracil. Source: Wikipedia.

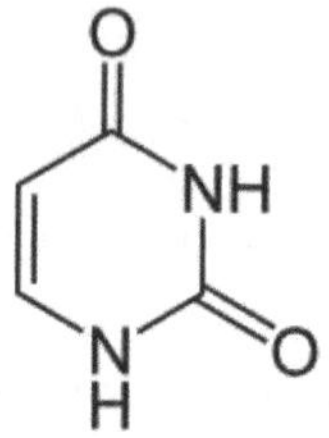

Figure 2.21   Uracil ($C_4H_4N_2O_2$). Source: Wikipedia.

Uracil (U), $C_4H_4N_2O_2$, is one of the four nucleobases in the nucleic acid of RNA that are represented by the letters A (adenine), G (guanine), C (cytosine), and U (Fig. 2.21). In RNA, U binds to adenine (A) via two hydrogen bonds. In DNA, the uracil nucleobase is replaced by thymine (see further below).

Thymine (T) (Fig. 2.22) is a nucleobase and one of the two pyrimidines in DNA, the other one being cytosine (C).

Thymine is also known as 5-methyluracil, a pyrimidine nucleobase obtained by methylation of uracil's carbon situated at position 5.

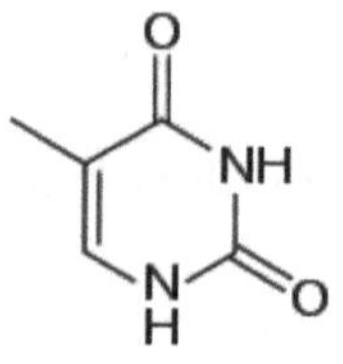

Figure 2.22   Thymine ($C_5H_6N_2O_2$). Source: Wikipedia.

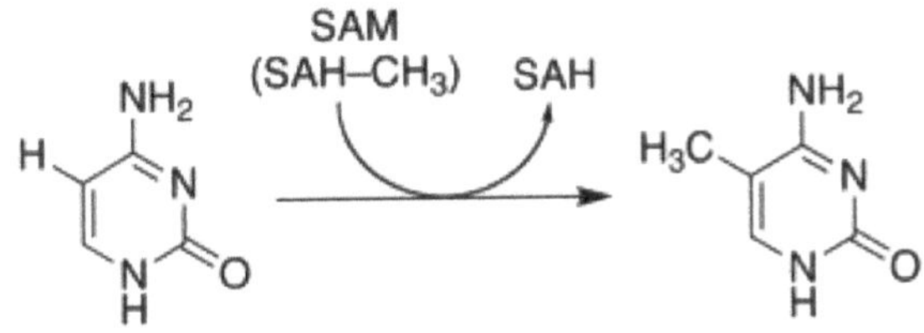

Figure 2.23 Methylation of cytosine, resulting in 5-methylcytosine. Source: Wikipedia.

Methylation denotes the addition of a methyl group (CH_3) to a substrate or the substitution of an atom or group by a methyl group. In our case, CH of uracil in position 5 is replaced by $C-CH_3$.

Frequently, cytosine is methylated to form 5-methylcytosine. When cytosine is methylated, the DNA will maintain the same sequence, but the expression of methylated genes can be altered (see our discussion of epigenetics in Section 3.10). Figure 2.23 shows the methylation of cytosine by means of an enzyme, DNA methyltransferase. DNA methyltransferases use S-adenosyl methionine (SAM) as the methyl donor.

In DNA, T binds to adenine (A) via two hydrogen bonds, thus stabilizing the nucleic acid structures. T may be derived by methylation of uracil at the 5th carbon and is therefore also called 5-methyluracil. Let's recap: Cytosine can turn into uracil via deamination, and methylation of uracil results in thymine.

On the other hand, methylation of cytosine creates 5-methylcytosine, which turns into thymine via deamination. This conversion of a DNA base from cytosine (C) to thymine (T) can result in a transition mutation, a main reason for single-nucleotide polymorphisms (SNPs), to be discussed further in Section 3.4.

Ribose is an organic compound with the formula $C_5H_{10}O_5$ (Fig. 2.24).

Ribose is a monosaccharide (a carbohydrate/sugar—see above) containing five carbon atoms. In the conventional numbering scheme for monosaccharides, the carbon atoms are numbered from C-1' (in the aldehyde group CH_2OH at the bottom in Fig. 2.24) to C-5'.

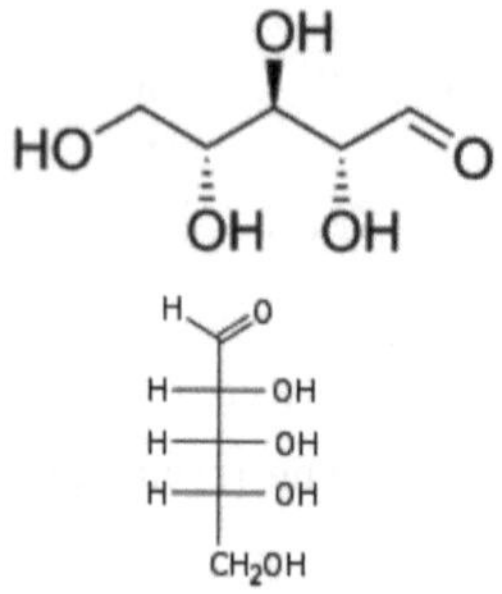

Figure 2.24 Ribose ($C_5H_{10}O_5$). Source: Wikipedia.

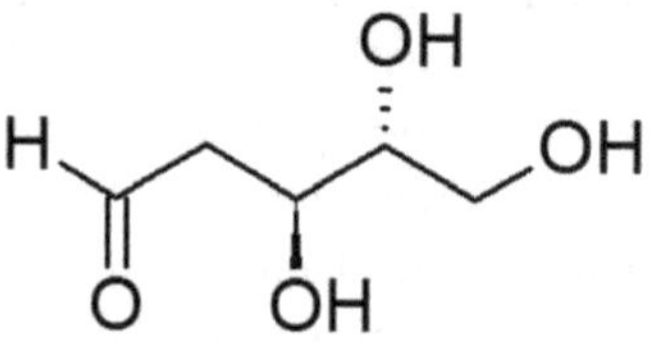

Figure 2.25 Deoxyribose ($C_5H_{10}O_4$). Source: Wikipedia.

Deoxyribose is a monosaccharide that is derived from ribose by a deoxy sugar, meaning that it is derived from the sugar ribose by loss of an oxygen atom. Its structure is shown in Fig. 2.25.

Ribose and deoxyribose derivatives play an important role in biology. The DNA molecule, the repository of genetic information in life, consists of a long chain of deoxyribose-containing units linked via phosphate groups. Before discussing further details, let's define the terms "nucleosides" and "nucleotides."

Nucleosides consist of a nucleobase (A, G, T, U, C) bound to a ribose or deoxyribose sugar via a glycosidic bond, defined as a type of covalent bond that joins a carbohydrate (sugar) molecule to the hydroxyl group of another organic compound such as an alcohol.

Nucleotides are composed of a nucleobase (nitrogenous base), a five-carbon sugar (either ribose or deoxyribose), and one or more phosphate groups. Without the phosphate group, the nucleobase and sugar compose a *nucleoside*.

Figure 2.26 illustrates the formation of nucleotides.

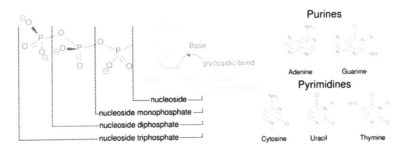

Figure 2.26 Overview of common nucleic acid constituents. Purines A and G and pyrimidines C, U, and T are connected to ribose or deoxyribose via a glycosidic bond and then phosphorylated to form nucleotides. Source: Wikipedia.

In the standard nucleic acid nomenclature, a DNA nucleotide consists of a deoxyribose molecule with an organic base (purine or pyrimidine) attached to the 1′ ribose carbon. The 5′ hydroxyl of each deoxyribose unit is replaced by a phosphate (forming a nucleotide) that is attached to the 3′ carbon of the deoxyribose in the preceding unit.

Ribonucleotides are nucleotides in which the sugar is ribose. Deoxyribonucleotides are nucleotides in which the sugar is deoxyribose.

In DNA and RNA, the *phosphodiester bond* $(PO_4)^{3-}$, as shown in Fig. 2.27, is the linkage between the 3′ carbon atom of one sugar molecule and the 5′ carbon atom of another, deoxyribose in DNA and ribose in RNA. Being groups of covalent bonds, phosphodiester bonds are very strong. They are central to most life on earth, as they make up the "backbone" of the strands of DNA.

Frequently, the four nucleotides A, T, G, and C making DNA are schematically depicted as in Fig. 2.28.

The four nucleotides all have the same pair of "hooks," namely, 5 prime, or 5′ phosphoryl, and 3 prime, or 3′ hydroxyl, depending on their respective positions in the deoxyribose sugar molecule shown in Fig. 2.28. Those hooks are used when nucleotides are linked with each other in a strand of DNA.

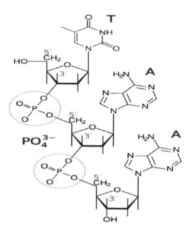

Figure 2.27 The phosphodiester bond between nucleotides. Source: Wikipedia.

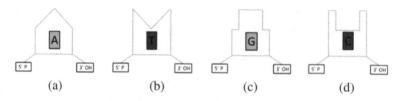

Figure 2.28 (a–d) DNA nucleotides A (a purine), T (a pyrimidine), G (a purine), and C (a pyrimidine).

Figure 2.29 shows that after linkage, the resulting DNA strand TGACT, internally connected by means of phosphodiester bonds, exhibits an unused phosphoryl group $(PO_3)^{2-}$ at the 5′ end and an unused hydroxyl group $(OH)^-$ at the 3′ end.

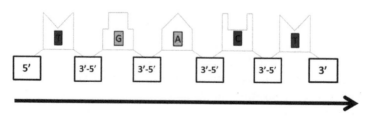

Figure 2.29 The DNA sequence TGACT is defined as the succession of nucleotides from the 5′ to the 3′ terminus.

Finally, Fig. 2.30 shows how double-stranded DNA is formed by adding A to T, C to G, T to A, G to C, etc.

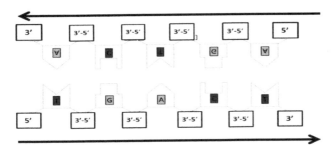

Figure 2.30 Formation of double-stranded DNA by complementarity of nucleotides.

Ligation is defined as the covalent linking of two ends of DNA or RNA molecules, most commonly done using enzymes such as DNA ligase or RNA ligase, respectively.

To summarize, nucleic acids are polymeric macromolecules made from (monophosphate) nucleotide monomers. In DNA, the purine bases are adenine and guanine, while the pyrimidines are thymine and cytosine. RNA uses uracil in place of thymine. *Adenine always pairs with thymine by two hydrogen bonds, while guanine pairs with cytosine through three hydrogen bonds*, each due to their unique structures.

Table 2.1 shows what we just discussed.

ATP $(C_{10}H_{16}N_5O_{13}P_3)$ is a nucleoside triphosphate used in cells as a coenzyme, a "helper molecule" that assists in biochemical transformations (Fig. 2.31). It is often called the "molecular unit of currency" of intracellular energy transfer. ATP transports chemical energy within cells for metabolism.

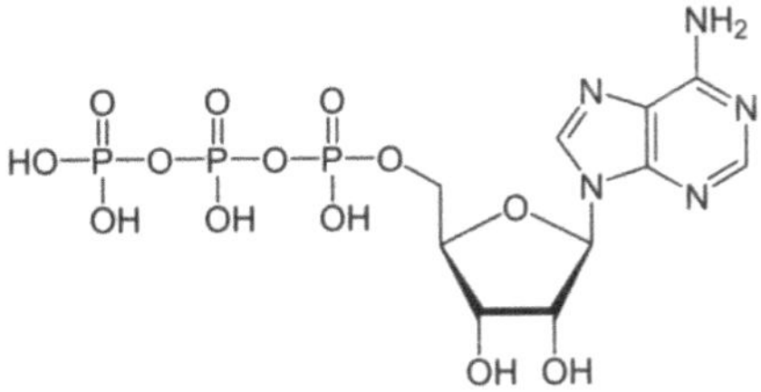

Figure 2.31 ATP $(C_{10}H_{16}N_5O_{13}P_3)$. Source: Wikipedia.

Table 2.1 RNA's and DNA's purine and pyrimidine bases, nucleosides, and nucleotides nucleoside triphosphates (such as ATP-Adenosine-5'-Tri-Phosphate) play an important role in metabolism

Base	RNA		DNA	
	Nucleoside	Nucleotide	Nucleoside	Nucleotide
PURINES:				
Adenine	Adenosine	Adenosine monophosphate (AMP)	Deoxyadenosine	Deoxyadenosine monophosphate (dAMP)
Guanine	Guanosine	Guanosine monophosphate (GMP)	Deoxyguanosine	Deoxyguanosine monophosphate (dGMP)
PYRIMIDINES:				
Cytosine	Cytidine	Cytidine monophosphate (CMP)	Deoxycytidine	Deoxycytidine monophosphate (dCMP)
Uracil	Uridine	Uridine monophosphate (UMP)	Not in DNA	
Thymine	Not in RNA		Deoxythymidine	Deoxythymidine monophosphate (dTMP)

Metabolic processes that use ATP as an energy source convert it back into its precursors. ATP is therefore continuously recycled in organisms: the human body, which on average contains only 250 g of ATP, turns over its own body weight equivalent in ATP each day.

In living cells, ATP is generated within the *mitochondrion*, a membrane-enclosed structure found in most eukaryotic cells (the cells that make up plants, animals, and many other forms of life). Mitochondria range from 0.5 to 1.0 μm (or 10^{-6} meters) in diameter and are acting as "cellular power plants" because they generate most of the cell's supply of ATP, used as a source of chemical energy. In addition to supplying cellular energy, mitochondria are involved in other tasks such as signaling, cellular differentiation, and cell death, as well as the control of the cell cycle and cell growth.

2.4 Amino Acids

Amino acids are biologically important organic compounds made from amine ($-NH_2$) and carboxylic acid ($-COOH$) functional groups, along with a side chain specific to each amino acid. A side chain is a chemical group that is attached to a core part of the molecule called main chain or backbone. The placeholder R is often used as a generic placeholder for alkyl (saturated hydrocarbon) group side chains in chemical structure diagrams. In Fig. 2.32, the central carbon atom to which the

- amine functional group ($-NH_2$)
- carboxylic acid functional group ($-COOH$)
- side chain R

are all attached is often referred to as the α-carbon.

Elements other than C,H,O, and N may be found in the side chains of certain amino acids. There are about 500 known amino acids but only 22 are of relevance as building blocks of proteins (also called proteinogenic amino acids).

In the form of proteins, amino acids comprise the second largest component (after water) of human muscles, cells, and other tissues. Outside proteins, amino acids perform critical roles in processes such as neurotransmitter transport and biosynthesis.

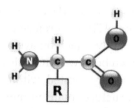

Figure 2.32 The generic structure of an amino acid, with R denoting the side chain. Source: Wikipedia.

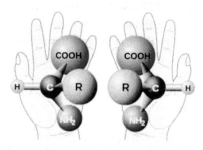

Figure 2.33 Enantiomers of amino acids. Source: Wikipedia.

Proteinogenic amino acids are produced by cellular machinery and guided by the genetic code (to be discussed below) of organisms.

At this point it is necessary to introduce the concept of *chirality*: Amino acids exist in two forms called enantiomers (Fig. 2.33) and including both left-handed and right-handed structures. A chiral molecule is a type of molecule that has a nonsuperposable mirror image caused by the presence of an asymmetric carbon atom.

Proteinogenic amino acids are all L-stereoisomers ("left handed" isomers) as compared to glyceraldehyde, a monosaccharide with the chemical formula $C_3H_6O_3$ that is used as the reference structure in the D/L system. Glyceraldehyde is the simplest of all common aldose sugars. Its respective "D" (dextrorotatory) and "L" (levorotatory) forms are shown in Fig. 2.33 on the left and right sides, respectively.

There are 22 standard amino acids, but only 21 are found in eukaryotes.[31] Of the 22, selenocysteine and pyrrolysine are incorporated into proteins by distinctive biosynthetic mechanisms. The other 20 are directly encoded by the universal genetic code. Humans can only synthesize 11 of these 20 from each other or

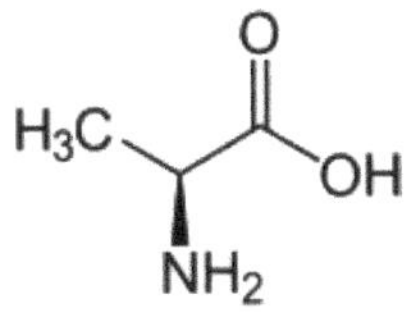

Figure 2.34   Alanine. Source: Wikipedia.

from other molecules of intermediary metabolism. The other 9 must be consumed (usually as their protein derivatives) in the diet and so are thus called essential amino acids. The *essential amino acids* are *histidine, isoleucine, leucine, lysine, methionine, phenylalanine, threonine, tryptophan, and valine.*

Here are brief descriptions of the proteinogenic amino acids, in alphabetical order of their respective single letter codes (see also Table 2.2).

Alanine (Ala/A): Molecular formula $C_3H_7NO_2$ (hydrophobic side chain $-CH_3$). Ala (Fig. 2.34) is one of the 22 amino acids that are used by cells to synthesize proteins. Its codons[32] are GCU, GCC, GCA, and GCG. It is classified as a nonpolar amino acid. Alanine is a nonessential amino acid, that is, it can be manufactured by the human body and does not need to be obtained directly through the diet. Alanine is found in a wide variety of foods but is particularly concentrated in meats.

Cysteine (Cys/C): Molecular formula $C_3H_7NO_2S$ (thiol side chain $-CH_2SH$). Although classified as a nonessential amino acid, in rare cases, cysteine may be essential for infants, the elderly, and individuals with certain metabolic disease. It is found in high-protein foods, including some plants like red peppers, broccoli, etc. Cysteine is coded for by the UGU and UGC (translation of messenger RNA [mRNA] molecules to produce polypeptides). Cysteine is required by sheep in order to produce wool and has been proposed as a preventative or antidote for some of the negative effects of alcohol because it counteracts poisonous effects of acetaldehyde (C_2H_4O), the major by-product of alcohol metabolism.

Aspartic acid (Asp/D): Molecular formula $C_4H_7NO_4$ (side chain $-CH_2COOH$). Asp is one of the 22 amino acids that are used by cells

Table 2.2 Proteinogenic amino acids

Proteinogenic amino acid	3-letter abbrev.	1-letter form	side chain	Polar	pH
Alanine	Ala	A	–CHS	–	–
Cysteine	Cys	C	–CH2SH	–	acidic
Aspartic acid	Asp	D	–CH2COOH	X	acidic
Glutamic acid	Glu	E	–CH2CH2COOH	X	acidic
Phenylalanine	Phe	F	–CH2C6H5	–	–
Glycine	Gly	G	–H		
Histidine	His	H	–CH2–C3H3N2	X	weak basic
Isoleucine	lie	1	–CH(CH3)CH2CH3	–	–
Lysine	Lys	K	–(CH2)4NH2	X	basic
Leucine	Leu	L	–CH2CH(CH3)2	–	–
Methionine	Met	M	–CH2CH2SCH3	–	–
Asparagine	Asn	N	–CH2CONH2	X	weak basic
Pyrrolysine	Pyl	O	–(CH2)4NHCOC4H5NCH3	X	weak basic
Proline	Pro	P	–CH2CH2CH2–	–	–
Glutamine	Gln	Q	–CH2CH2CONH2	X	weak basic
Arginine	Arg	R	–(CH2)3NH–C(NH)NH2	X	strongly basic
Serine	Ser	S	–CH2OH	X	weak acidic
Threonine	Thr	T	–CH(OH)CH3	–	weak acidic
Selenocysteine	Sec	U	–CH2SeH	–	acidic
Valine	Val	V	–CH(CH3)2	–	–
Tryptophan	Trp	w	–CH2C8H6N	X	weak basic
Tyrosine	Tyr	Y	–CH2–C6H40H	X	weak acidic

to synthesize proteins. Its codons are GAU and GAC. It is classified as acidic with a negatively charged side chain. The salt of aspartic acid is known as aspartate, a metabolite in the urea cycle.

Glutamic acid (Glu/E): Molecular formula $C_5H_9NO_4$ (side chain –CH_2CH_2COOH). Glu is one of the 22 amino acids that are used by cells to synthesize proteins. Its codons are GAA and GAG. It is classified as acidic with a negatively charged side chain. Glutamate (a salt of glutamic acid) is an important neurotransmitter that plays a key role in long-term potentiation and is important for learning and memory.

Phenylalanine (Phe/F): Molecular formula $C_9H_{11}NO_2$ (side chain –$CH_2C_6H_5$). Phe is one of the 22 amino acids that are used by cells to synthesize proteins. Its codons are UUU and UUC. It is classified as nonpolar because of the hydrophobic nature of the benzyl side chain. Phenylalanine is a precursor for tyrosine, to be discussed below, in addition to the signaling molecules dopamine, norepinephrine (noradrenaline), and epinephrine (adrenaline), and the skin pigment melanin. Phenylalanine is found naturally in the breast milk of mammals.

Glycine (Gly/G): Molecular formula $C_2H_5NO_2$ (hydrophobic side chain –H). Gly is the smallest of the 20 amino acids commonly found in proteins and is one of the 22 amino acids that are used by cells to synthesize proteins. Its codons are GGU, GGC, GGA, and GGG. It is unique among amino acids as not being chiral and hence not optically active.

Histidine (His/H): Molecular formula $C_6H_9N_3O_2$ (imidazole side chain $-CH_2-C_3H_3N_2$). His is one of the 22 amino acids that are used by cells to synthesize proteins. Its codons are CAU and CAC. It is classified as an amino acid with a positively charged side chain. It is a precursor to histamine, a compound involved in local immune responses, regulating physiological function in the gut and acting as a neurotransmitter.

Isoleucine (Ile/I): Molecular formula $C_6H_{13}NO_2$ (side chain $-CH(CH_3)CH_2CH_3$). Ile is one of the 22 amino acids that are used by cells to synthesize proteins. Its codons are AUU, AUC, and AUA. It is classified as an essential amino acid with hydrophobic side chain. Isoleucine must be ingested, usually as part of a protein. It was first discovered in hemoglobin.

Lysine (Lys/K): Molecular formula $C_6H_{12}N_2O$ (side chain $-(CH_2)_4$ NH_2). Lys is one of the 22 amino acids that are used by cells to synthesize proteins. Its codons are AAA and AAG. It is an essential amino acid for humans and is classified as an amino acid with a positively charged side chain, along with His and Arg. Lysine is not synthesized in animals; hence it must be ingested as lysine or lysine-containing proteins. In plants and bacteria, it is synthesized from aspartic acid (aspartate). Good sources of lysine are high-protein foods such as eggs, meat, soy, beans and peas, cheese, and certain fish (e.g., cod).

Leucine (Leu/L): Molecular formula $C_6H_{11}NO$ (side chain $-CH_2CH(CH_3)_2$). Leu is one of the 22 amino acids that are used by cells to synthesize proteins. Its codons are UUA, UUG, CUU, CUC, CUA, and CUG. It is classified as an amino acid with hydrophobic side chain. Leucine is utilized in the liver, adipose[33] tissue, and muscle tissue. Leucine is the only dietary amino acid that has the capacity to stimulate muscle protein synthesis. Leucine is an essential amino acid, meaning that the human body cannot synthesize it, and it therefore must be ingested. Sources are soy, beef, peanuts, fish, etc.

Methionine (Met/M): Molecular formula C_5H_9NOS (side chain $-CH_2CH_2SCH_3$). Met is one of the 22 amino acids that are used by cells to synthesize proteins. Its codon is AUG. This essential amino acid is classified as nonpolar. Together with cysteine, methionine is one of two sulfur-containing proteinogenic amino acids. Improper conversion of methionine can lead to atherosclerosis.

Asparagine (Asn/N): Molecular formula $C_4H_6N_2O_2$ (polar uncharged side chain $-CH_2CONH_2$). Asn is one of the 22 amino acids that are used by cells to synthesize proteins. Its codons are AAU and AAC. It is classified as a polar amino acid. It is not essential; hence it can be generated by humans. It can also be obtained from both animal and plant sources. It was first found in asparagus juice (1806) and was the first amino acid to be isolated.

Pyrrolisine (Pyl/O): Molecular formula $C_{12}H_{21}N_3O_3$ (side chain $-(CH_2)_4NHCOC_4H_5NCH_3$). It is considered the 22nd proteinogenic amino acid. Genetic code: UAG is normally the stop codon but encodes pyrrolysine if a pyrrolysine insertion sequence (PYLIS) element is present. Pyl is used by some methanogenic archaea and one known bacterium in enzymes that are part of their methane-producing metabolism. It is similar to lysine but with an added pyrroline ring linked to the end of the lysine side chain. It forms part of an unusual genetic code in these organisms and is considered the 22nd proteinogenic amino acid.

Proline (Pro/P): Molecular formula C_5H_7NO (side chain $-CH_2CH_2CH_2$). Pro is one of the 22 amino acids that are used by cells to synthesize proteins. Its codons are CCU, CCC, CCA, and CCG. Pro can be synthesized by the human body. It is unique among the protein-forming amino acids in that the amine nitrogen is bound to not one but two alkyl groups, thus making it a secondary amine.

Glutamine (Gln/Q): Molecular formula $C_5H_8N_2O_2$ (polar uncharged side chain $-CH_2CONH_2$). Glu is one of the 22 amino acids that are used by cells to synthesize proteins. Its codons are CAA and CAG. It is classified as a polar amino acid. It is not essential, but may become conditionally essential in certain situations, including intensive athletic training or certain gastrointestinal (GI) disorders. In human blood, glutamine is the most abundant free amino acid. The most eager consumers of glutamine are the cells of intestines, the kidney cells for the acid–base balance, activated immune cells, and many cancer cells. Glutamine is also known to have various effects in reducing healing time after operations. Dietary sources of L-glutamine include beef, chicken, fish, eggs, milk, dairy products, wheat, cabbage, beets, beans, spinach, and parsley.[34]

Arginine (Arg/R): Molecular formula $C_6H_{14}N_4O_2$ (positively charged side chain $-(CH_2)_3NH-C(NH)NH_2$). Arg is one of the 22 amino acids that are used by cells to synthesize proteins. Its codons are CGU, CGC, CGA, CGG, AGA, and AGG. In mammals, arginine is classified as a semiessential or conditionally essential amino acid, depending on the developmental stage and health status of the individual. Preterm infants are unable to synthesize or create arginine internally, making the amino acid nutritionally essential for them. There are some conditions that put an increased demand on the body for the synthesis of L-arginine, including surgical or other trauma, sepsis, and burns.[35]

Serine (Ser/S): Molecular formula $C_3H_5NO_2$ (side chain $-CH_2OH$). Ser is one of the 22 amino acids that are used by cells to synthesize proteins. Its codons are UCU, UCC, UCA, UCG, AGU, and AGC. It is classified as a polar amino acid. It is not essential to the human diet, since it is synthesized in the body from other metabolites (including glycine). Serine was first obtained from silk protein and its name is derived from *sericum* (Latin for *silk*). Serine is important in metabolism. It participates in the biosynthesis of purines and pyrimidines and plays an important role in the catalytic function of many enzymes.

Threonine (Thr/T): Molecular formula $C_4H_7NO_2$ (side chain $-CH(OH)CH_3$). Thr is one of the 22 amino acids that are used by cells to synthesize proteins. Its codons are ACU, ACC, ACA, and ACG.

It is classified as an essential amino acid with polar uncharged side chain. Along with serine, threonine is bearing an alcohol group. Threonine was discovered as the last of the 20 common proteinogenic amino acids in the 1930s. Foods high in threonine include cottage cheese, poultry, fish, meat, lentils, and sesame seeds.

Selenocysteine (Sec/U): Molecular formula C_3H_5NOSe (side chain $-CH_2\mathbf{SeH}$). Unlike other amino acids present in biological proteins, selenocysteine is not coded for directly in the genetic code. UGA is normally the stop codon, but encodes selenocysteine if a SECIS (selenocysteine insertion sequence) element is present. It is the 21st proteinogenic amino acid and exists naturally in all kingdoms of life as a building block of selenoproteins.

Valine (Val/V): Molecular formula C_5H_9NO (side chain $-CH(CH_3)_2$). Val is one of the 22 amino acids that are used by cells to synthesize proteins. Its codons are GUU, GUC, GUA, and GUG. It is classified as an essential nonpolar amino acid with hydrophobic side chain. Dietary sources are meats, dairy products, soy products, beans, and legumes. Valine is named after the perennial flowering plant valerian. In sickle-cell disease, the substitution of hydrophylic glutamic acid in hemoglobin by valine is leading to abnormal aggregation, due to Val's hydrophobic side chain.

Tryptophan (Trp/W): Molecular formula $C_{11}H_{10}N_2O$ (side chain $-CH_2\mathbf{C_8H_6N}$). Trp is one of the 22 amino acids that are used by cells to synthesize proteins. Its codon is UGG. It is classified as an essential amino acid with hydrophobic side chain. Tryptophan functions as a biochemical precursor to serotonin (a neurotransmitter), niacin (also known as vitamin B_3 or nicotinic acid), and auxin. Auxins have a cardinal role in the plant's life cycle and are essential for plant body development.

Tyrosine (Tyr/Y): Molecular formula $C_9H_9NO_2$ (side chain $-CH_2-C_6H_4OH$). Tyr is one of the 22 amino acids that are used by cells to synthesize proteins. Its codons are UAU and UAC. It is classified as a nonessential amino acid with a polar hydrophobic side chain. The name tyrosine is derived from the Greek word for *cheese*, where it was first found in the protein casein. Casein is making up 80% of the proteins in **cow milk** and between 20% and 45%

of the proteins in human milk. Tyrosine occurs in proteins that are part of cellular signal transduction processes. It functions as a receiver of phosphate groups that are transferred by way of protein kinases (enzymes that modify other proteins by chemically adding phosphate groups to them via phosphorylation). Tyrosine is a precursor to neurotransmitters and is also the precursor to the pigment **melanin**.

2.5 Proteins

The name "protein" (Greek for *primary* or *in the lead*) was coined by the eminent Swedish chemist Jöns Jacob Berzelius (1779–1848), who is also credited with originating the chemical terms "catalysis," "polymer," "isomer," and "allotrope." He proposed the name "protein" because the material seemed to be the primary substance of animal nutrition.[36]

Proteins are large biological molecules consisting of one or more chains of amino acids. Proteins perform a great number of important functions in biology. Proteins (as enzymes) are catalyzing metabolic reactions, they are involved in the replication of DNA, they control cellular processes and are transporting molecules from one location to another. In blood and in all tissues of the body, in muscles, nerves, and skin, proteins form an essential and functional constituent. It is the chemical individuality of proteins that is responsible for the species' differences among all living things. Proteins differ from one another primarily in their sequence of proteinogenic amino acids. The sequence of amino acids is then giving rise to the folding of the protein into a specific 3D structure that determines its activity.

As discussed above (Fig. 2.32), all proteinogenic amino acids possess common structural features. They include an α-carbon to which an amino group, a carboxyl group, and a variable side chain are all bonded.

When linked together in a protein, the amino acids are forming (covalent) peptide bonds among each other. The carboxyl group of an amino acid is reacting with the amino group of a second amino acid, forming a peptide bond and releasing a water molecule (Fig. 2.35).

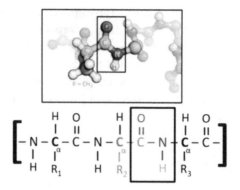

Figure 2.35 Peptide bond between adjacent amino acids. If R = CH$_3$, the amino acid is alanine. Source: Wikipedia.

Figure 2.36 Resonance structures—single and double (C–N) bond options of a peptide bond. Source: Wikipedia.

Once linked in the protein chain, an individual amino acid is called a residue, and the linked series of carbon, nitrogen, and oxygen atoms are known as the main chain or protein backbone. Note that the peptide bond has two resonance forms (Fig. 2.36) that limit rotation around its axis, so that the α-carbons are roughly coplanar.

The end of the protein with a free carboxyl group is known as the *C-terminus*, or carboxy terminus, whereas the end with a free amino group is known as the *N-terminus*, or amino terminus.

For clarity, let's look at the (sometimes overlapping) definitions of the terms "protein," "peptide," and "polypeptide":

- The term "protein" is generally used to refer to the complete biological molecule in a stable conformation.
- The term "peptide" is generally reserved for short amino acid oligomers that may lack a stable 3D structure. Peptides are assumed to contain no more than 20–30 residues.

- The term "polypeptide" can refer to any single linear chain of amino acids, usually regardless of length. It usually implies an absence of a defined conformation.

To understand the role and function of proteins we need to understand their structure. There are four aspects to a protein's structure:

1. *Primary structure*: The sequence of amino acids listed as a linear chain with letters or as three-letter abbreviations, as shown in Table 2.3. A protein is a polyamide. The sequence of amino acids was discovered by Frederick Sanger, who won his first Nobel Prize in 1958 for his work on the structure of proteins, especially that of insulin.[37] When studying insulin, a physiologically important hormone that is used to break down glucose in our body, Sanger had to sort out how the 51 amino acids of this protein are linked together. After 15 years of painstaking work he was able to solve the amino acid sequence problem. We now know that the sequence of a protein is unique to that protein, and that it defines the structure and function of the protein.

2. *Secondary structure*: Regularly repeating local structures stabilized by hydrogen bonds, as first discovered by Linus Pauling.[38] Pauling received the Nobel Prize for chemistry in 1954 for his research into the nature of the chemical bond and its application to the elucidation of the structure of complex substances. In 1948, he discovered how a polypeptide chain can fold into an α-helix while maintaining planar peptide bonds. His discovery was compatible with the results of X-ray diffraction studies, described in more detail below. A good example of a secondary structure is the α-helix (Fig. 2.37).

Another example of a secondary structure is the beta sheet consisting of beta strands connected laterally by at least two or three backbone hydrogen bonds, forming a generally twisted, pleated sheet. A beta strand (also β-strand) is a stretch of typically 3 to 10 amino acids with backbone in an almost fully extended conformation, as illustrated below (Fig. 2.38).

Figure 2.37 An α-helix with hydrogen bonds (yellow dots). Source: Wikipedia.

3. *Tertiary structure*: The overall shape of a single protein molecule, including the spatial relationship of the secondary structures to one another. The tertiary structure is generally stabilized by nonlocal interactions, and other modifications that may occur after the initial formation. The term "tertiary structure" is often used as synonymous with the term "fold." The tertiary structure is what controls the basic function of the protein.

Four factors are responsible for the tertiary structure of proteins:

(a) Disulfide linkages: If the folding of a protein brings two cysteine residues together, the two –SH side chains can be oxidized to form a covalent S–S bond. These disulfide bonds cross-link the polypeptide chain.

(b) Hydrogen bonding: In addition to the hydrogen bonds between peptide bonds that gives rise to the secondary structure of the protein, hydrogen bonds can form between amino acid side chains.

(c) Electrostatic interactions/ionic bonding: The structure of a protein can be stabilized by the force of attraction between

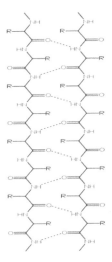

Figure 2.38 Illustration of the hydrogen-bonding patterns, represented by dotted lines, in a parallel beta sheet. Oxygen atoms are colored red and nitrogen atoms colored blue. Source: Wikipedia.

amino acid side chains of opposite charge, such as the $-NH3+$ side chain of Leu and the $-CO2-$ side chain of Asp.

(d) Hydrophobic interactions: Proteins often fold so that the hydrophobic side chains of the amino acids Gly, Ala, Val, Leu, Ile, Pro, Met, Phe, and Trp are buried within the protein, where they can interact to form hydrophobic pockets. These hydrophobic interactions stabilize the structure of the protein.

4. *Quaternary structure*: The structure formed by several protein molecules (polypeptide chains), usually called protein subunits in this context, which function as a single protein complex (Fig. 2.39).

Proteins are not entirely rigid molecules. In addition to these levels of structure, proteins may shift between several related structures while they perform their functions. In the context of these functional rearrangements, these tertiary or quaternary structures are usually referred to as conformations, and transitions between them are called *conformational changes*. Such changes are often induced by the binding of a substrate molecule to an enzyme's

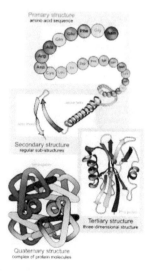

Figure 2.39 Protein structure, from primary to quaternary. Source: Wikipedia.

active site, or the physical region of the protein that participates in chemical catalysis. In solution proteins also undergo variation in structure through thermal vibration and the collision with other molecules.

The key experimental technique used to elucidate complex 3D (protein) structures is *X-ray crystallography*:

This method is based on the diffraction of X-rays caused by atoms in a crystalline structure: the atoms cause a beam of X-rays to diffract into many specific directions. By measuring the angles and intensities of these diffracted beams, it is possible to reconstruct a 3D picture of the density of electrons within the crystal. From this electron density, the mean positions of the atoms in the crystal can be determined, as well as their chemical bonds, their disorder and various other information. The basic idea of X-ray crystallography was first developed in 1912 by Max von Laue.[39] However, the mathematical foundation of the analysis of crystal structure by means of X-rays was first fully worked out by William Bragg and his son Lawrence in 1913–1914. This was recognized by the award of the Nobel Prize jointly to father and son in 1915.[40]

About 100 years later, the scientific field created by von Laue and by father and son Bragg has yielded a number of significant discoveries, most recently the elucidation of the structure and function of the ribosome by Venkatraman Ramakrishnan, Thomas A. Steitz, and Ada E. Yonath in 2009.[41] We will return to this topic when discussing the genetic code and its application to the fabrication of proteins by translation of DNA sequences in the ribosome.

What sometimes prevents X-ray crystallography from being applied to protein structure elucidation is the need to generate crystals, a tricky experimental activity.[42] Frequently, conditions unique to each protein must be obtained for a successful crystallization. Still, the number of solved protein structures in the Protein Data Bank (PDB)[43] has grown from seven in 1971 to over 20,000 in 2003 and about 84,000 in 2013. On May 13, 2014, a new milestone, 100,000 entries, was reached!

Let's discuss a few examples of beautiful protein structures that are essential for life.

Insulin is a peptide hormone, produced by beta cells in the pancreas, and is central to regulating carbohydrate and fat metabolism in the body. It causes cells in the skeletal muscles and fat tissue to absorb glucose from the blood. Insulin is a tiny protein, composed of only 51 amino acids. It moves quickly through the blood and is easily captured by receptors on cell surfaces, delivering its message. However, small proteins such as insulin pose a challenge to cells because they do not fold easily into a stable structure. Insulin in its more stable form is produced and stored in the body as a hexamer (a unit of six insulin molecules, see Fig. 2.40), while the active form is the monomer.

The hexamer is an inactive form with long-term stability, which serves as a way to keep the highly reactive insulin protected, yet readily available. The hexamer–monomer conversion is one of the central aspects of insulin formulations for injection.

Insulin is a very old protein that may have originated more than 1 billion years ago. Within vertebrates, the amino acid sequence of insulin is strongly conserved. Porcine insulin differs from human in only one amino acid residue, bovine insulin in three. Even insulin from some species of fish is similar enough to the human form to be

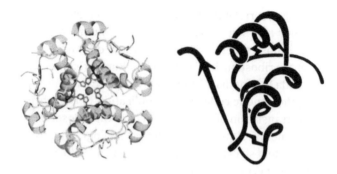

Figure 2.40   Insulin hexamer (stable form) and sketch of insulin monomer. Source: Wikipedia.

clinically effective. The strong homology seen in the insulin sequence of diverse species suggests that it has been conserved across much of animal evolutionary history.

Glucagon is a peptide hormone, produced by alpha cells of the pancreas, that raises blood glucose levels. Its effect is opposite that of insulin, which lowers blood glucose levels. The pancreas releases glucagon when blood sugar (glucose) levels fall too low. Glucagon causes the liver to convert stored glycogen into glucose, which is then released into the bloodstream.

Hemoglobin is the iron-containing oxygen transport metalloprotein in the red blood cells of all mammals. Hemoglobin in the blood carries oxygen from the respiratory organs (lungs or gills) to the rest of the body (i.e., the tissues) where it releases the oxygen to burn nutrients to provide energy to power the functions of the organism in the process called metabolism. In mammals, the protein makes up about 96% of the red blood cells' dry content (by weight) and around 35% of the total content (including water). Hemoglobin has an oxygen-binding capacity that increases the total blood oxygen 70-fold compared to dissolved oxygen in blood.

In 1959, Max Perutz determined the molecular structure of myoglobin (similar to hemoglobin) by X-ray crystallography. Together with John Kendrew, he was awarded the 1962 Nobel Prize for chemistry.[44]

Figure 2.41 Hemoglobin. The α and β subunits in red and blue; heme groups in green. Source: Wikipedia.

The role of hemoglobin in the blood was elucidated by French physiologist Claude Bernard. The name hemoglobin is derived from the words *heme* and *globin*, reflecting the fact that each subunit of hemoglobin is a globular protein with an embedded heme group. Each heme group contains one iron atom, which can bind one oxygen molecule through ion-induced dipole forces. The most common type of hemoglobin in mammals contains four such subunits. Hence, human hemoglobin molecule can bind up to four oxygen molecules. The hemoglobin structure is shown in Fig. 2.41.

Ribonuclease is an RNA-cleaving enzyme stabilized by four disulfide bonds. It was used in Anfinsen's seminal research on protein folding, which led to the concept (Anfinsen's dogma[45]) that a protein's 3D structure was determined by its amino acid sequence. Ribonuclease (commonly abbreviated RNase) catalyzes the degradation of RNA into smaller components. RNases play a critical role in many biological processes, including angiogenesis, the physiological process through which new blood vessels form from pre-existing vessels. They also play a role in flowering plants. The structure is shown in Fig. 2.42.

Immunoglobulin (also referred to as *antibody*) is a large Y-shaped protein produced by plasma cells that is used by the immune system to identify and neutralize foreign objects such as bacteria and viruses. The antibody recognizes a unique part of the foreign

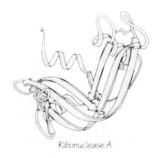

Ribonuclease A

Figure 2.42   Structure of ribonuclease A. Source: Wikipedia.

target, called an *antigen*. Each tip of the "Y" of an antibody contains a paratope (a structure analogous to a lock) that is specific for one particular epitope (similarly analogous to a key) on an antigen, allowing these two structures to bind together with precision, as illustrated in Fig. 2.43. The terms "antibody" and "immunoglobulin" are often used interchangeably (although there is a more general immunoglobulin superfamily that includes cell surface and soluble proteins involved in the recognition, binding, or adhesion processes of cells). Antibodies are typically made of basic structural units— each with two large heavy chains and two small light chains. The membrane-bound form of an antibody may be called a surface immunoglobulin (sIg) or a membrane immunoglobulin (mIg). It is part of the B cell receptor (BCR), which allows a B cell to detect when a specific antigen is present in the body and triggers B cell activation.

Glycoproteins are proteins that contain oligosaccharide chains (glycans) covalently attached to polypeptide side chains. An oligosaccharide (from the Greek *oligos*, for *a few*, and *sacchar*, for *sugar*) is a saccharide polymer containing a small number (typically three to nine simple sugars—monosaccharides). Oligosaccharides are commonly found on the plasma membrane of animal cells. When consumed, the undigested portion serves as food for the intestinal microflora, stimulating (suppressing) specific bacterial groups. The carbohydrate is attached to the protein in a process known as glycosylation. Secreted extracellular proteins are often glycosylated. Glycoproteins are important integral membrane proteins, where

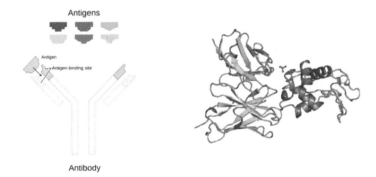

Figure 2.43   Schematic diagram of an antibody and antigens. Light chains are in lighter blue and orange, heavy chains in darker blue and orange. Antibody with an enzyme (hen egg lysozyme) shown to the right. Source: Wikipedia.

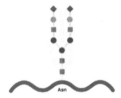

Figure 2.44   *N*-linked protein glycosylation at asparagine residues. Source: Wikipedia.

they play a role in cell–cell interactions. There are two types of glycosylation:

- In *N*-glycosylation, shown in Fig. 2.44, the addition of sugar chains can happen at the amide nitrogen on the side chain of the amino acid asparagine.
- In O-glycosylation, the addition of sugar chains can happen on the hydroxyl oxygen on the side chain of hydroxylysine, hydroxyproline, serine, or threonine.

While about 90% of all structures in the PDB have been determined by X-ray crystallography, about 9% of the structures were determined by another method, namely, NMR spectroscopy of proteins.

NMR involves the quantum mechanical properties of the atom's nucleus. These properties depend on the local molecular environment, and their measurement provides a map of how the atoms are linked chemically, how close they are in space, and how rapidly they move with respect to each other. The technology used is similar to magnetic resonance imaging (MRI), but the molecular application is different, appropriate to the change of scale from millimeters (of interest to radiologists using MRI in vivo, that is, in patients) to nanometers, since bonded atoms in a protein are typically a fraction of a nanometer apart. Both MRI and NMR take advantage of the nuclear magnetic moment of atomic nuclei, defined as the total spin of protons and neutrons. The nuclear magnetic moment varies from isotope to isotope of an element. It can only be zero if **both** the numbers of protons and of neutrons are even. An MRI scanner is a device in which the patient lies within a large, powerful magnet (typically with a field strength of at least 1.5 Tesla—more than 25,000 times the earth's magnetic field) where the magnetic field is used to align the magnetization of atomic nuclei in the body, and radio frequency magnetic fields are applied to systematically alter the alignment of this magnetization. On the basis of the fact that all soft tissues contain water and the principle that a hydrogen atom's nucleus behaves like a spinning dipole bar magnet with a north and a south pole, a radio frequency force field that resonates with the target protons will transfer them to a higher-energy state. When the radio frequency pulse is turned off, the excited protons relax back to their lower-energy state and release a small amount of energy that can be measured as a change of electric current. MRI provides good contrast between the different soft tissues of the body, which makes it especially useful in imaging the brain, muscles, the heart, and cancers, as compared to other medical imaging techniques such as computed tomography (CT) or X-rays. To summarize, MRI machines make use of the fact that body tissue contains lots of water, and hence protons (hydrogen nuclei), which will be aligned in a large magnetic field.

NMR is a physical phenomenon in which nuclei in a magnetic field absorb and re-emit electromagnetic radiation. This energy is at a specific resonance frequency (RF) that depends on the strength of the magnetic field and the magnetic properties of the isotope of the atoms. The most commonly studied nuclei are ^{1}H and ^{13}C, although nuclei from isotopes of many other elements (e.g., ^{2}H, ^{6}Li, ^{10}B, ^{11}B, ^{14}N, ^{15}N, ^{17}O, ^{19}F, ^{23}Na, ^{29}Si, ^{31}P, ^{35}P, ^{35}Cl) have been studied as well. A key feature of NMR is that the RF of a particular substance is directly proportional to the strength of the applied magnetic field. It is this feature that is exploited in imaging techniques; if a sample is placed in a nonuniform magnetic field then the resonance frequencies of the sample's nuclei depend on where in the field they are located.

NMR was first described and measured in molecular beams by Isidor Rabi[46] and later applied to liquids and solids by Felix Bloch and Edward Mills Purcell.[47] Rabi, Bloch, and Purcell observed that magnetic nuclei, like ^{1}H and ^{31}P, could absorb RF energy when placed in a magnetic field and when the RF was of a frequency specific to the identity of the nuclei. When this absorption occurs, the nucleus is described as being in resonance. Different atomic nuclei within a molecule resonate at different frequencies for the same magnetic field strength. The observation of such magnetic resonance frequencies of the nuclei present in a molecule allows NMR specialists to discover essential chemical and structural information about the molecule. NMR has a big advantage in that it is sensitive to chemical changes by catalysis. Because it looks at a very-low-energy interaction, NMR is intrinsically hard to do. NMR instruments are large and expensive. Most samples are examined in a solution in water, but methods are being developed to also work with solid samples.

Chapter 3

Genetics and DNA Sequencing

Grau, teurer Freund, ist alle Theorie
und grün des Lebens goldner Baum.
(All theory, dear friend, is gray,
but the golden tree of life springs ever green.)

—says Mephisto in *Faust* by Johann Wolfgang von Goethe

3.1 DNA and the Genetic Code

The year 1953 was a great year for the British Empire: Queen
Elizabeth II was coronated, Edmond Hillary and Tenzing Norgay
reached the top of Mount Everest as part of a successful British
expedition, and two scientists at the University of Cambridge were
able to uncover the structure of the most important "molecule of
life," DNA.

In 1953, on the basis of their deep analysis of X-ray crystallography data, provided first by an image of DNA, labeled "Photo 51," from
Rosalind Franklin[48] in 1952, the young American James Watson and
his British colleague Francis Crick came up with perhaps the most
significant scientific discovery of the 20th century: the 3D molecular

Nanomedicine: Science, Business, and Impact
Michael Hehenberger
Copyright © 2015 Pan Stanford Publishing Pte. Ltd.
ISBN 978-981-4613-76-7 (Hardcover), 978-981-4613-77-4 (eBook)
www.panstanford.com

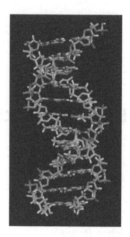

Figure 3.1   DNA double-helix structure. Source: Wikipedia.

structure of DNA is of such great importance because it dictates the general and individual properties of all living matter. DNA is the substance that is the carrier of heredity in higher organisms.

The DNA molecule can also be looked upon as two interwoven spiral staircases, forming one staircase, a double helix. The outside of this staircase consists of phosphate and sugar molecules. The steps are formed by the paired nucleobases adenine (A), guanine (G), thymine (T), and cytosine (C). The purines A and G are paired with the pyramidines T and C, respectively.

As to details regarding their chemical structure and their formation of nucleosides and nucleotides, we refer back to Section 2.3. As illustrated in Table 2.1, nucleic acids are polymers containing a sugar–phosphate "backbone" with the purines and pyrimidines sticking off to the side.

What Watson and Crick discovered was the way the two polymers of DNA are linked together in a double-helix "staircase" (Fig. 3.1).

If it were possible for a person to climb this staircase, this person would discover that A always was coupled to T and C was always coupled to G. The climber, who in molecules of (human) DNA had to ascend 3 billion steps, would see an endless variation in the sequence of A-T, T-A, G-C, and C-G steps. Watson and Crick realized

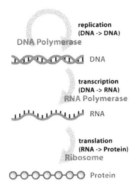

replication
(DNA -> DNA)
DNA Polymerase

DNA

transcription
(DNA -> RNA)
RNA Polymerase

RNA

translation
(RNA -> Protein)
Ribosome

Protein

Figure 3.2   DNA makes RNA makes protein.

that the 3 billion base pairs contained in human DNA determined the genetic code.

The code contained in the DNA of all living organisms—plants, animals and humans—is transferred in cell division, that is, in the normal growth of the organism, and also in the fusion of the sexual cells. In this way the code of the DNA can start and control the development of a new individual that has striking similarities with its parents.

Five years later, Francis Crick took another important step by stating the *central dogma* of molecular biology: "DNA makes RNA makes Proteins" (Fig. 3.2). Genetic information is transcribed into an intermediate molecule called RNA, which is then used to make a protein. Crick stated the dogma as a hypothesis and the scientific community has been working ever since on figuring out the details.

In 1966, as described in the excellent book *Francis Crick: Discoverer of the Genetic Code* by Matt Ridley,[49] Crick and a number of distinguished scientists, including Sidney Brenner,[50] Robert W. Holley, H. Gobind Khorana, and Marshall W. Nirenberg,[51] succeeded in "cracking the genetic code," the secret of life.

In addition to the language of four nucleic acids in DNA that describe our inherited traits (such as the color of our eyes), there also exists a second language in our cells—the language of proteins, written in the 22-letter (amino acid) alphabet of proteins. A single cell contains many thousands of proteins that perform all the

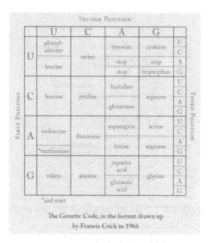

The Genetic Code, in the format drawn up
by Francis Crick in 1966

Figure 3.3   The genetic code. Source: Wikipedia.

chemical reactions required for the normal life of the organism. The synthesis of each protein is governed by rules summarized in the (triplet) genetic code that defines the rules used to translate the DNA alphabet into the alphabet of proteins. While Nirenberg[52] had been the dominant experimenter, Crick had been the dominant theorist and coordinator of research.

Figure 3.3 below shows the genetic code, as drawn by Crick in 1966[53] for the then known 20 amino acids.

Note that U stands for uracil, one of the four nucleobases in RNA. As already explained in Chapter 2, uracil takes the place of thymine in RNA and differs from thymine by lacking a methyl (CH_3) group on its ring.

Evolution has produced 22 types of amino acids that are assembled in the cell from DNA via transcription or derived from food. Every living organism needs those amino acids, but human and animal cells cannot make them all: for example, leucine and other essential amino acids—as discussed in Chapter 2—have to be acquired through food.

It is noteworthy that the genetic code is the first decoded natural language. Cracking this code remains an incredible achievement by the human race.

3.2 From DNA to Proteins and Cells

The human genome is distributed across 23 separate pairs of chromosomes. Of these, 22 pairs are numbered from 1 to 22, in approximate order of size, chromosome #22 being the smallest. The remaining pair includes one X and one Y chromosome for males and two X chromosomes for females. In size, sex chromosome X is larger than #7 but smaller than #8. Sex chromosome Y is very small, even smaller than #22.

There are ~22,000 genes distributed across the 23 chromosomes and the small but important mitochondrial genome (mtDNA), which is tasked with the conversion of chemical energy from food into a form that cells can use.

In most species, including humans, mtDNA is inherited solely from the mother.

Within each gene, there are coding regions, called exons, and large noncoding regions, called introns:

- Exons can be compared to paragraphs in a book, made up of words that are called codons.
- Each codon is composed of letters called nucleotides or bases. There are 3 billion bases in the human genome.
- Human genes are not unique—we share many of our genes with other mammals.
- What makes us uniquely human are small variations in the codons. The deviations are small percentage-wise but highly significant.

At the early stages of our understanding of the human genome, scientists focused entirely on exons and considered introns as irrelevant, sometimes even referring to them collectively as "junk DNA." We have since learned that introns play a much more important role but geneticists are still far away from a complete understanding of their function during the various stages of life processes.

Despite being the most evolved and complex species, our human genome is not the largest of genomes. Plants such as the onion have larger genomes as measured by the total number of included bases.

However, primitive organisms such as bacteria and viruses have much smaller genomes.

For example, the small virus first studied by Sanger[54] during his pioneering work on DNA sequencing, only had 5375 building blocks (or base pairs) in its DNA.

As mentioned above and illustrated in Fig. 3.2, DNA achieves its mission to perpetuate and control life by three mechanisms: *replication, transcription,* and *translation.*

Because of DNA's unique pairing properties (A-T, T-A; G-C, C-G), a single strand of DNA can copy itself by assembling a complementary strand: When the original (single) strand of DNA is intertwined with its complement, the double helix is formed. This property enables DNA to replicate indefinitely while retaining the original information.

Translation is achieved first *transcribing* DNA into RNA, a messenger molecule that is similar to single-stranded DNA (ssDNA) but uses uracil (U) instead of thymine (T). Messenger RNA (mRNA) is then edited by removal of all introns and the splicing together of all exons. The next important step is performed by the *ribosome*, a large and complex molecular machine that incorporates and moves along mRNA and *translates* each 3-letter codon into one letter of a different alphabet, the alphabet of 22 different amino acids, according to the genetic code shown below in an updated version of Fig. 3.4.

During the replication process, each amino acid is attached to the last forming a linear sequence in the same order as determined

	U	C	A	G
U	UUU = phe UUC = phe UUA = leu UUG = leu	UCU = ser UCC = ser UCA = ser UCG = ser	UAU = tyr UAC = tyr UAA = stop UAG = stop	UGU = cys UGC = cys UGA = stop UGG = trp
C	CUU = leu CUC = leu CUA = leu CUG = leu	CCU = pro CCC = pro CCA = pro CCG = pro	CAU = his CAC = his CAA = gln CAG = gln	CGU = arg CGC = arg CGA = arg CGG = arg
A	AUU = ile AUC = ile AUA = ile AUG = met	ACU = thr ACC = thr ACA = thr ACG = thr	AAU = asn AAC = asn AAA = lys AAG = lys	AGU = ser AGC = ser AGA = arg AGG = arg
G	GUU = val GUC = val GUA = val GUG = val	GCU = ala GCC = ala GCA = ala GCG = ala	GAU = asp GAC = asp GAA = glu GAG = glu	GGU = gly GGC = gly GGA = gly GGG = gly

Figure 3.4 Translation of codons into amino acids.[55]

by mRNA. Finally, this chain of amino acids is folding into a 3D distinctive shape known as protein.

As already discussed in Chapter 2, proteins are large and important molecules—almost everything in our body is made of proteins. Ribosomes are made from proteins and from RNA and are therefore often referred to as ribonucleoproteins.

Albert Claude, Christian de Duve, and George Emil Palade were jointly awarded the Nobel Prize for physiology or medicine, in 1974, for the discovery of the ribosomes.[56] Their pioneering work launched a fundamental new field of life sciences, namely, cell biology. Thirty-five years later, the Nobel Prize for chemistry, 2009, was awarded for determining the detailed structure and mechanism of the ribosome.[57]

3.3 History of Genetics

Born Johann Mendel, the father of modern genetics, took the name Gregor upon entering religious life as an Augustinian monk in Brno, now part of the Czech Republic, around 1850. Brno was then part of the Austro–Hungarian empire, located between Vienna and Prague. Mendel first started his studies of heredity with mice and bees but focused on his famous peas when his bishop expressed concern about doing research with animals. Between 1856 and 1863 Mendel cultivated and tested some 29,000 pea plants and found experimentally that one in four pea plants had pure-bred *recessive* alleles, two out of four were hybrid, and one out of four were pure-bred *dominant* (Fig. 3.5). His experiments led him to make two important generalizations that later came to be known as **Mendel's laws of inheritance**. He published his results in 1865.

Mendel concluded that organisms inherit traits (such as the color of phenotypes) via discrete units of inheritance that we now refer to as *genes*. Applied to his peas, each plant has two *alleles* (or versions) of each gene, one each inherited from its two parents. It is a matter of chance which gene from each parent is received. If the offspring receives a gene for the (dominant) yellow seed together with the (recessive) green seed, the offspring will have yellow color but the recessive gene for green is still present and may surface in the

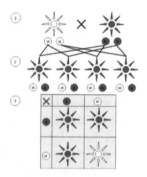

Figure 3.5   Dominant (red) and recessive (white) phenotypes. (1) Parental generation. (2) F1 (first) generation. (3) F2 (second) generation: dominant and recessive phenotype look alike in the F1 generation and show a 3:1 ratio in the F2 generation. Source: Wikipedia.

next generation if combined with another pea that carries a green gene.

Mendel's first law, the **law of segregation**, states that every individual possesses a pair of alleles (assuming diploidy) for any particular trait and that each parent passes a randomly selected copy (allele) of only one of these to its offspring. The offspring then receives its own pair of alleles for that trait. Interactions between alleles at a single locus are termed dominance and these influence how the offspring expresses that trait (e.g., the color and height of a plant, or the color of an animal's fur).

Mendel's second law, the **law of independent assortment**, also known as the **law of inheritance**, states that separate genes for separate traits are passed independently of one another from parents to offspring. That is, the biological selection of a particular gene in the gene pair for one trait to be passed to the offspring has nothing to do with the selection of the gene for any other trait.

Of the 46 chromosomes in a normal *diploid* (containing two homologous copies of chromosomes) human cell, half are maternally derived (from the mother's egg) and half are paternally derived (from the father's sperm). This occurs as sexual reproduction involves the fusion of two *haploid gametes* (the egg and sperm) to produce a new organism having the full complement of chromosomes.

Mendel's laws were ignored until they were rediscovered and validated around 1900 by Hugo de Vries, Carl Correns, and Erich von Tschermak-Seysenegg. All three were working independently on different plant hybrids and came to the same conclusions about inheritance as Mendel. By the early part of the 20th century, *chromosomes* were discovered and it became clear that heredity and development of organisms were dependent on information residing in genes contained in chromosomes.

Mendel's laws continue to serve as the foundation of genetics. It was Mendel who introduced the term "gamete" (Greek for *husband/wife*) to describe the fusion of cells during fertilization in organisms that reproduce sexually. Although human sex cells differ significantly in size (the female ovum, or egg, is about hundred thousand times larger than the male tadpole-like sperm), his insight that each parent gamete is carrying half the genetic information of the offspring turned out to be correct.

In humans, an ovum can carry only an X chromosome (of the X and Y chromosomes), whereas a sperm may carry either an X or a Y; thus the male sperm determines the sex of any resulting zygote. A *zygote* (Greek for *to join*) is the initial cell formed when two gamete cells are joined by means of sexual reproduction. In multicellular organisms, it is the earliest developmental stage of the embryo. If the zygote has two X chromosomes it will develop into a female; if it has an X and a Y chromosome, it will develop into a male.

Another important term from the vocabulary of genetics is *meiosis* (Fig. 3.6), a special type of cell division necessary for sexual reproduction where *diploid* cells are transformed into *haploid* gametes. Meiosis also leads to the *crossing-over* of genetic material, resulting in genetic variation. Although the resulting haploid gametes have only one set of chromosomes, that set is a shuffled mixture of genetic material from both parents. Every single egg or sperm may have a different selection of alleles (forms of genes) from the parental chromosomes. When two gametes later fuse during fertilization, the number of sets of chromosomes in the resulting zygote is restored to the original number.

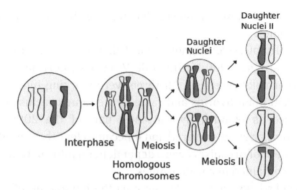

Figure 3.6 Events involving meiosis, showing chromosomal cross-over. The two gray chromosomes represent one set and the two red chromosomes another set, each set originally from one of the two parents. Source: Wikipedia.

3.4 Molecular Basis of Genetics

Molecular genetics is the field that studies the structure and function of genes at a molecular level. The methods of genetics and molecular biology are combined to arrive at a deeper understanding of developmental biology, the study of processes that cause organisms to grow and develop. By studying the genetic control of cell growth, differentiation, and morphogenesis we are trying to learn what gives rise to tissues, organs and body anatomy. Let us quickly describe what cells are about and then define and explain the associated terms introduced above:

The *cell* (from Latin *cella*, meaning *small room*) is the basic structural, functional, and biological unit of all known living organisms. Cells are the smallest unit of life. They can replicate independently and are often called the building blocks of life.

There are two basic types of cells, *eukaryotes* (which contain a nucleus) and *prokaryotes* (which do not contain a nucleus). Prokaryotic cells are usually single-celled organisms (most bacteria), while eukaryotic cells can be either single-celled or part of multicellular organisms, including plants and animals.

Cells consist of protoplasm enclosed within a membrane, which contains many biomolecules such as proteins and nucleic acids.

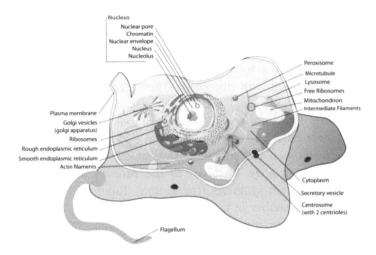

Figure 3.7 Major structures inside a eukaryotic animal cell. Source: Wikipedia.

While the number of cells in plants and animals varies from species to species, humans contain about 100 trillion (10^{14}) cells. Most plant and animal cells are visible only under the microscope, with dimensions between 1 and 100 μm (Fig. 3.7).

The cell was discovered by Robert Hooke in 1665, and cell theory (as first developed in 1839 by Matthias Jakob Schleiden and Theodor Schwann) states that all organisms are composed of one or more cells, that all cells come from preexisting cells, that vital functions of an organism occur within cells, and that all cells contain the hereditary information necessary for regulating cell functions and for transmitting information to the next generation of cells. Cells emerged on earth at least 3.5 billion years ago.

A few important definitions of cell structures depicted in Fig. 3.7 are given below:

- *Cytoplasm:* Comprises cytosol, the gel-like substance enclosed within the cell membrane, and organelles, the cell's internal substructures
- *Endoplasmic reticulum* (ER): Is a type of organelle in the cells of eukaryotic organisms that forms an interconnected network of flattened, membrane-enclosed sacs or tubes known

as cisternae. The ER occurs in most types of eukaryotic cells but is absent from red blood cells and spermatozoa. There are two types of ER, rough endoplasmic reticulum (RER) and smooth endoplasmic reticulum (SER). The RER is studded with ribosomes. The SER lacks ribosomes and functions in lipid metabolism, carbohydrate metabolism, and detoxification. SER is especially abundant in mammalian liver cells.

- *Ribosomes:* Is from ribonucleic acid and the Greek *soma*, meaning *body*. It serves as the primary site of biological protein synthesis (translation). Ribosomes link amino acids together in the order specified by mRNA molecules. Ribosomes consist of two major components, the small ribosomal subunit that reads the RNA and the large subunit that joins amino acids to form a polypeptide chain. Each subunit is composed of one or more ribosomal RNA molecules and a variety of proteins. The ribosomes and associated molecules are also known as the translational apparatus.

- *Lysosome:* Is derived from the Greek words *lysis*, meaning *to loosen*. It is a membrane-bound cell organelle found in animal cells but absent in red blood cells. They are spherical vesicles (see below) containing more than 50 enzymes capable of breaking down virtually all kinds of biomolecules, including proteins, nucleic acids, carbohydrates, lipids, and cellular debris. They act as the waste disposal system of the cell by digesting unwanted materials in the cytoplasm, both from outside the cell and obsolete components inside the cell. Further, lysosomes are involved in secretion, plasma membrane repair, cell signaling, and energy metabolism. They were discovered and named by Belgian biologist Christian de Duve.[58]

- *Golgi vesicles* (apparatus): Is named after 1906 Nobel laureate Camillo Golgi.[59] The Golgi apparatus (or complex) is an organelle that is part of the cellular endomembrane system and packages proteins inside the cell before they are sent to their destination; it is particularly important in the processing of proteins for secretion.

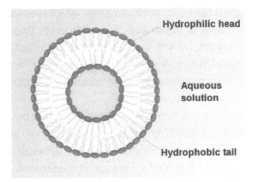

Figure 3.8 Scheme of a liposome formed by phospholipids in an aqueous solution. Source: Wikipedia.

- *Vesicle*: Is a small organelle within a cell, consisting of fluid enclosed by a lipid bilayer membrane. Vesicles can form naturally, for example, during the processes of transport of materials within the cytoplasm. Alternatively, they may be prepared artificially, in which case they are called *liposomes* (Fig. 3.8).

If there is only one phospholipid bilayer, they are called unilamellar liposome vesicles; otherwise they are called multilamellar. Vesicles can fuse with the plasma membrane to release their contents outside of the cell. Vesicles can also fuse with other organelles within the cell.

Prokaryotic genetic material is organized in a simple circular DNA molecule (the bacterial chromosome). All their intracellular water-soluble components (proteins, DNA, and metabolites) are located together in the same volume enclosed by the cell membrane, rather than in separate cellular compartments.

Eukaryotic genetic material is divided into different, linear molecules called chromosomes inside a discrete nucleus, usually with additional genetic material in some organelles like mitochondria and chloroplasts.

The main distinguishing feature of eukaryotes as compared to prokaryotes is the presence of membrane-bound compartments in which specific metabolic activities take place. Most important among these is a cell nucleus that houses the eukaryotic cell's DNA.

This nucleus gives the eukaryote its name, which means *true nucleus*. The eukaryotic DNA is organized in one or more linear molecules, called chromosomes, which are associated with histone proteins. All chromosomal DNA is stored in the cell nucleus, separated from the cytoplasm by a membrane. Some eukaryotic organelles such as mitochondria also contain some DNA.

Mitochondria are self-replicating organelles that occur in various numbers, shapes, and sizes in the cytoplasm of all eukaryotic cells. Mitochondria play a critical role in generating energy in the eukaryotic cell. The cell's energy is generated by oxidative phosphorylation, using oxygen to release energy stored in cellular nutrients (typically pertaining to glucose) to generate adenosine-5'-triphosphate (ATP). Chloroplasts can only be found in plants and algae, and they capture the sun's energy to make ATP through photosynthesis.

To understand cell replication, it is important to understand the respective roles of DNA, RNA, and proteins. To replicate, DNA needs the proteins because proteins/enzymes catalyze the replication process. On the other hand, the proteins cannot exist without DNA molecules because they contain the recipes (genetic code) for making proteins. What may have come first in the history of evolution was probably RNA, essentially a single-stranded version of DNA with uracil instead of DNA's thymine and an extra oxygen on each sugar. RNA comes as mRNA carrying the genetic information from the DNA in the nucleus to the ribosomes in the cell's cytoplasm. RNA also comes as transfer RNA (tRNA), small (~75 nucleotides) folded molecules that transport amino acids from the cytoplasm to the ribosome. In the ribosome, tRNA amino acids get strung together to form a protein, in the sequence dictated by mRNA and the rules set by the genetic code.

The term *cell growth* is used in the contexts of cell development (often studied in model organisms such as yeast) and cell division (reproduction). The process of cell division, called *cell cycle*, has four major parts called phases. The first part, called the G1 phase, is marked by synthesis of various enzymes that are required for DNA replication. The second part of the cell cycle is the S phase, where DNA replication produces two identical sets of chromosomes. The third part is the G2 phase. Significant protein synthesis occurs

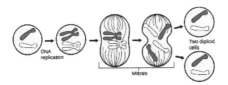

Figure 3.9   Mitosis is dividing the chromosomes in a cell nucleus. Source: Wikipedia.

during this phase, mainly involving the production of microtubules, which are required during the process of division, called *mitosis.* Mitosis should not be confused with meiosis (generation of haploid gametes, as discussed above) and is defined as the process by which a cell, which has previously replicated each of its chromosomes, separates the chromosomes in its cell nucleus into two identical sets of chromosomes, each set in its own new nucleus (Fig. 3.9).

The fourth phase (or M phase) consists of nuclear division (karyokinesis) and cytoplasmic division (*cytokinesis*), accompanied by the formation of a new cell membrane.

We define *cellular differentiation* as the process by which a less specialized cell becomes a more specialized cell type. It occurs numerous times as the organism evolves from a simple zygote to a complex system of tissues and cell types. Differentiation not only occurs as embryos develop into adults, it is also responsible for tissue repair and during normal cell turnover. On the basis of modifications in *gene expression,* differentiation changes a cell's size, shape, membrane potential, metabolic activity, and responsiveness to signals. Normally, cellular differentiation does not involve a change in the DNA sequence itself. Different cells can have very different physical characteristics despite having the same genome.

We define *morphogenesis* as the biological process that causes an organism to develop its shape. Morphogenesis arises because of changes in the cellular structure or how cells interact in tissues, resulting in tissue elongation, thinning, folding, or separation into distinct layers of tissue (also referred as cell sorting). Chemically, morphogenesis is associated with *morphogens*—soluble molecules that can diffuse and carry signals controlling cell differentiation based on their concentration and acting through binding to specific *protein receptors. Transcription factor proteins* determine the fate of cells by interacting with DNA and are typically coded for by master regulatory genes. They either activate or deactivate the transcription of other genes, which, in turn, can regulate the expression of still other genes in a regulatory cascade. At the end of this cascade, another class of molecules involved in morphogenesis is molecules that control cellular behaviors (e.g., cell migration) or, more generally, their properties, such as *cell adhesion*. Tissues can also change their shape and separate into distinct layers via *cell contractility*. Myosins, a family of ATP-dependent motor proteins that are best known for their role in muscle contraction, can contract different parts of the tissue to change its shape or structure.

In biochemistry, *metabolic pathways* are series of chemical reactions occurring within a cell. In each pathway, a principal chemical is modified by a series of chemical reactions. Enzymes catalyze these reactions and often require dietary minerals, vitamins, and other cofactors in order to function properly. Because of the many chemicals (metabolites) that may be involved, metabolic pathways can be quite elaborate. Numerous distinct pathways coexist within a cell. Dr. Gerhard Michal who started his career at the German diagnostics company Boehringer Mannheim (acquired by Roche in 1997) developed a "Biochemical Pathways" poster that was first published in 1968[60] and has since been updated and distributed to more than a million scientists around the globe. Roche is continuing the tradition until this day.[61]

The 2013 Nobel Prize for physiology or medicine was awarded to James E. Rothman, Randy W. Schekman, and Thomas C. Südhof for their discoveries of machinery regulating *vesicle traffic* and for their elucidation of the *molecular transport system in eukaryotic cells*.[62] In eukaryotes, specific cellular functions are compartmentalized into the cell nucleus and organelles surrounded by intracellular membranes. This compartmentalization improves the efficiency of cellular functions and prevents potentially dangerous molecules from moving freely inside the cell. However, when different compartments need to exchange specific molecules, or when certain molecules need to be exported to the cell's exterior, a transport mechanism is required to protect the molecules and to move them to the right place at the right time. Here is an extract of the very detailed description of the discoveries by the three Nobel laureates, as can be found at www.nobelprize.org. It is included because of its relevance for nanomedicine, in particular with respect to the *host–cargo concept* and its role in drug delivery, to be discussed in Chapter 5.

Randy Schekman (Stanford University and UC Berkeley University) used yeast genetics to dissect the mechanism involved in membrane and vesicle trafficking. He realized that baker's yeast secretes glycoproteins and decided to study vesicle transport and fusion. He used a genetic screen for identification of the 23 genes—named *sec1, sec2, ... sec23*—that regulate intracellular transport. He was further able to classify them as controlling traffic from the ER, the Golgi complex, or the cell surface (gene *sec1*). He systematically unraveled the events along secretory pathways involved in vesicle traffic and in the interaction of vesicles with target membranes.

James Rothman (Stanford University and Yale University) developed an in vitro reconstitution assay to study events involved in intracellular vesicle transport. He started by reconstituting the intracellular transport of the vesicular stomatitis virus (VSV)–G protein within the Golgi complex and then purified proteins from the cytoplasm that were required for transport. The first protein to be purified, *N*-ethylmaleimide-sensitive factor (NSF), paved the way for identification of other proteins responsible for vesicle fusion, among them soluble NSF attachment protein (SNAP). Together, Rothman and Schekman discovered that one of the yeast mutants, *sec18*, corresponds to NSF and that *sec17* was functionally equivalent to SNAP. They were further able to show that the vesicle fusion machinery is evolutionary ancient: what can be learned by investigating yeast cells does apply to human cells as well.

Rothman then turned to brain tissue from which he purified soluble NSF attachment protein receptors (SNAREs). He subsequently proposed the SNARE hypothesis, stipulating that target and vesicle SNAREs (t-SNAREs and v-SNAREs, respectively) were critical for vesicle fusion through a set of sequential steps for synaptic docking, activation, and fusion.

Thomas Südhof (University of Goettingen, Germany, and University of Texas, Dallas) set out to study how synaptic vesicle

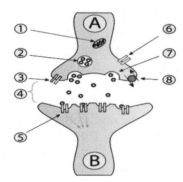

Neuron A (transmitting) to Neuron B (receiving)

1. Mitochondrion
2. Synaptic vesicle with neurotransmitters
3. Autoreceptor
4. Synapse with neurotransmitter released (serotonin)
5. Postsynaptic receptors activated by neurotransmitter (induction of a postsynaptic potential)
6. Calcium channel
7. Exocytosis of a vesicle
8. Recaptured neurotransmitter

Figure 3.10 Communication between nerve cells (neurons). Source: Wikipedia.

fusion between neurons of the central nervous system (CNS) was controlled. Figure 3.10 shows a synapse.

Although Schekman and Rothman had discovered the fundamental machinery for vesicle fusion, it was still unknown how it was controlled temporally: Vesicular fusions in the body must be kept carefully in check, and vesicle fusions must be executed frequently with high precision in response to specific stimuli. For example, neurotransmitter release in the brain and insulin secretion from the pancreas require such precision. Südhof was intrigued by the rapid process (exocytosis) by which a cell directs the contents of secretory vesicles out of the cell membrane and into the extracellular space. This process is regulated by changes in the free calcium concentration and by studying how calcium (Ca^{2+}) regulates neurotransmitter release, he discovered that the proteins complexin and synaptotagmin are play a critical role. By means of animal experiments (with mice) he discovered that complexin acts at a late step in synaptic fusion as a clamping mechanism that prevents constitutive (i.e., continual) fusion and allows regulated exocytosis to occur. He also discovered synaptotagmin-1, which couples calcium to neurotransmitter release by interacting with phospholipids as well as with syntaxin-1 and SNAREs. Südhof established a role for synaptotagmin-1 as a calcium sensor for rapid synaptic fusion by demonstrating that calcium binding to synaptotagmin-1 is triggering neurotransmitter release at the synapse. He further characterized Munc18-1, which corresponds to Schekman's *sec1* and is therefore also called a Sec/Munc (SM) protein. Deletion of Munc18-1 in mice leads to a complete loss of neurotransmitter secretion from synaptic vesicles. SM proteins along with SNARE proteins are now known to be an integral part of the vesicle fusion protein complex.

Vesicle transport and fusion are essential for physiological processes ranging from synapses in the brain to immunological responses and hormone secretion. Whenever there is a disruption in the machinery for the routing of cargo in cells in organisms based on eukaryotic cells, there will be risk for disease such as metabolic disorders, malfunctioning of immune responses, epilepsy, or other cognitive dysfunction.

The goal of developmental biology is not only to understand how organisms develop but also to apply the new knowledge to regeneration and perhaps eventually to slowing down of aging.

A useful way to study the function of genes is to relate genotype to phenotype. The *genotype* is the genetic makeup of a cell or organism, its internally coded, inheritable information, but usually associated with a specific characteristic under consideration. The *phenotype*, on the other hand, is the outward, physical manifestation of the organism, including not only appearance but also metabolism, energy utilization, tissues, organs, reflexes, and behaviors. The definition of phenotype includes anything that is part of the observable structure, function, or behavior of a living organism.

Whereas an individual's genomic sequence is an absolute measure of DNA base composition, genotype typically implies a measurement of how an individual differs or is specialized within a group of individuals or a species. In humans, particular attention is paid to what combination of alleles the individual carries. The words *homozygous* and *heterozygous* are used to describe the genotype of a diploid (such as human or plant) organism at a single locus on the DNA. Homozygous describes a genotype consisting of two identical alleles at a given locus; heterozygous describes a genotype consisting of two different alleles at a locus.

A cell is said to be homozygous for a particular gene when identical alleles of the gene are present on both homologous chromosomes. The cell or organism in question is called a *homozygote*.

When breeding plants or animals, the *phenotypic* traits to be held constant should be homozygous.

A diploid organism is heterozygous at a gene locus when its cells contain two different alleles of a gene. The cell or organism is called a *heterozygote* specifically for the allele in question.

A gene for a particular trait may exist in two allelic forms, dominant (denoted by capital letters such as A, B, etc.) or recessive (denoted by lower case letters a, b, etc.). For a given phenotypic trait, there could be three possible genotypes, namely, AA (homozygous dominant), Aa (heterozygous), and aa (homozygous recessive).

If the trait in question is determined by simple (complete) dominance, a heterozygote will express only the trait coded by the dominant allele. The trait coded by the recessive allele will not be present in the phenotype.

A main goal of molecular genetics is to discover and understand how genotypes relate to phenotypes, how traits are carried on, and how and why some may mutate.

In genetics, a *mutation* is a (discrete) change of the nucleotide sequence of the genome of an organism. Mutations result from unrepaired damage to DNA or to RNA genomes (typically caused by radiation or chemical mutagens), from errors in the process of replication, or from the insertion or deletion of segments of DNA by mobile genetic elements. As pointed out by Schrödinger in his influential lecture series and book *What Is Life*,[63] both Max Planck's discovery of quantum theory[64] and the rediscovery of Mendelian genetics happened in the same year, 1900. It took then more than a quarter of a century for Schrödinger to discover wave mechanics and for Heitler and London[65] to apply it to molecular chemistry, and another two decades to begin to explain mutations on a molecular level (see the Luria–Delbrück experiment, 1943)[66] and to provide an improved understanding of the role of mutations in Darwin's theory of evolution.

It is important to distinguish between DNA *damage and mutation*, the two major types of errors in DNA. DNA damages and mutation are fundamentally different. Damages are physical abnormalities in the DNA, such as single- and double-strand breaks. DNA damages can be recognized by enzymes and often correctly repaired as long as redundant information (such as the undamaged sequence in the complementary DNA strand or in a homologous chromosome) is available for copying. If a cell retains DNA damage, transcription of a gene and translation into a protein can often be prevented. Replication may also be blocked or the cell may die.

In contrast to DNA damage, a mutation is a change in the base sequence of the DNA. A mutation cannot be recognized by enzymes once the base change is present in both DNA strands, and thus, a mutation cannot be repaired. At the cellular level, mutations can cause alterations in protein function and regulation. Mutations are replicated when the cell replicates. However, the majority of mutations have a negative effect to a cell's survival. In a population of cells comprising a tissue with replicating cells, mutant cells will tend to be lost. Infrequent mutations that provide a survival advantage will tend to expand at the expense of neighboring cells in the tissue. This advantage to the cell may have a negative effect on the whole organism because such mutant cells can give rise to cancer. In contrast, DNA damages in infrequently dividing cells are likely a prominent cause of aging.

Nonlethal mutations accumulate within the gene pool and increase the amount of genetic variation. The abundance of some genetic changes within the gene pool can be reduced by natural selection, while other more favorable mutations may accumulate and result in adaptive changes.

For example, a butterfly may produce offspring with new mutations. The majority of these mutations will have no effect; but one might change the color of one of the butterfly's offspring, making it harder/easier for predators to see. If the color change is advantageous, it improves the chance of this butterfly's surviving and producing its own offspring. Over time the number of butterflies with this mutation may form a larger percentage of the population.

Here are two examples of (more or less) beneficial mutations in humans:

1. A specific 32-base-pair deletion (Delta 32) in the human C-C chemokine receptor type 5 (CCR5 or CD195, a protein on the surface of white blood cells that is involved in the immune system as it acts as a receptor for chemokines—a process by which T cells are attracted to specific tissue and organ targets) confers

human immunodeficiency virus (HIV) resistance to homozygotes (identical alleles of the gene on both homologous chromosomes) and delays acquired immunodeficiency syndrome (AIDS) onset in heterozygotes. One possible explanation of the etiology of the relatively high frequency of CCR5 Delta 32 in the European population is that it also conferred resistance to the bubonic plague in mid-14th century Europe. People with this mutation were more likely to survive infection, hence its frequency in the population increased. However, this mutation is not found in southern Africa, which remained untouched by bubonic plague. Alternatively, another theory suggests that the selective pressure on the CCR5 Delta 32 mutation was caused by smallpox instead of the bubonic plague.[67]

2. Another example is sickle-cell disease, a blood disorder in which the body produces an abnormal type of the oxygen-carrying substance hemoglobin in the red blood cells. Sickling decreases the cells' flexibility and may cause various life-threatening complications, including anemia. One-third of all indigenous inhabitants of sub-Saharan Africa carry the gene because there is a malaria survival value in carrying only a single sickle-cell gene.

A third interesting example supporting Darwin's theory of evolution via natural selection[68] is provided by the adaptation of humans to low-oxygen environments in areas of extremely high altitude.[69] Globally, there are three such regions: Tibet, the South American Andes (in particular Peru), and Ethiopia. It turns out that human adaptation strategies are different in these three areas, with the Tibetan adaptation probably optimal for very high altitudes.

More than 4 million Tibetans live at an altitude over 3500 m, of which 600,000 are at an altitude exceeding 4500 m, at oxygen levels of less than 60% of that at sea level. After 3000 years of adaptation, the Tibetans cope with oxygen deficiency by inhaling more air with each breath and breathing more rapidly than sea-level populations. Tibetans already have better oxygenation at birth, enlarged lung volumes throughout life, and a higher capacity for exercise. They show a sustained increase in cerebral blood flow, lower hemoglobin concentration, and less susceptibility to chronic mountain sickness. Although there seems to be a contradiction between living with less oxygen and low hemoglobin concentrations in the blood, it is actually a biological disadvantage to increase hemoglobin levels at very high altitudes: Too much hemoglobin limits blood circulation and increases the risk of blood clots and stroke. Instead, Tibetans have high levels of nitric oxide in their blood (double compared to lowlanders), which helps their blood vessels dilate for enhanced blood circulation.

The Andes highlanders in Peru have developed a different high-altitude adaptation strategy. Though their physical growth in body size is delayed, growth in lung volumes is accelerated. In contrast to the Tibetans, the Andeans exhibit the same elevated hemoglobin concentrations that lowlanders exhibit at high elevations. In addition, they have increased oxygen levels in their hemoglobin, that is, more oxygen per blood volume than other people. This confers an ability to carry more oxygen in each red blood cell and a more effective delivery system of oxygen throughout their bodies, while breathing essentially at the same rate as lowlanders. It enables the Andeans to overcome hypoxia and normally reproduce without risk of death for the mother or baby. However, Andeans are not able to withstand mountain sickness (at extreme altitudes) as well as the Tibetans.

Finally, the Amhara of Ethiopia also live at very high altitudes, around 3000 to 3500 m. Among healthy individuals, the average hemoglobin concentrations are lower than normal, almost similar to the Tibetans, but with a higher-than-average oxygen content of hemoglobin, similar to the Andeans. Ethiopian highlanders do not exhibit any significant change in blood circulation of the brain, makes them more susceptible to mountain sickness than the Tibetans. Still, the Ethiopian highlanders seem to be immune to the extreme dangers posed by high-altitude environment.

Recent genetic studies show significant differences between genetic makeups of the Tibetan, Andean, and Ethiopian populations.[70]

In fact, the statement made above about 3000 years of adaption by the Tibetans has to be re-examined on the basis of the recent discovery by an international research team including UC Berkeley population geneticist Rasmus Nielsen, his postdoc Emilia Huerta-Sanchez, and a large number of scientists at BGI (see Section 3.9 below for a discussion of BGI) that the *EPAS1* gene by the

Tibetans had a unique sequence variant only to be found in the gene of a Denisovan girl, found in the Denisova Cave in the Altai Mountains of Siberia. The Denisovan and Tibetan gene segments matched so closely that the inevitable conclusion seems to be that Denisovans, an extinct relative of humans who disappeared about 40,000 years ago, transferred the high-altitude *EPAS1* gene by mating with ancestors of the Tibetans. Although most other human groups who mated with Denisovans (like Han Chinese and Melanesians in Papua New Guinea) lost the Denisovans' version of the *EPAS1* gene because it wasn't particularly beneficial al low sea levels, Tibetans who settled on the high-altitude Tibetan plateau retained it because it helped them adapt to life there.[71]

Conclusion: The "superathlete" *EPAS1* gene that helps Sherpas and other Tibetans breathe easier and work harder at high altitudes was inherited from an ancient species of human relatives but was kept largely unchanged only in populations exposed to high-altitude environments with 40% less oxygen.

Genotyping is the process of elucidating the genotype of an individual with a biological assay. Genotyping is frequently used to find particular single-nucleotide polymorphisms (SNPs) that may be associated with genetic risk factors.

The term "single-nucleotide polymorphism," or "snip," refers to a DNA sequence variation occurring when a single nucleotide in the genome (or other shared sequence) differs between members of a biological species. For example, two sequenced DNA fragments from different individuals, AAGCCTA to AAGCTTA, contain a difference in a single nucleotide (boldfaced in our example).

Genotypic assay techniques include polymerase chain reaction (PCR), DNA fragment analysis, allele-specific oligonucleotide (ASO) probes, DNA sequencing, and nucleic acid hybridization to DNA microarrays or beads. Let's briefly discuss these techniques below in more detail.

PCR was invented in 1983 by Kary Mullis, then a scientist working at the Biotech company CETUS. In his 1993 Nobel lecture,[72] Mullis explained how it works and how the idea occurred to him on one of his weekly Friday night trips from San Francisco to his cabin in Mendocino, a beautiful village with spectacular ocean views that can be reached driving up North along the coast and through

California wine country. While his girlfriend was sleeping, he was thinking about oligonucleotides—short, single-stranded bits of DNA or RNA molecules that can be synthesized in chemical labs such as the one he then used at CETUS. While driving on, he had discovered the concept of PCR and it led him to exclaim the following: "Dear Thor! I may have just solved the most annoying problems in DNA chemistry in a single lightning bolt. Abundance and distinction. With two oligonucleotides, DNA polymerase, and the four nucleoside triphosphates I could make as much of a DNA sequence as I wanted and I could make it on a fragment of a specific size that I could distinguish easily." Mullis would receive a $10,000 bonus from CETUS for the invention. A few years later, CETUS sold the PCR patent to ROCHE for an estimated $330 M. As it turned out, it was a good business decision by the Swiss life sciences company. PCR is now a common and often indispensable technique used in medical and biological research labs for important applications such as DNA cloning for sequencing, functional analysis of genes, the diagnosis of hereditary diseases, the identification of genetic fingerprints (used in forensic sciences and paternity testing), and the detection and diagnosis of infectious diseases.

The PCR method relies on thermal cycling, consisting of cycles of repeated heating and cooling of the reaction for DNA melting and enzymatic replication of the DNA. Primers (short DNA fragments) containing sequences complementary to the target region along with a DNA polymerase (after which the method is named) are key components to enable selective and repeated amplification.

A *DNA polymerase* is a cellular or viral polymerase enzyme that synthesizes DNA molecules from their nucleotide building blocks. In 1956, Arthur Kornberg[73] and colleagues discovered the enzyme DNA polymerase I in *Escherichia coli*, a rod-shaped bacterium that is commonly found in the lower intestine of warm-blooded organisms. In 1959, Severo Ochoa and Arthur Kornberg received the Nobel Prize for physiology or medicine for their discovery of the mechanisms in the biological synthesis of RNA and DNA.

As PCR progresses, the DNA generated is itself used as a template for replication, setting in motion a chain reaction in which the DNA template is exponentially amplified. Today, almost all PCR applications employ a heat-stable DNA polymerase, such as Taq

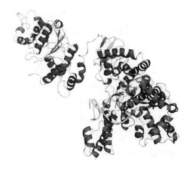

Figure 3.11   Taq polymerase bound to a DNA octamer. Source: Wikipedia.

polymerase, an enzyme originally isolated from the bacterium *Thermus aquaticus*, which lives in hot springs and acts as an enzyme able to withstand the protein-denaturing conditions (high temperature) required during PCR (Fig. 3.11).

DNA polymerase enzymatically assembles a new DNA strand from DNA-building blocks, the nucleotides, by using ssDNA as a template and DNA oligonucleotides (also called DNA primers), which are required for initiation of DNA synthesis. A key concept in PCR is *thermal cycling*, the repeated heating and cooling of the PCR sample through a defined series of temperature steps. In the first step, the two strands of the DNA double helix are physically separated at a high temperature in a process called *DNA melting*. In the second step, the temperature is lowered and the two DNA strands become templates for DNA polymerase to selectively amplify the target DNA.

DNA fragments are created by *restriction enzymes*. A restriction enzyme (or *restriction endonuclease*) is an enzyme that cuts DNA at or near specific recognition nucleotide sequences known as restriction sites. Restriction enzymes recognize a specific sequence of nucleotides and produce a double-stranded cut in the DNA. The recognition sequences usually vary between four and eight nucleotides, and many of them are *palindromic*, meaning the base sequence reads the same backward and forward. To cut DNA, all restriction enzymes make two incisions, once through each sugar–phosphate backbone (i.e., each strand) of the DNA double helix. These enzymes are found in bacteria and archaea (single-celled microorganisms without cell nuclei or membrane-bound organelles) and provide a defense mechanism against invading viruses. Thousands of restriction enzymes have been studied in detail, and several hundred are available commercially. For their work in the discovery and characterization of restriction enzymes, the 1978 Nobel Prize for physiology or medicine was awarded to Werner Arber, Daniel Nathans, and Hamilton O. Smith.[74] Their discovery led to the development of *recombinant DNA (rDNA) technology*, which allowed, for example, the large-scale production of human insulin for diabetics using *E. coli* bacteria. Restriction enzymes can also be used to distinguish gene alleles by specifically recognizing single-base changes in DNA (SNPs). This is only possible if a SNP alters the restriction site present in the allele. In this method, the restriction enzyme can be used to genotype a DNA sample without the need for (more expensive) gene sequencing.

Restriction enzymes are routinely used for DNA modification in laboratories, and are a vital tool in *molecular cloning*. The word cloning is used to describe the replication of a single DNA molecule starting from a single living cell to generate a large population of cells containing identical DNA molecules. Molecular cloning generally uses DNA sequences from two different organisms: the species that is the source of the DNA to be cloned, and the species that will serve as the living host for replication of the rDNA.

Cloning in biotechnology (BT) refers to processes used to create copies of DNA fragments (molecular cloning), cells (cell cloning), or organisms.

Cloning is commonly used to amplify DNA fragments containing whole genes. It is used in a wide array of biological experiments and practical applications ranging from genetic fingerprinting to large-scale protein production.

To amplify any DNA sequence in a living organism, that sequence must be linked to an origin of replication, which is a sequence of DNA capable of directing the propagation of itself and any linked sequence. However, a number of other features are needed and a variety of specialized cloning vectors (small pieces of DNA into which a foreign DNA fragment can be inserted) exist that allow protein expression, tagging, single-stranded RNA (ssRNA) and ssDNA production, and a host of other manipulations.

Cloning of any DNA fragment proceeds in four steps:

1. Fragmentation—breaking apart a strand of DNA
2. Ligation—gluing together pieces of DNA in a desired sequence
3. Transfection—inserting the newly formed pieces of DNA into cells
4. Screening/selection—selecting the cells that were successfully transfected with the new DNA

Cloning a cell means to derive a population of cells from a single cell. In the case of unicellular organisms such as bacteria and yeast, this process only requires the inoculation of the appropriate medium. However, in the case of cell cultures from multicellular organisms, cell cloning is an arduous task.

Somatic cell nuclear transfer, known as SCNT, can also be used to create embryos for research or therapeutic purposes. The most likely purpose for this is to produce embryos for use in *stem cell research*. This process is also called research cloning or therapeutic cloning. The goal is not to create cloned human beings (reproductive cloning), but rather to harvest stem cells that can be used to study human development and to potentially treat disease. Therapeutic cloning is achieved by creating embryonic stem cells (ESCs) with the goal of treating diseases such as diabetes and Alzheimer's disease (AD).

The process of cloning an animal using SCNT is relatively the same for all animals.

A good illustration of nuclear transfer cloning is provided by the story of Dolly, the famous sheep that in 1996 was cloned into existence at the Roslin Institute in Edinburgh, Scotland.

As illustrated in Fig. 3.12, Dolly was produced from the frozen udder cell of a six-year-old Finn–Dorset ewe that had died three years prior to Dolly's birth.

The cell nucleus of this ewe was physically transferred into an egg cell whose nucleus had previously been removed. Dolly's embryo was created by implantation of the egg cell into the uterus of an animal of the same species, where it developed into the fully formed live clone.

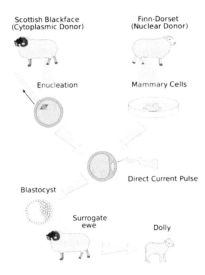

Figure 3.12   The cloning of Dolly the sheep. Source: Wikipedia.

Cloning Dolly the sheep had a low success rate per fertilized egg: she was born after 277 eggs were used to create 29 embryos, which only produced 3 lambs at birth, only 1 of which lived.

However, Dolly was a nearly exact genetic duplicate of the deceased ewe from which she had been cloned.

Since Dolly the sheep, scientists have used the SCNT technique to clone mice, pigs, dogs, and other animals. However, human cells proved trickier to work with. In 2007, the Oregon National Primate Research Center (ONPRC) in Beaverton, OR, was able to clone monkey embryos and derived ESCs from them. In 2013, Shoukrat Mitalipov, a Kazakhstan-born researcher at the ONPRC, cloned the first human ESCs, a highly controversial project from an ethical point of view. Mitalipov uses cell manipulation machines to strip the nucleus from an egg and replace it with DNA from a skin cell. His team activates the cell using chemicals, electricity, and caffeine. The result is what biologists call an early embryo, formed using cloning techniques, although unfertilized. Then stem cells are extracted from the embryo and grown into bunches. Thousands of stem cells are stored in a translucent plastic vial and then frozen in liquid nitrogen.

The stem cells produced are an identical match to the skin cell donor, making them potentially ideal for transplants and personalized therapies—to be discussed in Section 5.6.

The majority of biologists agree that human cloning is taboo. However, in the United States, despite the ban on federal funding for ESC research, no federal law or rules exist to prevent it (unlike in countries like the UK).

3.5 DNA Sequencing and the Human Genome Project

Along with Bach's music, Shakespeare's sonnets, and the Apollo Space Programme, the Human Genome Project is one of those achievements of the human spirit that makes me proud to be human.

—Richard Dawkins (born 1941; author of *The Selfish Gene*)

DNA sequencing technologies are used to determine the order of the nucleotide bases A, G, C, and T in a molecule of DNA. The knowledge of DNA sequences—of all organisms, not just humans— has fundamental importance for basic biological research as well as in numerous applied fields of medicine, BT, plant, and animal science. The decision in 1990 by the Department of Energy (DOE) and the National Institutes of Health (NIH), two US government agencies, to initiate a Human Genome Project (HGP) has both accelerated sequencing technology and significantly stimulated biological research and life sciences discoveries. At the NIH, it was first James Watson, codiscoverer of the genetic code, who was responsible at the launch of the project. In 1993, Francis Collins took over as the head of the National Human Genome Research Institute (NHGRI) of the NIH. At the DOE, the project was initiated a few years earlier by Charles DeLisi and then, after creation of the Office of Biology and Environmental Research, headed by David Galas and later by Aristides Patrinos. See Robert Cook–Deegan's excellent book[75] for a detailed chronology of the genesis of the HGP, starting with a first meeting in May 1985 at the University of California, Santa Cruz, convened by Robert Sinsheimer.[76] In 1998, Celera Corporation launched a private research project led by Craig Venter.[77] Both public and private human genome projects produced drafts of the human genome in 2000 and announced completion of the project by 2003, only 13 years after its start.

Most of the government-sponsored research was conducted at universities in the United States and in other participating countries, among them the UK, Japan, France, Germany, and China.[78] The 20-member consortium of HGP sequencing centers had the following members:

- The Whitehead Institute/MIT Center for Genome Research, Cambridge, MA, USA
- The Wellcome Trust Sanger Institute, The Wellcome Trust Genome Campus, Hinxton, Cambridgeshire, UK
- The Washington University School of Medicine Genome Sequencing Center, St. Louis, MO, USA
- The United States DOE Joint Genome Institute, Walnut Creek, CA, USA
- Baylor College of Medicine Human Genome Sequencing Center, Department of Molecular and Human Genetics, Houston, TX, USA
- The RIKEN Genomic Sciences Center, Yokohama, Japan
- Genoscope and CNRS UMR-8030, Evry, France
- The GTC Sequencing Center, Genome Therapeutics Corporation, Waltham, MA, USA
- The Department of Genome Analysis, Institute of Molecular Biotechnology, Jena, Germany
- The Beijing Genomics Institute/Human Genome Center, Institute of Genetics, Chinese Academy of Sciences, Beijing, China
- The Multimegabase Sequencing Center, The Institute for Systems Biology, Seattle, Wash.
- The Stanford Genome Technology Center, Stanford, CA, USA
- The Stanford Human Genome Center and Department of Genetics, Stanford University School of Medicine, Stanford, CA, USA
- The University of Washington Genome Center, Seattle, WA, USA
- The Department of Molecular Biology, Keio University School of Medicine, Tokyo, Japan

- The University of Texas Southwestern Medical Center at Dallas, Dallas, TX, USA
- The University of Oklahoma's Advanced Center for Genome Technology, Dept. of Chemistry and Biochemistry, University of Oklahoma, Norman, OK, USA
- The Max Planck Institute for Molecular Genetics, Berlin, Germany
- The Cold Spring Harbor Laboratory, Lita Annenberg Hazen Genome Center, Cold Spring Harbor, NY, USA
- GBF, German Research Centre for Biotechnology, Braunschweig, Germany

Although focused primarily on a complete mapping of the human genome, the project also investigated several other nonhuman organisms such as *E. coli*, the fruit fly, and the laboratory mouse. It was an enormous undertaking, one of the largest single investigative projects in modern science, but it was completed 2 years ahead of the anticipated 15-year schedule. As to the HGP budget number, it may be difficult to fully analyze, given the global participation. Estimated contributions by the participating countries have been the United States, 54%; the UK, 33%; Japan, 7%; France, 2.8%; Germany, 2.2%; and China, 1%.[79] The two contributing US government agencies spent a total of ~$3.8B on the project, split between the DOE ($1,015 B) and the NIH ($2.786 B).[80]

The objectives of the HGP were stated as follows:

- Identify all protein-coding genes in human DNA. Initial estimates of their number were much too high—estimated at ~100,000 genes. By 2014, the number has dropped below 20,000.
- Determine genome sequences in human DNA.
- Store all found information into databases.
- Improve the tools used for data analysis.
- Transfer technologies to the private sector/industry.
- Address the ethical, legal, and social issues (ELSI).

The main reasons for the project's success were the evolution of sequencing technology and a wise decision of the funding agencies to set aside a considerable fraction of the overall budget for ELSI.

By doing so, possible objections by the media and by political and religious interest groups could be contained.

3.6 Sequencing Technologies

The science of today is the technology of tomorrow.

—Edward Teller (1908–2003)

The first DNA sequences were obtained in the early 1970s by academic researchers. Early forms of nucleotide sequencing were based on chromatography, laboratory techniques for the separation of mixtures that were invented in 1900 by Michail Tsvet[81] and first applied by him to the extraction of plant pigments such as chlorophyll and carotene. RNA sequencing was done first because it was easier to deal with a single strand of a helical molecule than working with the full double helix. Between 1972 and 1976, Walter Fiers and his coworkers at Ghent (Belgium) were able to sequence the first complete gene and the complete genome of a viral genome, bacteriophage MS2.[82]

During the 1970s, Sanger at the Medical Research Council in Cambridge, UK, developed his method of DNA sequencing with chain-terminating inhibitors,[83] and Walter Gilbert and Allan Maxam at Harvard in Cambridge, Massachusetts, developed DNA sequencing by chemical degradation.[84]

In 1980, Sanger and Gilbert shared one half of the Nobel Prize[85] for their contributions concerning the determination of base sequences in nucleic acids, with Paul Berg of Stanford University, capturing the other half for his fundamental studies of the biochemistry of nucleic acids, with particular regard to rDNA.

3.6.1 *Chemical Sequencing*

Maxam–Gilbert sequencing is based on chemical modification of DNA and subsequent cleavage at specific bases. Also known as *chemical sequencing*, this method's interesting history is described in detail by Gilbert in his Nobel lecture.[86] The basic idea is to find a repressor protein that can protect a short stretch of

DNA from degradation, caused, for example, by digestive enzymes. Subsequently, they found that protein binding would block not only enzymes but also chemical methylation, the addition of CH_3 groups to the bases in DNA. When comparing DNA fragments in the presence and absence of the repressor, there would be a stretch of DNA that would be methylated without repressor but be protected with it. By finding four separate chemical ways to fracture DNA selectively at each of the four bases A, G, C, and T, Maxam and Gilbert were able to deduct DNA sequences. However, the method's use of radioactive labeling and its technical complexity discouraged extensive use.

3.6.2 *Sanger Sequencing, NGS, and the $1000 Genome Challenge*

In his Nobel lecture[87] in 1959, Arthur Kornberg stated that "what Sanger has done for protein sequence remains to be done for nucleic acids. The problem is more difficult, but not insoluble." Amazingly, Sanger accepted the challenge and succeeded. His method, called the *chain termination method*,[88] developed by him and a team of coworkers in 1977, soon became the method of choice, owing to its relative ease and reliability.

Sanger's DNA sequencing method was based on his deep understanding of DNA's chemistry: the outer backbone of DNA is a monotonous repeat of sugar–phosphate–sugar–phosphate, etc., the only variables being which base (A, C, T, G) is inserted in the "rungs" of the DNA "ladder." Sanger's initial idea was to supply all of the components needed to produce the DNA sequence but to "starve" the reaction of one of the four bases needed to make DNA. For instance, in the sequence ACGTCGGTGC,[89] starving for T would produce ACGTCGG (blank) and ACG (blank). By separating the resulting molecules by length, it could be found that the eighth and fourth position of ACGTCGGTGC should be T.

Likewise, starving for G would produce AC (end), ACGTC (end), and ACGTCGGT (end), meaning G was in positions 3, 6, 7, and 9.

The whole sequence would follow directly from starving for each of the four nucleotide precursors and separating the molecules by length.

Sanger's breakthrough idea[90] was to find chemicals that were inserted in place of As, Cs, Ts, and Gs, but caused a growing DNA chain to end. For our example, the fragments obtained via chain termination would be ACGT* and ACGTCGGT*. With one terminator for each base type, the sequence could be read just by measuring the respective lengths of the fragments.

He divided the DNA sample to be sequenced into four separate sequencing reactions, containing all four of the standard deoxynucleotides (dATP, dGTP, dCTP, and dTTP) and the DNA polymerase, but added to each reaction only one of the four dideoxynucleotides ddATP, ddGTP, ddCTP, or ddTTP, respectively. Dideoxynucleotides are chain-terminating inhibitors of DNA polymerase. The dideoxyribonucleotides do not have a 3′ hydroxyl group, hence no further chain elongation can occur once this dideoxynucleotide is on the chain as no phosphodiester bond (see Fig. 2.24) can be created.

Sanger sequencing is the method that prevailed from the 1980s until the mid-2000s. Over that period, great technical advances were made and Sanger sequencing was automated. Important contributions were made by Prof. Leroy Hood's team at the California Institute of Technology and by the start-up company Applied Biosystems (ABI), which was founded in 1981 by two Hewlett–Packard engineers to build instruments for the new BT. ABI soon became the leading manufacturer of both protein and DNA sequencers, thanks to a very close cooperation with Prof. Hood's molecular biology group at CalTech. Prof. Hood created not only an academic environment that attracted top talent but also had a gift for fundraising and entrepeneurship. Key members of his instrumentation team were Henry Huang, Lloyd Smith, and Tim and Mike Hunkapiller.[91] Mike Hunkapiller joined ABI in 1983, where he moved from an initial technical leadership role all the way to the position of president, which he held from 1995 until his retirement in 2004. He went back into the field of DNA and RNA sequencing in 2005, first as board member and since 2012 as CEO of Pacific Biosciences,[92] one of the third-generation sequencing companies to be discussed below.

The automated DNA sequencing instruments that were commercialized by ABI were still based on Sanger's original concept but

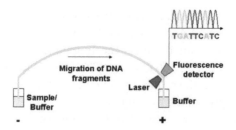

Figure 3.13   Automated Sanger sequencing. Source: Wikipedia.

included several important modifications, leading to the workflow described below and illustrated by Fig. 3.13:

- PCR-based cloning is used to create the desired number of copies of a given DNA sequence
- Double-stranded DNA is first heated to separate the strands. The single strands are then cut at various points to generate fragments.
- The fragments are added to a mixture of fluorescent dyes in which strands that end in a specific letter (A, C, G, or T) are all tagged with the same fluorescent dye
- The mixture is run through an agarose gel. Agarose (often extracted from seaweed) is a linear polymer that is frequently used for the separation of large molecules, especially DNA, by electrophoresis. Electrophoresis is an analytical technique used to separate DNA or RNA fragments by size and reactivity. In an electric field, nucleic acids are migrating toward the anode, due to the net negative charge of the sugar–phosphate backbone of the nucleic acid chain.
- The agarose gel is next passed through a light amplification by stimulated emission of radiation (laser) beam that identifies fragments by length and by fluorescent color. Laser devices differ from other sources of light because they emit light coherently, resulting in accurate and reproducible results of the fluorescence detection.
- The DNA sequence can be inferred from the sequence of fluorescing colors emitted as the fragments in the gel go past the laser beam. An important part of the instrumentation is

performed by computer algorithms that analyze the signals and perform better than visual examination by human operators.

The DNA fragments sequenced by automated Sanger sequencing instruments have to be reassembled in order to reconstruct the original sequence to be analyzed. The size of Sanger sequencing fragments is limited to about 800–1000 base pairs, which has to be compared to a full human genome consisting of 3 billion base pairs. Assembling a full genome is therefore a highly complex task requiring sophisticated computer software and expert bioinformatics skills. More about it will be given below in Section 3.8.

ABI dominated the DNA sequencing market until the completion of the HGP.

Although alternative sequencing technologies were pursued in Japan by RIKEN Institute and Hitachi and in Europe by the European Molecular Biology Lab (EMBL) in Heidelberg and by the Swedish company Pharmacia, only ABI had the commercial success required to continue supplying the global need for DNA sequencing instruments.

After three decades of gradual improvement, the Sanger biochemistry can be achieve per base "raw" accuracies as high as 99.999%, higher than all next-generation sequencing (NGS) technologies to be covered below. The common NGS goal is to significantly increase sequencing speed and to cut cost with an ultimate objective of achieving the "$1000 ($1 K) human genome."

In September 2003, the J. Craig Venter Science Foundation promised $500,000[93] for the achievement of a $1 K genome. To attract even more resources to this goal, Venter later joined forces with the X Prize Foundation, wrapping his competition and prize purse into the $10 M Archon Genomics X Prize, announced in October 2006, to the first team to sequence 100 human genomes of centenarians within 30 days at a cost of less than $1000 per genome.[94]

But already in 2004 the NIH launched a $70 M grant program to support researchers working to sequence a complete mammalian genome initially for $100,000 and ultimately for $1000.

That program dramatically accelerated progress, encouraging researchers to pursue a wide variety of new ideas that led to an explosion of start-up companies, each pursuing its own angle on the technology.

In fact, progress has been so fast that the "Archon Genomics X Prize" competition was canceled on August 22, 2013. The reason quoted was the fact that only two competitors (George Church, Harvard, and Ion Torrent) had signed up and that the $10 M prize was no longer a sufficient incentive for sequencing companies. In other words, the sponsors felt that the competition had been outpaced by actual innovation.

Let's review what happened.

3.6.3 *Second-Generation DNA Sequencing*

Automated Sanger sequencing, or first-generation sequencing, can be characterized as a low-throughput operation with high accuracy, relying on fluorescent tags, electrophoresis, and optical detection via laser. The next step in the history of DNA sequencing was to significantly improve throughput, while sacrificing accuracy but still trying to retain good-enough accuracy.

Sequencing by synthesis involves taking a single strand of the DNA to be sequenced and then synthesizing its complementary strand.

The *pyrosequencing* method, invented at Stockholm's Royal Institute of Technology by Mostafa Ronaghi, Pål Nyrén, and Mathias Uhlén et al.,[95] is based on detecting the activity of DNA polymerase with another chemiluminescent enzyme. Essentially, the method allows sequencing of a single strand of DNA (ssDNA) by synthesizing the complementary strand along it, one base pair at a time, and detecting which base was actually added at each step. Solutions of A, C, G, and T nucleotides are sequentially added to ssDNA and then removed from the reaction. Light is produced only when the nucleotide solution complements the first unpaired base of the template, that is, enters into a chemical reaction that releases energy leading to a visible signal that can be detected. As shown in Fig. 3.14, both pyrophosphate and a hydrogen ion are released when the fitting nucleotide is added. The proposed pyrosequencing method relies on the detection of pyrophosphate and was first

Polymerase integrates a nucleotide.

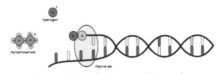

Hydrogen and pyrophosphate are released.

Figure 3.14 Pyrosequencing leads to release of pyrophosphate $(P_2O_7)^{4-}$ and hydrogen (H^+) ions. Source: Wikipedia.

commercialized by the company 454 Life Sciences founded in Branford, CT, by Jonathan Rothberg.[96]

An enzyme, ATP sulfurylase, quantitatively converts pyrophosphate to ATP in the presence of adenosine-5′-phosphosulfate. This ATP acts as fuel to the luciferase-mediated conversion of luciferin to oxyluciferin that generates visible light in amounts that are proportional to the amount of ATP. The light produced in the luciferase-catalyzed reaction is detected by the charged-coupled device (CCD) sensor of a camera and then analyzed. Unincorporated nucleotides and ATP are degraded by other chemicals and the reaction can restart with another nucleotide.

A limitation of the pyrosequencing method is that the lengths of individual reads of DNA sequence are somewhat shorter than the 800 base pairs obtainable with Sanger sequencing. This can make the process of genome assembly more difficult.

The Swedish inventors started a company, Pyrosequencing AB in Uppsala, to commercialize the technology but also licensed it to 454 Life Sciences, initially part of Curagen, a genomics company founded by J. Rothberg in 1993 and acquired by Celldex[97] in 2009.

Pyrosequencing AB was listed on the Stockholm Stock Exchange in 1999, later renamed to Biotage in 2003 and acquired by Qiagen, a German–Dutch life sciences company, in 2008.

454 Life Sciences subsequently developed an array-based pyrosequencing technology that has emerged as a platform for large-scale DNA sequencing and the first sddecond-generation sequencing company reaching the market in 2005.

To make it a high-throughput method, 454 Life Sciences added a number of innovations to turn basic pyrosequencing into a large-scale parallel sequencing instrument. The system relies on fixing nebulized and adapter-ligated DNA fragments to small DNA capture beads in a water-in-oil emulsion. The DNA fixed to these beads is then amplified by PCR. Each DNA-bound bead is placed into a ~29 μm well on a PicoTiterPlate, a fiber-optic chip. A mix of enzymes such as DNA polymerase, ATP sulfurylase, and luciferase are also packed into the well. A microfluidics subsystem delivers sequencing reagents (containing buffers and nucleotides) across the wells of the plate. The four DNA nucleotides are added sequentially in a fixed order across the PicoTiterPlate device during a sequencing run.

During the nucleotide flow, millions of copies of DNA bound to each of the beads are sequenced in parallel. When a nucleotide complementary to the template strand is added into a well, the polymerase extends the existing DNA strand by adding nucleotide(s). Addition of one (or more) nucleotide(s) generates a light signal that is recorded by the CCD camera and data is stored for downstream analysis.

Despite being the first commercial manufacturer of second-generation sequencing instruments and despite being acquired in 2007 by leading diagnostics company Roche, the technology developed by 454 Life Sciences has not been able to conquer enough market share to survive the next wave of innovation. In 2013, Roche decided to discontinue 454 Life Sciences operations by 2016.[98]

The winning second-generation sequencing technology, also based on sequencing by synthesis (SBS), was developed by Solexa, later (in 2007) to be acquired by Illumina,[99] based in San Diego, CA, the DNA sequencing market leader.

Solexa was founded by Shankar Balasubramanian and David Klenerman in 1998, and developed a sequencing method based on *reversible dye terminators technology*[100] and engineered polymerases. The reversible terminator (RT) chemistry was developed internally at Solexa, based on a concept invented by the founders at Cambridge University's chemistry department. In 2004, Solexa acquired the company Manteia Predictive Medicine in order to gain a massively parallel sequencing technology based on DNA clusters, which involves the clonal amplification of DNA on a surface.

Sample preparation is a very important first step: DNA molecules and primers are first attached on a slide and amplified with polymerase so that local clonal DNA colonies, later coined "DNA clusters," are formed. This method for in vitro clonal amplification is also referred to as a bridge PCR. Then, to determine the sequence, four types of RT bases are added and nonincorporated nucleotides are washed away. After having taken images of the fluorescently labeled nucleotides, the dye is chemically removed from the DNA, allowing the next cycle to begin. The DNA fragments are extended one nucleotide at a time, allowing image acquisition to be performed at a delayed moment. A key Solexa advantage over 454's pyrosequencing is provided by the decoupling of the enzymatic reaction from the image capture, allowing optimal throughput and theoretically unlimited sequencing capacity.

Here are 12 detailed steps in the workflow of a Solexa/Illumina sequencer, where the first 6 are related to sample preparation and amplification, steps 7–11 are performing the core RT chemistry and image acquisition, and the final step 12 can be classified as data analysis.

1. Sample preparation: The sample double-stranded DNA to be sequenced is randomly fragment and ligate adapters added to both ends of the fragments.
2. Attachment of DNA to the surface: Single-stranded fragments are randomly bound to the inside surface of the flow cell channels.
3. Bridge amplification: Unlabeled nucleotides and an enzyme are added to initiate solid-phase bridge amplification.
4. Creation of double-stranded fragments: The enzyme builds double-stranded bridges on the solid-phase substrate.
5. Denaturing of the double-stranded molecules: Denaturization leaves single-stranded templates anchored to the substrate.

6. Complete amplification: Several million dense clusters of double-stranded DNA are generated in each channel of the flow cell.
7. Determination of first base: Four labeled RTs, primers, and DNA polymerase are added to start the first sequencing cycle.
8. Imaging of the first base: After laser excitation, the emitted fluorescence in each cluster is captured and the first base is identified.
9. Determination of the second base: Step 7 is repeated.
10. Imaging of the second base: After laser excitation, the image is captured and the second base is identified and recorded.
11. Sequencing-over of multiple chemistry cycles: Sequencing cycles are repeated to determine the sequence of bases in a fragment, one base at a time.
12. Alignment of data: The data are aligned and compared to a reference, and sequencing differences are identified.

A third second-generation sequencing technology, sequencing by oligonucleotide ligation and detection (SOLiD) was developed by ABI and brought to market in 2006. Sequencing by ligation uses the enzyme DNA ligase to identify the nucleotide present at a given position in a DNA sequence.

SOLiD has its origins in the Polony sequencing method, developed in the laboratory of George M. Church at Harvard.[101] The technology was licensed to Agencourt Biosciences, subsequently spun out into Agencourt Personal Genomics, and eventually incorporated into ABI's SOLiD platform.

The target molecule to be sequenced is a single strand of unknown DNA sequence, flanked on at least one end by a known sequence. A short anchor strand is brought in to bind the known sequence.

A mixed pool of eight- to nine-base probe oligonucleotides is then brought in and labeled with fluorescent dyes according to the position that will be sequenced. These molecules hybridize to the target DNA sequence, next to the anchor sequence, and DNA ligase preferentially joins the molecule to the anchor when its bases match the unknown DNA sequence. On the basis of the fluorescence produced by the molecule, one can infer the identity of the nucleotide at this position in the unknown sequence. More technical details may be found on the ABI website.[102]

The future of ABI's SOLiD technology appears similar to Roche's 454 Life Sciences pyrosequencing platform: both have lost out to Illumina's market dominance. Internally, SOLiD has given way to Ion Torrent's platform after acquisition (2010) of Ion Torrent by Life Technologies, the company in charge of ABI's legacy assets since 2008. In 2013, Thermo Fisher acquired Life Technologies and rebranded the company as Thermo Fisher Scientific.[103]

Finally, let's discuss the second-generation sequencing technology developed by Complete Genomics[104] (CG), a company that was founded in 2006 by Clifford Reid and Radoje (Rade) Drmanac. CG never sold sequencing instruments but developed its own technology to provide instead sequencing services. CG's proprietary sequencing by hybridization technology includes several interesting components:

1. DNA Nanoball Arrays: During library preparation, genomic DNA is fragmented and each fragment is then copied in a manner (also referred to as rolling circle replication) that results in a long single molecule containing hundreds of copies of the same fragment. Each long single molecule is then consolidated into a small DNA nanoball (DNB), approximately 200 nm in diameter. The DNBs are then packed together very tightly on a silicon chip. Established photolithography processes adapted from the semiconductor industry are then used to create a silicon chip that has a grid pattern of small spots, also referred to as sticky spots. CG's process ensures that over 90% of the sticky spots contain exactly one DNB, without adherence of the DNA to the areas between the spots. The silicon chip filled with DNBs is referred to as a DNB array. Each finished DNB array contains up to 180 billion bases of genomic DNA prepared for imaging.

2. Combinatorial Probe–Anchor Ligation (cPAL): The DNB contains both genomic DNA sequence and adapter sequence. During sequencing, an anchor probe binds to the adapter sequence. A ligase enzyme then attaches one of four possible fluorescent-labeled probes to the anchor, depending on the sequence being read in the fragment. By imaging the fluorescence during each

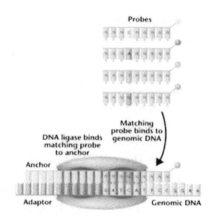

Figure 3.15 Complete Genomics's combinatorial probe–ancor ligation (cPAL).[105]

ligation step, the cPAL process, as illustrated in Fig. 3.15, can subsequently determine the sequence of nucleotides in each DNB.

By focusing on high-throughput processes and workflow automation, CG created a powerful human genome sequencing services capability. Instead of building instruments for external use by customers, CG spent all its resources on internal operations. CG's technology development has been relying on advances in semiconductor technology (photolithography) and digital camera technology (processing of images). CG technology goals were to miniaturize, to parallelize, and to focus heavily on advanced data analysis capability (including image processing) and information technology (IT) infrastructure.

In 2009, CG's cofounder and CEO Reid announced a price of $20,000 per human genome to be sequenced by his company. He further presented a vision of 100,000 human genomes to be sequenced by 2012. CG did not quite reach this ambitious goal but made a lot progress along the way. CG's whole human sequencing throughput increased to:

2010: ~300 genomes per quarter
2011: ~500 genomes per quarter
2012: ~2000 genomes per quarter

To buy DNA sequencing services market share, the cost per genome was reduced to $5000 in 2012 and the cost for reagents (chemicals) was—according to CG management—already reduced to <$1000 in 2010.

Still, by 2012, CG was not yet profitable and was running out of cash. As a consequence, CG agreed in December 2012[106] to be acquired by BGI,[107] the Shenzhen, China, based global sequencing services leader. For more details see below when we will discuss business aspects of DNA sequencing.

3.6.4 *Third-Generation DNA Sequencing*

Whereas second-generation sequencing technologies are relying on both fragment amplification and optical detection, third-generation technologies only use one of those two features.

Let us discuss two such technologies, first Pacific Biosystems and second Ion Torrent technology.

Heliscope single-molecule sequencing is another third-generation method of single-molecule sequencing developed by Helicos Biosciences. Helicos BioSciences was founded in 2003 by Stephen Quake of Stanford University, Stanley Lapidus, and Noubar Afeyan and was based in Cambridge, MA. The Helicos Genetic Analysis Platform was the first DNA-sequencing instrument to image individual DNA molecules. Helicos technology uses DNA fragments attached to a flow cell surface. The next steps involve extension-based sequencing with cyclic washes of the flow cell with fluorescently labeled nucleotides (one nucleotide type at a time, as with the Sanger method). Although successful technically and particularly suitable for RNA sequencing, the instruments were too expensive and eventually not competitive in a fast changing market with rapid price/performance improvements by competitors such as Illumina and Ion Torrent. On November 15, 2012, Helicos BioSciences filed for Chapter 11 bankruptcy.

Pacific Biosystems (PacBio)[108] was founded by Stephen Turner in 2004 with the goal of developing a single-molecule real-time (SMRT) approach for nucleic acid sequencing. The concept, to "eavesdrop on single molecules of DNA polymerase synthesizing virgin DNA in real time,"[109] was initially developed by Turner and his colleague Jonas

Korlach at Cornell University in the Laboratories of Watt Webb and Harold Craighead. In 2005, PacBio won a $6.6 M grant sponsored by the NHGRI of the NIH for the development of technologies leading to the "$1000 genome," the complete sequencing of a human genome at a cost of only $1000.

PacBio's proposal was based on single-molecule sequencing (avoiding the sample preparation and DNA chemistry needed for all second-generation sequencing technologies) and optical detection.

The DNA is synthesized in zero-mode wave guides (ZMWs), small well-like containers with the capturing tools located at the bottom of the well. A thin layer of aluminum on top of a silica slide has etched into it a grid of 20 zL (20 × 10^{-21} L) wells, all holding single DNA polymerase molecules. As fluorescently tagged bases are diffusing in and out of the ZMW's, they are registered by a camera. The camera needs a few milliseconds to detect the fluorescing dye before the DNA polymerase enzyme ratchets along the DNA to the next position. The wells are constructed in a way that only the fluorescence occurring at the bottom of the well is detected. Upon incorporation into the DNA strand, the fluorescent label is detached from the nucleotide, leaving an unmodified DNA strand.

The SMRT concept, watching DNA polymerase, the molecule described above in Section 3.4 that earned Arthur Kornberg of Stanford University a Nobel Prize in 1959, work in real time as it performs DNA synthesis, is indeed attractive. The PacBio start-up received strong interest by Silicon Valley investors[110] and Turner's presentation in 2008 at the yearly Advances in Genome Biology and Technology (AGBT)[111] conference in Marco Island, FL, was the "hit of the show." PacBio has since struggled to live up to high expectations but has found application areas where the SMRT concept has advantages over mainstream sequencers such as Illumina's. In September 2013, Roche and PacBio signed a $75 M collaboration[112] to develop diagnostic products, including sequencing systems and consumables. Led by DNA sequencing pioneer Mike Hunkapiller, Pacific Biosciences will partner with Roche to develop and manufacture products intended for clinical use in the field of human in vitro diagnostics (IVD), as well as continue to market its current and future products for all fields outside of human IVD, including research, plant, animal, and applied markets.

The Ion Torrent concept is based on a combination of DNA fragment amplification and electrical detection, replacing the optical detection of pyrophosphate used in pyrosequencing with electrical detection of the hydrogen ions (protons) which—as shown above in Fig. 3.15—are also released whenever a nucleotide is integrated.

Having sold his Branford, CT, company 454 Life Sciences to Roche Diagnostics on March 29, 2007, Jonathan Rothberg went ahead and started Ion Torrent in nearby Guilford, Connecticut. He licensed the electrical detection concept from DNA Electronics, an Imperial College, London, spinoff founded by Chris Toumazou and based on his 2001 invention of detecting protons released during DNA synthesis by means of ion-sensitive field-emission transistors (ISFETs).[113] An ISFET incorporates microfluidics, systems that manipulate small $(10^{-9}-10^{-18}$ L) amounts of fluids on microfabricated substrates, on top of a conventional transistor.

As already demonstrated with 454, Rothberg was again able to assemble and lead a strong team capable of turning concepts and ideas into a working prototype and soon successful products. Rothberg realized that advances in semiconductor technology could help him to quickly propel Ion Torrent toward faster and cheaper detection and therefore significantly increase sequencing throughput, while reducing the footprint of the sequencing instrument.[114]

On August 10, 2010, Life Technologies announced the acquisition of Ion Torrent for up to $725 M.[115]

In the press release, it was mentioned that ". . . unlike existing second-generation sequencers, Ion Torrent does not require lasers, cameras, or labels."

After Thermo Fisher's acquisition of Life Technologies in 2013, Rothberg left to pursue other interests.

Finally, Bionano Genomics (San Diego, CA) has launched a single-molecule sequencing technology and a desktop platform, IRYS, that is based on a nanochannel technology developed at Princeton University by co-inventor, founder, and chief scientific officer Han Cao.

Founded in 2003 as BioNanomatrix in Philadelphia, Bionano received $22 M in total funding by 2010,[116] received another $23.3 M and moved to San Diego in 2011 where it started distributing its first commercial product. Previously, in 2007, BioNano and CG received an NIST grant to combine BioNano's platform with CG's sequencing chemistry. In October 2013, BioNano Genomics received another $10 M round of funding.

Unamplified, native-state DNA molecules, each up to a megabase long, can be loaded into the IrysChip's nanochannels. Laser excitation is used to illuminate fluorescently labeled DNA in the nanochannels of the chip. The DNA can be labeled using a wide variety of methods, including IrysPrep reagents or user-defined methods. Molecules are uniformly stretched and optically separated in 40nm wide channels, enabling high-resolution imaging of single molecules that is performed by an onboard CCD camera. The camera has a proprietary autofocusing mechanism and control software. It rapidly scans the chip and detects the fluorescence labels. The big advantage of IRYS technology is the effective long read length, accomplished by piecing together the labeled short pieces of DNA.

IrysChips preserve the true architecture of the genome. Genome maps are generated from massively parallel, single-molecule visualization of extremely long DNA without amplification, providing long-range contiguity and eliminating PCR bias.

IRYS allows the generation of genome maps: DNA molecules labeled with IrysPrep reagents contain sufficient sequence uniqueness to unambiguously map to a reference map or can be assembled de novo. The result is a comprehensive and accurate view that lends itself to mapping the "dark matter of the genome," regions that have not been fully explored by conventional methods with short read lengths. Provides unique insights into genome structure and opens the door to countless applications, both with and without sequencing.

Figure 3.16 illustrates IRYS de novo genome map assembly.

3.6.5 *Fourth-Generation DNA Sequencing*

Following a classification introduced by Roche Diagnostics,[118] we define fourth-generation sequencing as single-molecule sequencing with electrical detection.

So far, only one fourth-generation sequencing instrument has reached the market, namely, products developed by Oxford Nanopore.[119]

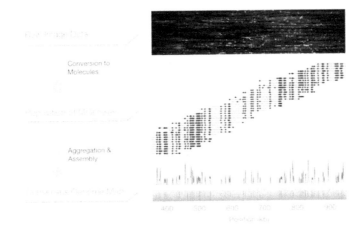

Figure 3.16 BioNano Genomics IRYS de novo genome map assembly.[117]

As already mentioned above, the AGBT conference that has been held annually at Marco Island, Florida, since 2000, and is considered the most important meeting for the DNA sequencing instrument community.

In 2008, PacBio made the strongest impression at AGBT by presenting its technology, in 2009 CG made the above-mentioned announcement to sequence a human genome for only $5000, in 2010 Jonathan Rothberg of Ion Torrent took center stage by introducing his first product and in 2012, it was Oxford Nanopore's turn to dazzle the participants with a demonstration of a working nanopore sequencing prototype.

On the basis of independent proposals by David Deamer (UC Santa Cruz) and George Church (Harvard), Deamer et al.[120] were able to demonstrate as early as 1996 that ssDNA and ssRNA molecules can be driven through a pore-forming protein and detected by their effect on the ionic current through this nanopore. This system used the *Staphylococcus aureus* toxin, α-hemolysin, the use of which as a biosensor had been pioneered by a research team led by Hagan Bayley.[121] Bayley was educated at Oxford but spent many years in the United States at leading academic institutions such as Harvard, MIT, and Texas A&M, before returning back to the UK in 2003 as professor of chemistry, Oxford University. Bayley and coworkers have shown that an α-hemolysin pore is remarkably stable and remains functional at high temperatures close to the boiling point of water. Because the inside diameter of the α-hemolysin pore is barely as large ($\sim$2–5 nm) as the diameter of a single nucleic acid strand, Deamer et al. were able to show that ssDNA and RNA nucleotides can be translocated through the pore in strictly single-file, sequential order by means of an ionic current. Because the current through the nanopore is partially blocked by the translocating molecule, each translocating molecule reduces the ionic current relative to that which flows through the open, unblocked pore. Assuming that each nucleotide causes a characteristic modulation of the ionic current during its passage through the nanopore, the sequence of current modulations would reflect the sequence of bases, holding the promise of nanopore sequencing.

Oxford Nanopore is headquartered in Oxford, UK, with offices in Cambridge, MA. It was cofounded by Gordon Sanghera, Spike Willcocks, and Hagan Bayley in 2005 to translate academic nanopore research into a commercial, electronics-based sensing technology.

A nanopore can be defined as a nanoscale hole and may be:

- biological: formed by a pore-forming protein in a membrane such as a lipid bilayer;
- solid state: formed in synthetic materials such as silicon nitride or graphene; or
- hybrid: formed by a pore-forming protein set in synthetic material.

Oxford Nanopore's technology advisory board includes representatives of Oxford, Harvard, and Boston University, as well as the University of California at Santa Cruz. In addition, the company was able to attract two key members of the successful Solexa team, namely, John Milton and Clive Brown, who are serving as Oxford Nanopore's chief scientific and chief technology officers, respectively.

Nanopore sequencing presents several challenges, including controlled fabrication, DNA/RNA translocation control, and sensing of the nucleotides. In addition to measurement of the modulated voltage through the nanopore, caused by differences in the chemical properties of A, C, T, and G, Oxford Nanopore is pursuing another technology named *strand sequencing*.[122] As an ssDNA polymer is passed through a protein nanopore, individual DNA bases on the strand are identified in sequence as the DNA molecule passes through. The strand-sequencing method promises to generate single-molecule sequencing read lengths of many tens of kilobases.

However, as a DNA polymer passes through a nanopore, a number of individual DNA bases occupy the aperture of the nanopore at any time. Therefore, whatever the sensing mechanism, it covers several individual bases within a given strand and must therefore be analyzed further to translate the characteristic electronic signals into DNA sequence data.

It is easily seen that reading bases in groups of three or more can lead to data analysis challenges.

If it's possible to read single bases, four signals must be distinguished corresponding to DNA's four bases A, C, T, and G. If two bases are read at a time, there will be $4 \times 4 = 16$ different patterns, namely, AA, CC, TT, GG, AC, AT, AG, CA, CT, CG, TA, TC, TG, GA, GC, and GT.

If three bases are read together, there will be $4 \times 4 \times 4 = 64$ patterns and they will all have to be processed and recognized in order to solve the nucleotide sensing problem.

Another requirement for strand sequencing to work is a precise method of controlled translocation of the strand through the nanopore. In fact, the above-mentioned data analysis will have to assume controlled ratcheting of DNA through the nanopore.

Note that nanopore sequencing can also be applied to the analysis of proteins. By combining a given protein with an aptamer—an oligonucleic acid that can bind specifically to a site on a target protein—the protein–aptamer complex creates a characteristic disruption of the current running through a nanopore, leading to a corresponding characteristic electric signal.

Time will tell how commercially successful the Oxford Nanopore company will be. It certainly is a beautiful example of nanotechnology applied to important biodiagnostic problems.

Another fourth-generation sequencing technology is developed by Genia, founded in March of 2009 and headquartered in Mountain View, CA. The heart of Genia's technology is the biological nanopore, a protein pore embedded in a lipid bilayer membrane. Electronic sensor technology enables highly efficient nanopore–membrane assembly and accuracy of current readings. Genia's NanoTag sequencing approach, developed in collaboration with Jingyue Ju of Columbia University, NY, and George Church of Harvard University, uses a DNA replication enzyme to sequence a template strand with single-base precision as base-specific engineered tags cleaved by the enzyme are captured by the nanopore. Genia identifies DNA sequences not by detecting the nucleotides themselves with the nanopore, but by measuring the current changes caused by the passage of each of four different tags that are released from the incorporated nucleotide during the polymerase reaction. As the cleaved tags travel through the pore, they attenuate the current flow across the membrane in a sequence-dependent manner. Genia's base recognition technology is based on measuring changes in electrical currents.

In 2013, Genia presented data from an alpha version of its Nanopore chip at the 8th International Conference on Genomics (ICG-8) in Shenzhen, China, and promised a beta version in the near future. The beta version of the chip, a circuit with 128,000 integrated sensors, was actually working in 2014 and caught the interest of Roche. On June 2, 2014, Genia was acquired by Roche.

Still another fourth-generation sequencing technology worth mentioning is IBM's *DNA transistor* concept, an idea conceived at IBM's T.J. Watson Research laboratory in Yorktown, NY, by Stas Polonsky, Steve Rossnagel, and Gustavo Stolovitzky.[123] The IBM scientists proposed the concept of a nanoelectromechanical device capable of controlling the position of DNA inside a nanopore with single-nucleotide accuracy. In analogy to solid-state transistors in which a small voltage controls the current between two electrodes, a voltage strategically located inside the nanopore could control the translocation of a single DNA molecule. As an immediate application of the device, the researchers proposed DNA sequencing and were able to first attract a grant from the NIH and, in July 2010, to enter into a significant partnership with Roche/454 Life Sciences.[124]

Figure 3.17 shows the result of successful computer simulations. The device consists of a multilayer (metal/dielectric/metal/dielectric/metal) nanostructure built into the membrane that contains

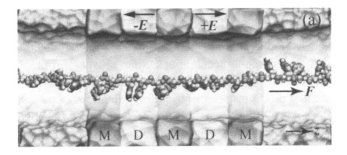

Figure 3.17 IBM Research concept of DNA translocation control through a multilayer metal–dielectric nanopore.[125]

the nanopore. Voltage biases between the electrically addressable metal layers will modulate the electric field inside the nanopore. This device utilizes the interaction of discrete charges along the backbone of a DNA molecule with the modulated electric field to trap DNA in the nanopore with single-base resolution. By cyclically turning on and off these gate voltages, it can be shown theoretically that DNA can move (and even ratchet back and forth) through the nanopore at a rate of one nucleotide per cycle. The device is called a DNA transistor because a DNA current is produced in response to modulation of gate voltages in the device.

By combining expertise in life sciences research, computational biology, semiconductor fabrication and nanotechnology, IBM and Roche hoped to come up with breakthrough sequencing technologies that would make DNA sequencing significantly faster and cheaper.

When trying to implement the concept, IBM and Roche experienced experimental setbacks caused by the very small dimensions of the device that were testing the limits of available fabrication technology. The setbacks did not invalidate the concept but caused delays in the perhaps overoptimistic plan to move from an idea (2007) to a working proof of concept in only six years. In April 2013, Roche decided to abandon the project due to high technical risks.[126]

It will be interesting to see if this promising but technically challenging project will be pursued further.

Another interesting piece of the nanopore/nanochannel sequencing puzzle could be the use of tunneling currents to distinguish the nucleotides. Stuart Lindsay and coworkers at Arizona State University have proposed *recognition tunneling*[127] between a biopolymer and a functionalized electrode as a way to solve the single-molecule nucleotide sensing problem. Both quantum chemical calculations and laboratory experiments seem to indicate that this approach has technical potential.

Before concluding, let us review the promise and challenges of fourth-generation sequencing technologies.

One of the most compelling advantages of nanopore sequencing is the prospect of inexpensive sample preparation requiring minimal chemistries or enzyme-dependent amplification. The components of an ideal commercial sequencing system using electrical measurements would consist of a disposable detector chip containing an array of nanopores having the required integrated microfluidics and electronic probes.

However, translocation control and nucleotide sensing remain serious challenges. How can single bases be detected accurately when they are only spaced by 0.4 nm in a pore of, say, 20 nm of depth?

Regarding the fabrication challenges of synthetic nanopores, it may be easier to choose nanochannel geometries instead. Fabricating a pore with a <10 nm hole is extremely difficult. Channel geometry will certainly be easier to fabricate and manufacture.

Given the significant global brainpower focused on single-molecule nanopore/nanochannel sequencing, we can expect to see more fourth-generation sequencing technologies that will keep alive the dream of eventually sequencing whole human genomes with inexpensive hand-held devices.

3.7 Sequencing Data Analysis

I'm fascinated by the idea that genetics is digital. A gene is a long sequence of coded letters, like computer information. Modern biology is becoming very much a branch of information technology.

—Richard Dawkins (born 1941)

Since the first decade of the 21st century, the term "big data analytics" has been applied to the analysis of huge volumes of data generated by (among others) retail and financial industries, to understand trends in customer behavior and to design successful marketing campaigns. In the life sciences, the dawn of "big data" already happened in 1953, when Watson and Crick discovered the structure of DNA and the associated role of DNA to store crucial information. Ever since, and accelerated by the HGP, we are developing big data analytics methods for sequencing data.

A typical DNA or RNA sequencing project can be broken down into four principal activities:

1. Sample preparation and experimental design
2. Sequencing
3. Data reduction and management
4. Downstream analysis

Before discussing steps 3–4, that is, the role of computers in DNA sequencing, let us define a few terms routinely used in the field of NGS.

- **Shotgun sequencing**: So far, there is no sequencing technology that can read DNA like a punch tape. Instead, the DNA to be sequenced is broken up into small pieces (fragments) that can be analyzed experimentally. After having done so, the pieces have to be put together again or reassembled, a task requiring significant computing resources.
- **Read length**: Typical length of DNA that a sequencing technology can handle before having to refresh all reagents and/or equipment. Note that first-generation Sanger sequencing, while slow and expensive, still provides higher read lengths and higher accuracy than second-generation sequencing techniques.
- **Coverage**: Number of times a single genome has to be sequenced to statistically reduce error rates. Typically, second-generation technologies require at least 5–30 times' coverage.

Data analysis has to be adapted to the respective strengths and weaknesses of the various sequencing technologies.

Comparisons published in 2012 showed the following results for the various commercially available sequencing technologies discussed above:[128]

- Read length: PacBio can handle read lengths of 10,000 base pairs; Oxford Nanopore could handle up to 9000 base pairs, as reported at AGBT 2014;[129] all other NGS technologies are working with a read length below Sanger's 1000 base pair limit; 454 can handle 700 base pairs. Accuracy: PacBio has the lowest accuracy; Sanger sequencing has still the highest. Other NGS technologies must compensate with increased coverage.
- Cost per million bases (US$): Sanger sequencing $2400; PacBio and IonTorrent <$1; Roche 454 ≈ $10; Illumina ≈ $0.05–$0.1.

Improving on read length and accuracy values is difficult as they are intrinsic properties of a given technology. Companies like Illumina and Ion Torrent (via ISFET chip density) have mostly focused on increasing throughput and have been able to cut the cost per million bases in each new generation of their instruments. Illumina and Ion Torrent are therefore the price/performance leader.

In January 2014, Illumina announced a new sequencing machine capable of delivering five human genomes in a day at a sequencing cost of ~$1000 per genome, not counting full amortization of the instrument.

Sequencing technology has come a long way since completion of the HGP!

It is interesting to compare the evolution over time of the relative roles of the four steps in a sequencing project. According to an analysis by a team led by Mark Gerstein of Yale University,[130] data management and downstream analysis are expected to require increasing amounts of resources and, as shown in Fig. 3.18, will dominate by 2020 as whole genome sequencing (WGS) will require <$1000.

Let's briefly discuss what is involved in the computational activities.

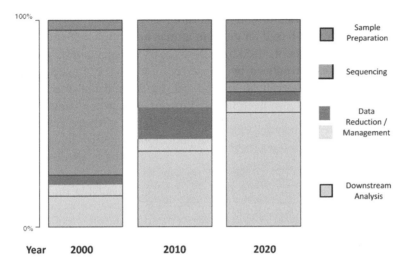

Figure 3.18 Contribution by activity (%) to a sequencing project (2000, 2010, 2020).

3.7.1 *Data Reduction and Management*

Although the number of nucleotides in the human genome is ~3 billion, WGS with NGS technology provided by industry leader Illumina will, due to required coverage, generate as much as 500–600 billion nucleotides, captured rather inefficiently in FASTQ files that store the nucleotide sequence in a text-based format along with corresponding quality scores. To reduce the data volumes, compression strategies such as binary sequence alignment/map (BAM)[131] are used to reduce the size of the files. Reference-based compression algorithms such as CRAM[132] can achieve a more than tenfold reduction of required storage space. Further improvements will certainly be possible as successful strategies from video compression will be applied.

In de novo sequencing the genome is not yet known, that is, there is no reference genome available.

Software packages dealing with the huge task of assembling a de novo genome require significant computing resources. Modern algorithms are typically relying on the mathematics of graph theory to deal with the challenge. The VELVET algorithm, developed at the European Bioinformatics Institute (EBI) by Daniel Zerbino and Ewan Birney, is based on deBrujn graphs and can handle both the short DNA fragments used by NGS sequencing instruments, and the complexities introduced by NGS error reduction strategies based on high coverage values. VELVET can be downloaded from EBI's website but has also been licensed to commercial companies such as BioNumerics by Applied Maths, a Belgian software company, founded in 1992, that specializes on Bioinformatics. Its website provides a good introduction and overview of the steps needed in sequencing data analysis.

If a *reference genome* for a given organism is available, the task of the data analysis software is to align the (re-)sequenced fragments with the reference. Although slightly less demanding than de novo sequencing data analysis, sequence alignment is a formidable task. A popular software package is BOWTIE, developed at Johns Hopkins University by Ben Langmead et al. and made available at JHU's SourceForge website.

Subsequent data reduction activities may have the goal to generate meaningful high-level summaries. For example, variant call format (VCF) files[133] describe genomic variants including SNPs, structural sequence variations such as deletions, duplications, inversions and translocations. The creation of such high-level summaries is similar to the encoding of protein structures in Protein Data Bank (PDB) files discussed already in Section 2.4.

In addition, sequencing instrument vendors such as Illumina and Thermo/Life Technologies/Ion Torrent are providing software to review the quality and accuracy of sequencing runs.

As NGS instruments are developed and improved at ever increasing speed, the software to analyze sequencing data is trying to catch up. Still, if the trends established since the completion of the HGP continue, data management and analysis may well be the bottleneck of the future.

3.7.2 *Downstream Analysis*

The goal of all downstream analysis projects is to unravel the complexity of biological mechanisms hidden in the sequence reads. WGS is expected to move into the mainstream of healthcare as sequencing costs are coming down, sequencing results are becoming increasingly reliable and trustworthy, and data analysis techniques are maturing. During the first decade after completion of the HGP, sequencing was mostly used for research and practiced by a small community of scientists working in dedicated sequencing centers, academic medical research centers with particular focus on diseases such as cancer.

What can be expected to happen by 2020 is

- the translation of WGS-based medical research results into diagnosis, prevention, and treatment of diseases, first of all focused on cancer but also metabolic disease (diabetes, etc.) and cognitive disorders such as AD, Parkinson's disease (PD), etc.
- WGS-based clinical care provided to patients that will include personalized diagnosis, prevention advice, and the most effective treatment plan

In addition, there will be important plant and animal genomics applications as the human population will continue to depend on a stable and sustainable food supply. We need to better understand the role of nutrition as related to our health and wellness. We are expecting to see a certain convergence of research conducted by food companies and what the biopharmaceutical industry has been working on for over a century. More about that topic will be given below in Chapter 6.

3.8 Bioethics

Whatever, in the course of my practice, I may see or hear (even when not invited), whatever I may happen to obtain knowledge of, if it be not proper to repeat it, I will keep sacred and secret within my own breast.

—Hippocrates (460–371 BC)

Bioethics is the study of (sometimes controversial) ethical issues brought about by advances in biology and medicine.

Debates in the field of bioethics range from the boundaries of life (including abortion and euthanasia), surrogacy (carrying of a pregnancy for intended parents), and the allocation of scarce healthcare resources (e.g., organ donation, healthcare rationing) to the right to refuse medical care for religious or cultural reasons.

The scope of bioethics did expand with recent advances in genetics and BT, in particular cloning, gene therapy, life extension, human genetic engineering, and manipulation of basic biology through altered DNA, RNA, and proteins.

An important milestone in the history of bioethics is *Belmont Report: Ethical Principles and Guidelines for the Protection of Human Subjects of Research*, issued by the US government in 1979.[134]

The report starts by defining the boundaries between medical practice and biomedical and behavioral research.

The term "practice" refers to interventions that are designed solely to enhance the well-being of an individual patient or client and that have a reasonable expectation of success (by following standard practice). By contrast, the term "research" designates an activity designed to test a new hypothesis, permit conclusions to be drawn and new knowledge to be developed. The general rule is that if there is any element of research in an activity, that activity should undergo review for the protection of human subjects.

Three basic ethical principles are then stated as follows:

1. Respect for persons (AUTONOMY): Respect for persons incorporates at least two ethical convictions: first, individuals should be treated as autonomous agents, and second, persons with diminished autonomy are entitled to protection. The principle of respect for persons thus divides into two separate moral requirements, the requirement to acknowledge autonomy and the requirement to protect those with diminished autonomy. In most cases of research involving human subjects, respect for persons demands that subjects enter into the research voluntarily and with adequate information. Essential to achieving and maintaining autonomy is access to information and counseling so that utilization is on the basis of INFORMED CONSENT. Confidentiality is required to ensure the privacy of that decision making.

2. BENEFICENCE: Beneficence is understood as an obligation to ensure that policies and practices (i) do not harm and (ii) maximize possible benefits and minimize possible harms. Beneficence is an obligation to treat persons such that they are protected from harm and that their well-being is secured. The Hippocratic maxim "Do no harm" has long been a fundamental principle of medical ethics. Claude Bernard, the great French physiologist (1813–1878) who suggested the use of blind medical experiments to ensure the objectivity of scientific observations and elucidated the roles of the pancreas and liver, extended this principle it to the realm of research, saying that one should not injure one person regardless of the benefits that might come to others. However, even avoiding harm requires learning what is harmful, and in the process of obtaining this information, persons may be exposed to risk of harm. Further, the Hippocratic Oath requires physicians to benefit their patients "according to their best judgment." Learning what will in fact benefit may require exposing persons to risk. The problem posed by these imperatives is to decide when it is justifiable to seek certain benefits despite the risks involved, and when the benefits should be foregone because of the risks. Following Bernard, there is a clear separation of human and animal experiments. He believed that "relief of human suffering justified the suffering of animals," still a practice in medical research.

3. JUSTICE: Within the context of healthcare, justice addresses two main themes:

   (i) The right to a minimum standard of care that includes values concerning public good and social justice. The interests of one individual or group of individuals should not disadvantage others. This notion implies respect for the disabled.

   (ii) Equity of access to services and information regardless of place of residence, ethnicity, gender, religion, age, or disability.

Until recently these questions have not generally been associated with research. However, the ethics of research involving human subjects is gaining in importance. In the 19th and early 20th centuries the burdens of serving as research subjects fell largely upon the poor and disadvantaged, while the benefits of improved medical care flowed primarily to the rich and wealthy. The exploitation of unwilling prisoners as research subjects in Nazi concentration camps is another example of a particularly flagrant injustice.

Others have added nonmaleficence, human dignity, and the sanctity of life to this list of cardinal values.

The HGP and related initiatives, have introduced powerful new methods to the study of genes. Genomics has laid the foundations for new approaches to the diagnosis and treatment of human disease, and introduced new possibilities for reproductive choices. This progress is accompanied by important ethical and social issues. Although many of these issues are not unique to genomics (such as confidentiality, informed consent, discrimination, and stigmatization), they require focused consideration in the context of genomics. Genomics is special in that gene-based approaches introduce a new language of *probability* and *susceptibility* to medical care, and furnish information about disorders; that often is of great interest to third parties—be they families, governments, insurance companies, law enforcement, or scientific researchers.

The ELSI program was founded in 1990 as an integral part of the HGP. The mission of the ELSI program was to identify and address issues raised by genomic research that would affect individuals, families, and society. The ELSI program focused on the possible consequences of genomic research in four main areas:

- Privacy and fairness in the use of genetic information, including the potential for genetic discrimination in employment and insurance
- The integration of new genetic technologies, such as genetic testing, into the practice of clinical medicine
- Ethical issues surrounding the design and conduct of genetic research with people, including the process of informed consent
- The education of healthcare professionals, policy makers, students, and the public about genetics and the complex issues that result from genomic research

Genetic testing extends ethical concerns to family members of a patient. A finding of inherited predisposition has implications for other family members (blood relatives). Consequently, recording an individual's information in a medical record may also create a record of the health status or risk of relatives. The doctor has therefore ethical responsibilities concerning privacy and confidentiality, and an obligation to inform the patient about shared nature of genetic information within families. Counseling before testing is therefore essential.

Predictive/presymptomatic testing introduces another set of ethical issues: Limitations of the test need to be discussed, including considerations regarding the accuracy of a given test. Also, *predisposition* for a medical condition does not guarantee the actual onset: the patient may never develop the condition.

Prenatal testing introduces another set of ethical issues. When a fetal abnormity is detected, support is essential for whatever difficult decision is made. The decision to either continue or terminate the pregnancy will depend on moral, religious, and cultural beliefs and values.

In 2002, the World Health Organization (WHO) made the following three main recommendations to its member states:

1. Individual member states need to develop detailed ethical frameworks to guide the conduct of genomic research and its medical application in their own unique social, cultural, economic, and religious context. The coordination of ethical guidelines should take place on the basis of internationally agreed upon fundamental principles. Capacity building in middle to low-income countries can take place and should be encouraged by virtue of information exchange and knowledge transfer between the high- and low-income countries. Capacity building will also include the development of human resources infrastructure and access to accurate information on advances in genomics.

2. The member states must also establish regulatory frameworks to monitor and control the commercial and medical application of genomics research, which present a number of potential risks and hazards to public safety and the environment.

3. The ethical and social issues raised by genomics research require open and comprehensive international debate to sift out the underlying commonalties and the concerns. Tapping on this debate, member states are encouraged develop collaborative approaches for capacity building in this field.

The establishment of national or disease-specific biobanks, biorepositories that store biological samples (usually human) for use in research, is another recent trend that is creating new ethical issues. An important requirement for responsible biobanking (and avoidance of possible negative reactions in the media) is the creation of a chain of trust between the research community and the public. The care and protection of human research participants must always be the dominant, foremost value. Protecting them from harms (physical injury) and wrongs (damage to self-esteem, dignity) is absolutely essential. Consent should be obtained in ways that ensure consistent definitions for anonymization and de-identification, explicit policies concerning when (if ever) subjects will be recontacted, a clear explanation of commercial issues, and perhaps even the right to withdraw.

With respect to consent, the main issue is that biobanks usually collect sample and data for multiple future research projects, and that it may therefore not be feasible to obtain specific consent for every single future project. A broad consent for various future research purposes may not satisfy strict ethical and legal requirements, but it could work in an environment where a chain of trust has been established, where the governance process is transparent and where there are no hidden commercial gains.

The patenting of genes is another controversial issue in terms of bioethics. There are three main concerns voiced about genetic patenting:

a. It is unethical to patent genetic material because it treats life as a commodity.
b. Living materials occur naturally, and therefore cannot be patented.
c. There is the fear that allowing patents on genetic material will undermine the dignity of people and other animals by subjecting their genes to ownership by other people.

Celera, a private company started by Venter in 2008, participated in the HGP by competing with the publicly financed project that was led by Francis Collins. Celera initially announced that it would seek patent protection on 200–300 genes, but later amended this to seeking intellectual property (IP) protection on fully characterized important structures amounting to an estimated 100–300 valuable drug targets. The firm eventually filed preliminary (placeholder) patent applications on 6500 whole or partial genes. Celera also promised to publish its findings in accordance with the terms of the 1996 "Bermuda Statement," by releasing new data annually (the HGP released its new data daily). Unlike the publicly funded project, Celera would not permit free redistribution or scientific use of the data. For this reason, publicly funded competitor UC Santa Cruz was compelled to publish the first draft of the human genome before Celera, on July 7, 2000. The scientific community downloaded about 500 GB of information from the UCSC genome server in the first 24 hours of free and unrestricted access to the first ever assembled draft of the human genome.

In March 2000, President Clinton announced that the genome sequence could not be patented, and should be made freely available to all researchers. The statement punished Celera's stock and dragged down the BT sector, which lost about $50 B in market capitalization in two days.

In a landmark ruling in June 2013, the US Supreme Court rattled the biotech world when it declared naturally occurring DNA sequences ineligible for patents. The Supreme Court ruled against Myriad Genetics, a Utah-based company that owned patents on the human breast cancer genes *BRCA1* and *BRCA2*. As a consequence, breast cancer diagnostic tests competing with Myriad's may now be offered by others.

In 2011, the European Union reconciled the legislation of biological patents among countries under the jurisdiction of the European Patent Organization (EPO) by allowing for the patenting of natural biological products, including gene sequences, as long as they are "isolated from [their] natural environment or produced by means of a technical process."

It is safe to assume that the controversies will continue as judges will be asked to clarify exactly how the law defines a product of nature.

In view of this ongoing debate, WHO's third recommendation appears to be highly relevant—international harmonization of bioethics and related IP issues are essential for continued progress in the field of nanomedicine.

Another important bioethical topic is the use of animals in medical research, estimated at 50 to 100 million worldwide. About 85% of all animal experiments are conducted on mice and rats.[135]

In view of ongoing research on the human brain and cognitive dysfunction (associated conditions PD, AD and dementia, schizophrenia, depression, autism, attention deficit hyperactivity disorder [ADHD], etc.), there is an increased interest in animal experiments involving nonhuman primates (NHPs). Around 65,000 NHPs are used every year in the United States, and around 7000 across the European Union. According to the Nuffield Council on Bioethics,[136] a highly respected organization that is funded by the Nuffield Foundation, the U.K. Medical Research Council and the Wellcome Trust, NHPs are used because their brains share structural and functional features with human brains, but "while this similarity has scientific advantages, it poses some difficult ethical problems, because of an increased likelihood that primates experience pain and suffering in ways that are similar to humans."

Human beings are recognized as *persons* and protected in law by the United Nations Universal Declaration of Human Rights.[137] NHPs are not classified as persons; hence their individual interests have no formal recognition or protection. The status of NHPs has generated much debate, particularly through the Great Ape Project (GAP),[138] which argues that gorillas, orangutans, chimpanzees, and bonobos be given limited legal status and the protection of three basic interests: (i) the right to live, (ii) the protection of individual liberty, and (iii) the prohibition of torture. In 2008, Spain became the first country to announce that it will extend rights to the great apes in accordance with GAP's proposals.

The use of NHPs in the European Union is regulated under a directive[139] that took effect in 2013. It permits the use of NHPs if no other alternative methods are available. Testing on NHPs is permitted for basic and applied research, quality and safety testing of drugs, food and other products, and research aimed on the preservation of the species. The use of great apes is generally not permitted, unless it is believed that the actions are essential to preserve the species or in relation to an unexpected outbreak of a life-threatening or debilitating clinical condition in human beings.

There is some hope that in silico modeling, to be discussed below in Section 5.1, will reduce the need for animal experiments.

3.9 Business Aspects of Sequencing Technology and the Human Genome Project

Entrepreneurs are simply those who understand that there is little difference between obstacle and opportunity and are able to turn both to their advantage.

—Nicolo Machiavelli (1469–1527)

The progression from promising new idea to the marketing of a commercial product is difficult and there are at least as many examples of failures as there are success stories, as already mentioned during our discussion of NGS.

The story of 454 Life Sciences provides a good example: In 2005, *The Wall Street Journal* awarded its Gold Medal for Innovation to 454 Life Sciences, the first NGS company able to bring a product to market. Compared to Sanger sequencing, 454 sequencing was 500 times faster and 50 times cheaper. James Watson's genome was sequenced in 2007 by 454 in two months using three instruments. The cost was less than $1 M and the accuracy was good enough to require only ~7 times coverage, quite similar to Sanger sequencing. Sequencing of the Neanderthal genome was started in 2006 and completed in 2009, in collaboration with the renowned Swedish anthropologist Svante Pääbo, who is heading the Max–Planck Institute for Evolutionary Anthropology in Leipzig, Germany, and has published the book *Neanderthal Man: In Search of Lost Genomes.*

In March 2007, Swiss Life Sciences giant Roche acquired 454 Life Sciences for ~$150 M. Together with Illumina's Solexa and Life Technologies/ABI's SOLiD technologies, Roche 454 was fighting for a market share in the competitive NGS arena.

Meanwhile, 454's founder and driving force Rothberg was working on the next iteration of 454's NGS concept, namely,

Ion Torrent's breakthrough way of replacing image capture of phosphates with electronic capture of hydrogen ions (protons), an idea licensed from DNA Electronics.

While 454 stagnated, Ion Torrent surprised the market with its new technology that clearly was heading toward the $1000 genome faster than 454's next generation of products. In 2010, Ion Torrent marketed their first generation of sequencing instruments as rapid, compact and economical sequencers that could be utilized in a large number of laboratories as a bench top machine.

Roche 454 pursued a similar DNA Electronics relationship but fell behind Ion Torrent while the pyrosequencing based instruments lost their competitive edge against Illumina's aggressive product improvement roadmap.

In 2013, after unsuccessful attempts to acquire Illumina, Roche decided to close down 454 Life Sciences by 2016, a disappointing development for 454 staff in Branford, CT.

On the other hand, Ion Torrent was acquired by Life Technologies in August 2010 for $725 M and soon displaced the ABI SOLiD platform as Life Technologies' leading NGS offering.

In November 2013, Thermo Fisher, a huge scientific instrument company with a market capitalization exceeding $45 B (in the 2nd quarter 2014), announced that it had reached an agreement with the European Commission to approve the purchase of Life Technologies and therefore Ion Torrent.

Meanwhile, after highly publicized unsuccessful bids for Illumina and a much less public attempt to acquire Life Technologies, Roche Diagnostics made a small $75 M deal with PacBio to "develop and supply DNA sequencing–based instruments and other products for clinical diagnostics."

While considering DNA sequencing a strategic area for cancer diagnostics and clinical care, high priorities for both Roche Diagnostics and Roche Pharma, the highly disciplined decision makers in Roche's corporate Swiss headquarters were not willing to overpay for Illumina's and Ion Torrent's sequencing assets. Time will tell if they made the right decision. As of September 2014, Illumina's market capitalization was $25 B, about three times the amount Roche was willing to pay in January 2013 for a hostile takeover eventually rejected by Illumina management. In June 2014, as already mentioned above, Roche acquired Genia's (still immature) nanopore technology for $125 M and additional milestone payments of $225 M.

Another interesting story is the story of *sequencing services*, with BGI and CG as the key players, along with genomic sequencing powerhouse Illumina and a few (at least currently) minor actors.

When Reid and Drmanac founded CG in 2005, they made three important decisions:

1. They would run the business as a pure services model, that is, not trying to sell sequencing instruments.
2. They would focus 100% on HUMAN Genome Sequencing (HGS) and try to become leaders in WGS throughput and accuracy.
3. They would focus both on proprietary sequencing technology and on sequencing data analysis, supported by a heavy investment in IT infrastructure (a Google-style data center in a warehouse in Santa Clara, CA) and highly qualified personnel.

For instance, they hired Bruce Martin, an early member of SUN's JAVA development team, to lead CG's IT infrastructure and software development efforts.

Reid had an interesting business background. In 1988, he cofounded Verity Inc., an enterprise text search engine company, and served as its vice president of engineering from 1988 to 1992 and as its executive vice president from 1992 to 1993. In those days, before the emergence of Yahoo! and Google, Verity was the emerging enterprise search engine leader.[140]

Drmanac has been described by many as a remarkable individual who stubbornly pursued his vision to realize the potential of sequencing-by-hybridization (SBH) technology. He gained the respect of outstanding pioneers in the field such as Prof. George Church, Harvard Genetics. Prof. Church is still engaged with CG as a member of CG's Scientific Advisory Board, along with another pioneer of sequencing, Prof. Hood, now leading the Institute for Systems Biology[141] (ISB) in Seattle, Washington. Both Profs. Church and Hood have repeatedly been advocates of CG's technology and business model. Prof. Hood's ISB is a preferred CG customer. Prof. Church has published several papers with Rade Drmanac and the CG team along with members of his genetics lab at Harvard.

Another CG Scientific Advisory Board member is Dr. Uta Francke, medical director of 23andMe, the well-known personal genomics company founded by Anne Wojcicki (ex-wife of Google cofounder Sergej Brin) and Linda Avey. Prof. Church has founded another personal genomics company, KNOME (based in Cambridge, MA). He advocates access to personal genetic information and is a leader in the new field of synthetic biology.

When discussing CG's technology, the CG business model and CG's aggressive pricing of WGS services were already mentioned. It was a risky strategy, leading to the events followed closely by business media (Forbes, Bloomberg, etc.) in 2012.

In February 2012, the influential Mayo Clinic, based in Rochester, MN, decided to pick CG as the supplier of sequencing services. In March, CG was featured in the science section of *The New York Times*.[142]

However, despite the positive news, it was also reported that BGI intended to acquire CG.

The status of CG's business and financials for the quarter ending September 30, 2012, was reported publicly as follows:

- 2200 genomes sequenced, of which 1900 revenue-generating genomes
- Total quarterly revenue $7.3 M (as compared to $4.2 M one year earlier)
- Total quarterly cost $24.6 M (compared to $25 M in 2011)
- Net quarterly loss $18 M (compared to $21.6 M in 2011)
- Cash position $34.7 M
- Backlog 3800 genomes expected to generate $18 M in revenue

In December 2012, *Bloomberg Business Week*[143] reported the following:

- The struggling genome sequencing services company CG put itself up for sale in June 2012.
- In December 2012, the leading Chinese genomics player BGI offered $118 M to buy CG (stock symbol GNOM), followed by an Illumina bid to offer $123.5 M.

- Prof. Church, Harvard Genetics, stated that "BGI and CG together would have a database of 30,000 human genomes, 10 times more than any of their nearest competitors."
- Illumina's bid was rejected by CG's board due to Illumina's market dominance.
- Subsequently, Illumina raised national security and privacy concerns that were rejected by CG.

On March 25, 2013, *Biotech Business Week*[144] announced the completion of the acquisition of CG by BGI-Shenzhen as follows: BGI-Shenzhen (BGI) and Complete Genomics, Inc. (Nasdaq: GNOM) announced that they have obtained approval from the State Administration of Foreign Exchange of the People's Republic of China. . . . all relevant regulatory approvals in connection with the proposed acquisition have been received.

By acquiring CG, BGI achieved a dominating position as the world's largest provider of genomic sequencing services.

BGI was created in 1999, cofounded by Prof. Huanming Yang, BGI's current President Jian Wang, and two more geneticists. BGI initially participated in the Human Genome Project as China's representative, contributing about 1% of the total effort. After completion of the HGP, BGI built its first branch in Hangzhou with funding from the local government. The Rice Genome was sequenced in 2002 which became a cover story in *Science*. In 2003 the severe acute respiratory syndrome (SARS) virus was sequenced and a detection kit quickly created. Later the first Asian human genome was sequenced. In 2007 the organization's headquarters relocated to Shenzhen and founded the first "citizen-managed, non-profit research institution in China". Since 2008, BGI-Shenzhen has been dramatically expanding its sequencing capacity.

In October 2003, The BGI Hangzhou branch and Zhejiang University founded a new research institute, the James D. Watson Institute of Genome Sciences, located at Zhejiang University. The Watson Institute was conceived as a major center for research and education in East Asia, modeled after the Cold Spring Harbor Laboratory in New York.

In 2010, BGI Americas was established and set up its main office in Boston. BGI Europe was established in Copenhagen, Denmark.

In 2011 the BGI had 4,000 scientists and technicians and those numbers were still predicted to increase.

BGI's sequencing platform had been standardized on Illumina's instruments.

After BGI's acquisition of Complete Genomics, BGI can provide sequencing services on several platforms. While Illumina has an edge in speed and throughput, CG technology may produce results that are (at least by some observers) considered more accurate. In addition, BGI will pay close attention to the emerging importance of Ion Torrent's platform. In fact, BGI announced the acquisition of about 50 new Ion Torrent sequencing instruments in October 2013.[145]

BGI has both a private and a public character. It receives funds from private investors and the Chinese government. The laboratory is also the National GeneBank of the country. Finally, a few words about BGI's management and staff.

Dr. Yang Huanming, also known (outside China) as Dr. Henry Yang, is one of China's leading genetics researchers. After studies in China, Yang earned his PhD in 1988 at the Institute of Medical Genetics, University of Copenhagen, Denmark. He then completed six years of post-doctoral training in Europe (INSERM/CNRS, France) and the United States (Harvard Medical School and UCLA). His research interests include the mapping and cloning of human genes, the sequencing and analysis of the human genome, human genome diversity and evolution, and the ethical, legal, and social issues related to genome research.

The study of the rice genome by Yang and his BGI collaborators made the cover of *Science* magazine (April 2002). Yang also led China's participation in the global Human Genome Project. Yang is a member of the Chinese Academy of Sciences, an Associate Fellow of the European Molecular Biology Organization (EMBO),[146] recipient of the 2005 Biology Prize of The World Academy of Sciences (TWAS),[147] a Member of the German Academy of Sciences,[148] and a Foreign Associate of the US National Academy of Sciences.[149] He was Coordinator-in-China of the International HapMap Consortium and Chief Coordinator of the Chinese Hybrid Rice Genome Consortium.

At BGI, Chairman Henry Yang and President Jian Wang have instilled a high performance culture. BGI is pursuing a wide range of scientific interests, including:

- genomics, cell biology, proteomics, transcriptomics, etc.;
- plant genomics and breeding (rice, drought tolerant hybrid millet, balsa wood);
- animal genomics, breeding and cloning (disease models, transgenic pigs);
- cloning of plants and animals;
- human translational research; and
- synthetic biology.

The management team includes a number of young and promising scientists. BGI tries to attract the best and brightest and even is engaged in new educational models. For example, BGI researcher Zhao Bowen had already published 17 papers in either *Nature* or *Science* at age 19. BGI's scientific "think-bank" includes Nobel Laureates James Watson, John Sulston, Harold Varmus, and Richard Roberts, as well as the eminent geneticists Francis Collins, Eric Lander, George Church, Fred Dubee and Lars Bolund.

BGI's "team building" exercises include climbs of some of the highest (>8000 m) peaks in the Himalayan mountains. In 2010, at age 56, BGI's President Jian Wang climbed Mount Everest.

Even before the CG acquisition, BGI was the global services leader in sequencing, covering not just the human genome but also that of animals and plants such as pandas and rice, respectively.

BGI's for-profit businesses broke even already in 2011, with $192 million in revenue, roughly ten times what Complete Genomics brought in that year. About 90% of its business is done with private clients, split equally between overseas and domestic. Among the most lucrative are long-term contracts, lasting two to three years, with at least 17 of the world's top 20 pharmaceutical companies.

BGI's acquisition of CG has already produced an interesting result: The NIFTY test, a genetic screening of human embryos that is performed with BGI's software on CG's sequencing platform, has been approved by China's FDA (CFDA) and appears to become a big success. The BGI-CG merger could become a first example of

US-China research (not just manufacturing!) partnerships in high technology.[150]

Finally, BGI is heavily engaged in Shenzhen's ambitious plan to turn the region into China's leading "Biotech Valley".

An interesting question is what economic *impact* the HGP has had so far on the overall economy. According to the already quoted report authored by S. Tripp and M. Grueber of the Battelle Memorial Institute,[151] the cumulative HGP impact on the US economy 1988–2010 has been:

- the creation of more than 3.8 million jobs across all 50 states over the 23-year period 1988–2010
- $796.3 B total economic output
- $78.4 B total tax revenue, of which $48.9 B is federal tax

Compared to direct expenditures in 1988–2003 of $3.8 B (or $5.6 B in constant 2010 dollars) by the US government (NIH + DOE), every $1 of the investment has helped to generate $141 in the US economy.

The various applications of Sequencing and their impact on society and various sectors of the economy are illustrated in Fig. 3.19 (adapted from the above-mentioned Battelle report).

The financial impact for the year 2010 alone was $67 B in US economic output, $20 B in personal income for US citizens, and 310,000 jobs.

So far we only covered the investments 1988–2003 and the impact of those investments. Beyond those specific HGP investments, both the NIH and the DOE (as well as other governments and research foundations) made and continue to make in the science, tools and applications required to successfully build on the groundwork laid by the HGP. The estimated US government funding 2004–2010 was $7.214 B in constant 2010 money (adjusted for inflation), and it generated 173,000 job-years and a total output exceeding $21.5 B, on top of the previously mentioned HGP output.

The impact of the HGP on the global economy should be even more impressive. It would be interesting to see the results of a study similar to what Battelle Institute did on behalf of the US government. A rough guess would be that the global impact up to 2010 should

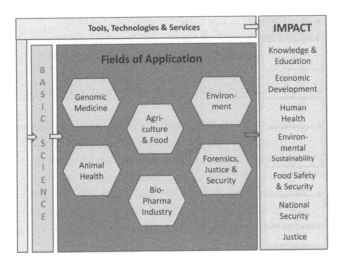

Figure 3.19 Applications of sequencing and impact of the Human Genome Project.

be about twice the impact on the US economy, trending even higher than twice in the future as the global share of the United States' contributions will decline.

The public HGP investment provides a shining example for the important role of governments in creating scientific and technological advances that stimulate economic growth and the creation of new jobs.

3.10 Post–Human Genome Project "-Omics"

The completion of the HGP was a major milestone for the life sciences, but in many ways it was just the beginning of our human quest to understand life. Instead of answering all biological questions, it created a basis for asking new questions such as how to bridge the gap between a complete genome and the cells and organs of a living organism.

The HGP's impact on biology is one of a revelation of complexity and will challenge scientists for decades to come. It is now known that:

- the same gene can code for multiple proteins;
- the expression of genes is less straightforward than assumed before the HGP, resulting in over 1 million different proteins;
- genes interact with one another and code for multiple RNAs;
- although structural genomic variation between organisms are relatively minor, the great variation across living organisms is a result of regulatory processes that need to be studied and better understood;
- portions of DNA between our ~20,000 (or less) genes that previously were called "junk DNA" play important roles in transcription and translational regulation of protein folding; and
- most common diseases including cancer, heart disease, and cognitive disorders cannot be simply traced to a few faulty genes and differ significantly across individual genomes from patient to patient.

Another important outcome of the HGP was the realization that biology is an information science, giving rise to new disciplines such as bioinformatics, computational biology and systems biology.

Genomics itself has previously been classified by the National Center for Biotechnology Information (NCBI)[152] at the NIH's National Library of Medicine (NLM) as:

- structural genomics—including mapping and sequencing
- comparative genomics—including genetic diversity and evolutionary studies
- functional genomics—the study of the roles of genes in biological systems

but has since been further broken down into smaller, specialized disciplines.

Here is a very brief overview of the new fields of investigation that sometimes are referred to as the "omics":

- *Genomics* is the genetic discipline that is focused on sequencing, assembling, and analyzing the function and structure of genomes, defined as the complete set of DNA within a single cell of an organism. Advances in genomics have triggered a revolution in discovery-based research to understand even the most complex biological systems such as the brain. The field includes efforts to determine the entire DNA sequence of organisms and fine-scale genetic mapping.

- *Transcriptomics* is the study of the *transcriptome,* defined as the set of all RNA molecules, including mRNA, ribosomal RNA (rRNA), tRNA, and other noncoding RNA produced in one or a population of cells. Because it includes all mRNA transcripts in the cell, the transcriptome reflects the genes that are expressed at any given time. The study of transcriptomics, also referred to as gene expression profiling, examines the expression level of mRNAs in a given cell population, often using high-throughput techniques based on DNA microarray technology. The transcriptomes of stem cells and cancer cells are of particular interest: Transcriptomics applied to those cells help us in the understanding of cellular differentiation and carcinogenesis.

- *Proteomics* is the large-scale study of proteins as encoded by a genome. Proteomics is different from protein science which is mostly focused on the 3D structure of proteins and their function. Proteomics generally refers to the large-scale experimental analysis of proteins, and it is often specifically used for protein purification and mass spectrometry, an analytical chemistry technique that helps identify the amount and type of chemicals present in a sample by measuring the mass-to-charge ratio and abundance of gas-phase ions. The *proteome* is the entire set of proteins, produced or modified by an organism or system. After genomics and transcriptomics, proteomics is the next step in the study of biological systems. While an organism's genome is more or less constant, the proteome differs from cell to cell and from time to time. Distinct genes are expressed in different cell types, which means that even the basic set of proteins that is produced in a cell requires identification.

- *Metabolomics* is the scientific study of chemical processes involving metabolites, the systematic study of the unique chemical fingerprints that specific cellular processes leave behind. The metabolome represents the collection of all metabolites in a biological cell, tissue, organ, or organism, which are the end products of cellular processes. While mRNA gene expression data and proteomic analyses do not tell the whole story of what might be happening in a cell, metabolic profiling can give an instantaneous snapshot of the physiology of that cell. A related term sometimes used and of great importance in toxicology is *Metabonomics*. Metabonomics extends metabolic profiling to include information about perturbations of metabolism caused by environmental factors (including diet and toxins), disease processes, and the involvement of extragenomic influences, such as gut microflora. As to instrumentation, the field of metabolomics is usually relying on mass spectrometry, whereas metabonomics is more associated with nuclear magnetic resonance (NMR) spectroscopy.

- *Metagenomics* (also named environmental genomics, ecogenomics, or community genomics) is the study of genetic material recovered directly from environmental samples. Recent studies use "shotgun" Sanger sequencing or second generation pyrosequencing to get largely unbiased samples of all genes from all the members of the sampled communities. Metagenomics offers a powerful lens for viewing the microbial world. However, a challenge that is present in human microbiome studies is to avoid including the host DNA in the study.

Epigenetics is defined as the study of heritable changes that are not caused by changes in the DNA sequence, but by functionally relevant changes to the genome that do not involve a change in the nucleotide sequence. Epigenetics addresses the question why many cell types of the body maintain drastically different gene expression patterns while sharing exactly the same DNA?

An example of an epigenetic mechanism would be the methylation of cytosine in DNA (see Fig. 2.23), illustrated also below in Fig. 3.20.

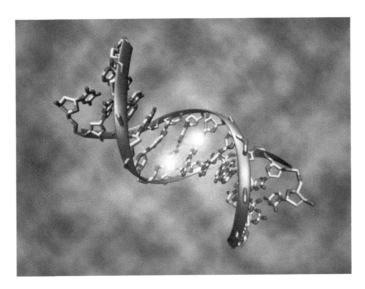

Figure 3.20 DNA methylation at the two center cytosines. Source: Wikipedia.

Methylation alters how genes are expressed without altering the underlying DNA sequence. DNA methylation may stably alter the expression of genes in cells as cells divide and differentiate from ESCs into specific tissues. The resulting change is normally permanent and unidirectional, preventing a cell from reverting to a stem cell or converting into a different cell type. For more stem cell details see Section 5.6.

A list of large post-HGP projects in biology is given below:

- *International HapMap Project*: Started in 2002, HapMap is a public resource catalog of common genetic variants that occur in human beings. It describes what these variants are, where they occur in human DNA and how they are distributed among people within populations in different parts of the world. Website: http://hapmap.ncbi.nlm.nih.gov/.
- *1000 Genomes Project*: Launched in 2008, the goal of the 1000 Genomes Project is to find most genetic variants that have frequencies of at least 1% in the populations studied. In October 2012, the sequencing of 1092 genomes

was announced in a *Nature* publication.[153] Website: http://www.1000genomes. org/about.

- *ENCODE Project*: The Encyclopedia of DNA Elements (EN-CODE) is a public research project launched by the US NHGRI in September 2003. Humans are estimated to have ~20,000 protein-coding genes (collectively known as the exome), which account for only about 1.5% of DNA in the human genome. The primary goal of the ENCODE project is to determine the role of the remaining component of the genome, much of which was traditionally regarded as junk (i.e., DNA that is not transcribed). Approximately 90% of SNPs in the human genome (that have been linked to various diseases by genome-wide association studies) are found outside of protein-coding regions. Website: https://www.encodeproject.org/.

- *Microbial Genome Project*: Started 1994 as HGP spinoff, the Microbial Genome Project (MGP) sequenced the genomes of nonpathogenic microbes useful in solving DOE's mission challenges in environmental waste cleanup, energy production, carbon cycling, and BT. Websites:

  - http://www.genome.gov/25520338
  - http://www.ncbi.nlm.nih.gov/genomes/MICROBES/ microbial_taxtree.html

- *Cancer Genome Anatomy Project*: Started in 1996 by the US National Cancer Institute (NCI), the Cancer Genome Anatomy Project (CGAP) was established to decipher the molecular anatomy of the cancer cell, to determine the gene expression profiles of normal, precancer, and cancer cells, leading eventually to improved detection, diagnosis, and treatment for the patient. Website: http://cgap.nci.nih.gov/.

- *Cancer Genome Atlas*: Since 2006; see Section 6.3. Website: http://cancergenome.nih.gov/.

- *Cancer Genome Project*: Since 2004; see Section 6.3. Website: https://www.sanger.ac.uk/research/projects/cancergenome/.

- *Human Microbiome Project (HMP)*: Since 2007; see Section 6.1. Website: http://commonfund.nih.gov/hmp/index.

Chapter 4

Biopharmaceutical R&D

Many have said of alchemy, that it is for the making of gold and silver. For me such is not the aim, but to consider only what virtue and power may lie in medicines.

—Paracelsus, born Philippus Aureolus Theophrastus Bombastus von Hohenheim (1493–1541)

Paracelsus was born in Einsiedeln, Switzerland, and died in Salzburg, Austria. He studied medicine in Basel and received his doctorate from the University of Ferrara, Italy. He led an eventful life that took him to Germany, France, Spain, Hungary, the Netherlands, Denmark, Sweden, Poland, and Russia. Paracelsus pioneered the use of chemicals and minerals in medicine. He used the name "zink" for the element zinc in about 1526 on the basis of the sharp pointed appearance of its crystals after smelting and the old German word *zinke* for pointed. As a contemporary of Copernicus and Leonardo da Vinci, he used experimentation in learning about the human body. It is said that Paracelsus was also responsible for the creation of laudanum, an opium tincture very common until the 19th century. As a physician and medical chemist at the time, he also sharply criticized apothecary practices that did not pay proper attention to dosage. He coined the phrase "Dosis facit venenum" or "The

Nanomedicine: Science, Business, and Impact
Michael Hehenberger
Copyright © 2015 Pan Stanford Publishing Pte. Ltd.
ISBN 978-981-4613-76-7 (Hardcover), 978-981-4613-77-4 (eBook)
www.panstanford.com

dose makes the poison." Paracelsus had many controversies with his contemporaries, not only because he was ahead of his time, but also because he had a tendency to be "bombastic" and sometimes excessively arrogant. Here is an example: "Let me tell you this: every little hair on my neck knows more than you and all your scribes, and my shoe buckles are more learned than your Galen and Avicenna, and my beard has more experience than all your high colleges."[154]

Galen of Pergamon (129–200) was a prominent Greek physician, surgeon, and philosopher in the Roman empire and arguably the most influential of all medical researchers of antiquity. Galen influenced the development of various scientific disciplines, including anatomy, physiology, pathology, pharmacology, and neurology, as well as philosophy and logic. Following the teachings by ancient Greek physicians such as Hippocrates, his theories dominated and influenced Western medical science for more than 1300 years. His anatomical reports, based mainly on dissection of monkeys, remained uncontested until 1543. His theory of the physiology of the circulatory system endured until 1628, when William Harvey established that blood circulates, with the heart acting as a pump. What is still accepted today, however, is Galen's correct assertion that the brain controls all the motions of the muscles by means of the cranial and peripheral nervous systems.

Ibn Sine (980–1037), better known by his Latinized name Avicenna, was a Persian sage, philosopher, and physician who wrote *The Book of Healing*, a vast philosophical and scientific encyclopedia, and *The Canon of Medicine*, a standard medical text at many medieval universities. His books provided an overview of all aspects of medicine according to the principles of Galen (and Hippocrates).

The point Paracelsus tried to make in his provocative way was that it was better to make one's own observations and experiments instead of relying on ancient books by Galen and Avicenna that had to be studied in Latin. When, as Professor of medicine at the University of Basel, Switzerland, he gave lectures in German instead of in Latin, he was making powerful enemies and had to leave after only one year. There are many stories about Paracelsus but one thing seems to be clear: he revolutionized the practice of medicine by questioning established procedures such as bloodletting, by advocating the cleaning of wounds, by understanding the inherited character of syphilis, by recognizing sepsis, and by trying to find antiseptic remedies. He even pioneered the insight that some diseases are rooted in psychological illness.

Today's methods of drug discovery still have one thing in common with the approach first advocated by Paracelsus: Use the most advanced chemical and biological knowledge, gain a deep understanding of disease mechanisms, and try to find substances that have a desirable therapeutic effect. If a biopharmaceutical company succeeds with the discovery and development of a new drug, the benefits to mankind and the commercial rewards for the company can be enormous. However, although less serendipitous and increasingly based on latest breakthroughs of science and technology, today's drug discovery process is still lengthy, expensive, difficult, and inefficient.

4.1 Stages of Biopharmaceutical Drug Discovery and Clinical Development

Success is often achieved by those who don't know that failure is inevitable.

—Gabrielle Bonheur "Coco" Chanel (1983–1971)

With the sequencing of the human genome and a huge volume of scientific publications on molecular biology discoveries, the field of life sciences has witnessed an explosion in the volume of electronic data generated by, and available to, drug discovery researchers in both the biopharmaceutical industry and in government/academic research. The expectation of this data revolution was that it would lead to a fundamental increase in the number of drugs discovered. However, quite the opposite has been true. Indeed, a recent review of the drug discovery process by the US Government Accountability Office (GAO) stated that "company researchers are still learning whether the data will lead to potentially valid drug candidates."[155]

On page 8 of this GAO report it is shown (Fig. 4.1) what steps are required in discovering and developing a new drug that,

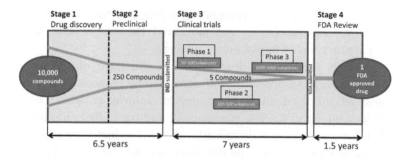

Figure 4.1 The drug discovery, development, and FDA review process.

after approval by the Food and Drug Administration (FDA), can be marketed by the biopharmaceutical industry.

Several independent studies have shown that of ~$100 B spent annually by global biopharmaceutical R&D organizations, more than 60% of all projects have to be terminated.[213] Most failures occur during the drug discovery and preclinical phases of the R&D process.

4.2 Stage 1: Drug Discovery

4.2.1 *Target Identification and Validation*

A critical first step in drug discovery is the identification and validation of a drug target, a protein (enzyme, ion channel, receptor, etc.) or DNA/RNA whose activity is modified by a drug, resulting in a desirable therapeutic effect. In other words, the biological target is "hit" by a signal and its behavior or function is then changed. The most common protein drug targets of currently marketed drugs include

- *G protein–coupled receptors (GPCRs)*: These are a large protein family of receptors that sense molecules outside the cell and activate inside signal transduction pathways and, ultimately, cellular responses. GPCRs are also called transmembrane receptors because they pass through the cell membrane. The ligands that bind and activate these receptors include light-sensitive compounds, odors, hormones, and neurotransmitters, varying in size from small molecules to peptides to large proteins. GPCRs are involved in many diseases and are therefore the target of approximately 40% of all modern medicinal drugs.[157] The 2012 Nobel Prize for chemistry was awarded to Robert J. Lefkowitz (Duke Univ) and Brian K. Kobilka (Stanford) for their groundbreaking discoveries that reveal the inner workings of G-protein–coupled receptors.[158] Lefkowitz started to use radioactivity in 1968 in order to trace cells' receptors. He attached an iodine isotope to various hormones, and thanks to the radiation, he managed to unveil several receptors, among those a receptor for adrenalin: beta-adrenergic receptor. During the 1980s, the newly recruited Kobilka accepted the challenge to find out which gene from the human genome is coding for the beta-adrenergic receptor. When the researchers analyzed the gene, they discovered that the receptor was similar to rhodopsin—the receptor in the eye that captures light. They realized that there is a whole family of receptors—GPCRs—that look alike and function in the same manner. We know today that about a thousand genes code for such receptors, including GPCRs for light, flavor, odor, adrenalin, histamine, dopamine, and serotonin.

- *Enzymes* (especially protein kinases, proteases, esterases, and phosphatases): As already discussed previously, enzymes are highly selective catalysts, greatly accelerating both the rate and specificity of metabolic reactions, from the digestion of food to the synthesis of DNA. A *protein kinase* is a kinase enzyme that modifies other proteins by chemically adding phosphate groups to them. This phosphorylation usually results in a functional change of the target protein (substrate) by changing enzyme activity, cellular location, or association with other proteins. The human genome contains about 500 protein kinase genes ($\sim$2% of all human genes). Up to 30% of all human proteins may be modified by kinase activity. Kinases are known to regulate the majority of cellular pathways, especially those involved in signal transduction. *Proteases* are enzymes performing proteolysis, the breakdown of proteins into smaller polypeptides or amino acids. In general, this occurs by the hydrolysis of the peptide bond. Proteolysis in organisms serves many purposes: For example, digestive enzymes break down proteins in food to provide amino acids for the organism. *Esterases* split esters into an acid and an alcohol in a chemical reaction with water called hydrolysis. Finally, *phosphatases* are enzymes that remove a phosphate group from its substrate, typically creating a phosphate ion and a molecule with a free hydroxyl group. This action (also called dephosphorylation) is directly opposite to that of phosphorylases and kinases, which attach phosphate groups to their substrates by using energetic molecules like adenosine-5′-triphosphate (ATP). A common phosphatase in many organisms is alkaline phosphatase. Another large group of proteins is associated with deoxyribonucleotide and ribonucleotide phosphatase or pyrophosphatase activities.

- *Ion channels*:

  - *Ligand-gated ion channels (LGICs)*: LGICs are a group of transmembrane ion channel proteins that open to allow ions such as Na^+, K^+, Ca^{2+}, or Cl^- to pass through the membrane in response to the binding of a chemical messenger (i.e., a ligand), such as a *neurotransmitter*. Neurotransmitters are endogenous (i.e., originating from within an organism, tissue, or cell) chemicals that transmit signals from a neuron to a target cell across a synapse, a structure in the nervous system that permits a nerve cell (neuron) to pass an electrical or chemical signal to another cell (neural or otherwise). Neurotransmitters are packaged into synaptic vesicles clustered beneath the membrane in the axon terminal, on the presynaptic side of a synapse. LGIC proteins are typically composed of at least two different domains, a transmembrane domain that includes the ion pore and an extracellular domain that includes the ligand binding location (an allosteric binding site). The function of such receptors located at synapses is to convert the chemical signal of presynaptically released neurotransmitter directly and very quickly into a postsynaptic electrical signal.

- *Voltage-gated ion channels (VGICs):* VGICs are a class of transmembrane ion channels that are activated by changes in electrical potential difference near the channel; these types of ion channels are common in many types of cells. They have a crucial role in excitable neuronal and muscle tissues. Examples include the sodium (Na^+) and potassium (K^+) voltage-gated channels of nerve and muscle and the voltage-gated calcium (Ca^{2+}) channels that play a role in neurotransmitter release in presynaptic nerve endings. VGICs directionally propagate electrical signals.

- *Nuclear hormone receptors*: These are a class of proteins found within cells that are responsible for sensing steroid and thyroid hormones and certain other molecules. In response, these receptors work with other proteins to regulate the expression of specific genes, thereby controlling the development, homeostasis, and metabolism of the organism. Some of these receptors such as farnesoid X receptor (FXR), expressed at high levels in the liver and intestine; liver X receptor (LXR), an important regulator of cholesterol, fatty acid, and glucose homeostasis; and peroxisome proliferator-activated receptors (PPARs), functioning as transcription factors regulating the expression of genes and playing essential roles in the regulation of cellular differentiation, development, metabolism (carbohydrate, lipid, protein), and tumorigenesis, may function as metabolic sensors.

- *Structural proteins such as tubulin*: These confer stiffness and rigidity to biological components. Most structural proteins are fibrous proteins; for example, collagen and elastin are critical components of connective tissue such as cartilage, and keratin is found in hard or filamentous structures such as hair, nails, feathers, hooves, and some animal shells. Some globular proteins can also play structural functions, for example, actin and tubulin are globular and soluble as monomers but polymerize to form long, stiff fibers that make up the cytoskeleton, which allows the cell to maintain its shape and size.

- *Membrane transport protein* (or simply transporter): This is a membrane protein involved in the movement of ions, small molecules, or macromolecules, such as another protein across a biological membrane. The proteins may assist in the movement of substances by facilitated diffusion or active transport. These mechanisms of action are known as carrier-mediated transport.

Another category of potential drug targets are "druggable genes," as defined by Hopkins and Groom:[159] A druggable gene contributes to a disease phenotype and can be modified by a small-molecule drug.

Having identified/selected a target, the biopharmaceutical organization will only move ahead with subsequent steps in the drug discovery process if the target can be *validated*. A target is considered validated if its inhibition reverts a disease phenotype in a model system that is as close to the disease as possible.

Target validation can therefore be defined as the investigation of the phenotype resulting from inhibition of the target. The investigation usually involves animal models chosen to recapitulate the pathophysiology of human disease. The model should further allow measurement of pharmacological efficacy and reveal mechanism based toxicity. Finally, there should be a path toward translation of the animal model response to the human clinical environment.

For a leading research-based pharmaceutical company, picking and validating a target may involve the following checklist:[160]

- Better pharmaceutical properties
- Refined understanding of mechanism
- Validated by drugs that failed that act on the same mechanism
- Validated genetically in humans
- Validated by the use of proteins as therapeutics
- Validated by drugs that act at a different point on the same pathway

- Animal models exist that correlate with human pathophysiology
- Transgenic animals

The concept of transgenic animals was introduced in the late 1980s as an effective way to use animal (mostly mouse) models to better understand and validate a target mechanism. *Transgenesis* is the process of introducing an exogenous gene—called a transgene— into a living organism so that the organism will exhibit a new property and transmit that property to its offspring.

A special case of transgenic animal is the "knockout mouse," a genetically engineered mouse in which researchers have inactivated, or "knocked out," an existing gene by replacing it or disrupting it with an artificial piece of DNA. The loss of gene activity often causes changes in a mouse's phenotype, which includes appearance, behavior, and other observable physical and biochemical character- istics. Knockout mice are important animal models for studying the role of genes that have been sequenced but whose functions have not been determined. By inactivating a specific gene and observing differences from normal behavior or physiology, researchers can infer its probable function. Mice are currently the most closely related laboratory animal species to humans for which the knockout technique can easily be applied. Gene knockout in rats is much harder and has only been possible since 2003. The first recorded knockout mouse was created by Mario R. Capecchi, Martin Evans, and Oliver Smithies in 1989. They won the 2007 Nobel Prize for physiology or medicine[161] for their discoveries of principles for introducing specific gene modifications in mice by the use of embryonic stem cells (ESCs).

During the early 1980s, Capecchi and Smithies discovered a way to experimentally modify genes. Independently of each other, they found that DNA molecules that resemble parts of normal genes yet differ from them in crucial respects can be inserted in the right place in the genome. Their discoveries made it possible to carry out targeted gene modifications in individual cells in a culture, but an important problem remained to be solved. Every cell in our bodies carries our entire genome, so if we want to understand the function of a given gene in its full context—in real life—the same genetic change must be introduced into all the cells of the body. Evans solved this problem through his discovery of ESCs—cells from early embryos that can be cultured, grown and genetically modified in a test tube. Like a fertilized egg cell, these ESCs can give rise to all the cells of the body, thereby passing their genes, including modified ones, onward to future generations.

Unless there is a strong medical reason, it is ethically impossible to make hereditary, targeted changes in the genes of humans. However, doing this in mice is much less controversial. Since humans share many genes with mice, the knockout mouse concept has been extremely useful.

Examples of medical research areas in which knockout mice have played a role include different kinds of cancer, obesity, heart disease, diabetes, arthritis, substance abuse, anxiety, aging, and Parkinson's disease (PD). Knockout mice offer a biological context in which drugs and other therapies can be developed and tested.

Target identification and validation activities in biopharmaceutical industry are largely driven by basic biomedical research into the understanding of disease mechanisms. Since a lot of this research is pursued by academic and government research teams, drug discovery benefits from close contacts, and interactions between the private and public sectors. The global public sector investments are adding up to more than $60 B per year, with the United States providing close to 50% of the funding via the National Institutes of Health (NIH).

Public–private partnerships (PPPs) are supporting those inter-actions and precompetitive collaboration among biopharmaceutical companies is emerging as a new trend and a possible answer to the above-mentioned R&D productivity crisis in the life sciences industry.

There is also an important tradeoff between industrial projects focused on new targets versus work on established validated receptors. Drug discovery is a high-risk activity and decision makers have to pick risk levels they feel comfortable with.

For the scientist, it is more interesting and challenging to work on new receptor subtypes and unvalidated animal models. However, to reduce risk, research management may instead focus on improvements over an existing therapy.

4.2.2 *Lead Identification and Optimization*

After target validation, drug companies try to find a new molecular entity (NME) that inhibits or blocks a given unfavorable biochemical pathway and therefore has a positive effect on the patient suffering from a given disease.

Whereas medical researchers and biologists dominate target identification and validation activities, the chemists—in particular synthetic organic and medicinal chemists—take over when it's time to find suitable NMEs. In terms of information technology (IT) support of drug discovery activities, we are moving from bioinformatics to cheminformatics. A key member of the cheminformatics toolbox is the chemical database and its associated data management and analysis tools. To be useful to researchers, chemical databases must be searchable by chemical structure. Research-based drug companies have all accumulated large libraries of chemical structures thought to be potential candidate NMEs that can be tested against chosen targets. Those chemical databases are highly valuable knowledge assets that are closely guarded and protected because chemical compounds with drug molecule potential can be patented and—contingent on approval by regulatory authorities such as the US FDA—can lead to significant and highly profitable revenue streams.

A standard way to generate large libraries of chemical compounds is *combinatorial chemistry*, a technique to first synthesize large numbers of compounds and then identify the useful components of the library. The idea is to find the compounds with the optimal chemical characteristics for a given drug target. A good drug compound would bind specifically and selectively to that drug target, thereby inhibiting the associated biomolecular pathway. Starting from a core structure or scaffold, various chemical groups (commonly denoted as R groups) are added to the scaffold. Although applied to small organic molecules in the case of conventional drug discovery, the experimental synthesis of large libraries of compounds has benefited from the groundbreaking work of Bruce Merrifield, who in 1984 received the Nobel Prize for chemistry for his development of methodology for chemical synthesis on a solid matrix.[162] Merrifield's simple and ingenious method for obtaining peptides and proteins has created completely new possibilities in the field of peptide and protein chemistry, as well as in the field of nucleic acid chemistry. Merrifield's method involves binding the first of the many amino acid residues of which a protein is composed to a polymer. After each subsequent synthetic step, by-products and remaining starting materials can then be removed by filtration and washing the polymer. Only when the entire synthesis of the desired peptide has been completed, the polymer is removed. The advantages of this method are very considerable. Through the replacement of a complicated isolation procedure for each intermediate product with a simple washing procedure much time is saved. In addition, it has proven possible to increase the yield in each individual step to 99.5% or better, a result that cannot be attained using conventional synthetic approaches. Finally, this method is also suitable for automation and automatic peptide synthesizers are now commercially available.

Having synthesized large libraries of variations of an identified lead compound, the drug discovery team still has to go through a number of necessary and time-consuming steps. The next significant experimental activity during lead identification is called *high-throughput screening* (HTS). Using robotics, sensitive detectors, control software and customized data processing, HTS allows a researcher to quickly conduct hundreds of thousands and even millions of experimental tests. The term "ultrahigh-throughput screening" (uHTS) refers to screening in excess of 100,000 compounds per day. The goal is to rapidly identify active chemical compounds that react with targets, modulate a particular biomolecular pathway, and can therefore serve as starting points for drug design. In parallel, computational chemists interact with their experimental colleagues by modeling the interactions of lead compounds with target proteins. The term "docking" is used to describe the interaction between a lead compound and the active site of a protein (enzyme) target. Flexible docking algorithms, based on a combination of classical mechanics and quantum chemistry, are used to perform in silico screening of drug targets that can complement the experimental HTS activities. They are highly compute intensive but increasingly feasible, accurate and predictive due to the rapid evolution of compute power and software development since the 1980s when molecular modeling was first introduced and embraced by the pharmaceutical industry. A less rigorous but highly useful empirical technique to screen proposed drug molecules is the use of *quantitative structure–activity relationships* (QSARs). QSAR models summarize a supposed relationship between chemical structures and biological activity in a data set of chemical compounds. They further predict the activities of new chemicals on the basis of molecular descriptors such as molecular weight, the number of hydrogen bonds, etc., as summarized concisely by Lipinski in his "rule of five" for orally active (small-molecule) drugs.[163]

After decades of successful drug discovery activities, synthetic chemists have learned about certain basic "rules" that lead to "druggable" compounds, molecules that are soluble in water, have no toxic effect on healthy tissues, could be taken by mouth, and stay intact through the stomach before making their way into the bloodstream.

Combinatorial hemistry and HTS have introduced an element of automation and certainly transformed the previous "artisan" character of drug discovery. The story of ibuprofen versus naproxen sodium may serve as an illustration of what combinatorial chemistry could have done to accelerate the discovery of popular painkillers:

Ibuprofen (also known as Motrin, Advil, Nuprin, etc.), $C_{13}H_{18}O_2$,[164] was introduced first in the UK in 1969 as a new painkiller and then launched by American Home Products (now merged into Pfizer) as Advil. Naproxen (also known as Naprosyn and Aleve), $C_{14}H_{14}O_3$, was introduced seven years later, in 1976, by Syntex (now part of Roche) as an alternative to ibuprofen with some advantages in treating inflammatory conditions. Both ibuprofen/Advil and naproxen/Aleve target the cyclooxygenase (COX) receptor, the central enzyme in the biosynthetic pathway that promotes the production of the natural mucus lining that protects the inner stomach. There are two isozymes of COX encoded by distinct gene products, COX-1 and COX-2, which differ in their regulation of expression and tissue distribution. COX-1 is inhibited by nonsteroidal anti-inflammatory drugs (NSAIDs) such as aspirin (acetylsalicylic acid, $C_9H_8O_4$, the pain medicine contained in the bark of the willow tree that was

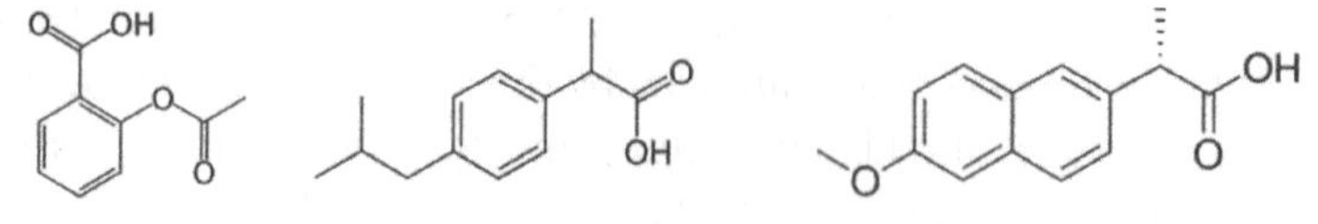

Figure 4.2 Chemical structures of (a) aspirin, (b) ibuprofen/Advil, and (c) naproxen/Aleve. Source: Wikipedia.

already known to Hippocrates). TXA2, the major product of COX-1 in platelets, induces platelet aggregation. Research has shown that the inhibition of COX-1 is sufficient to explain why aspirin is effective at reducing cardiac events. Both Advil and Aleve are nonselective COX inhibitors—they inhibit both COX-1 and COX-2. See Fig. 4.2.

Today's synthetic chemists would assert that if combinatorial chemistry had been used in developing ibuprofen, the competing compound naproxen could have been discovered the very same day!

Note that another widely used fever-reducing painkiller, acetaminophen or paracetamol ($C_8H_9NO_2$, also known as Tylenol, Panadol, Feverall, Doliprane, Alvedon, etc.) is not generally classified as an NSAID because it exhibits only weak anti-inflammatory activity. Although known since 1887, is popularity first surged around 1950 as new laboratory studies resulted in improved purity of the synthesized compound, thereby reducing negative side effects and confirming its fever-reducing and painkilling properties. Compared to aspirin, paracetamol is much gentler on the stomach but lacks aspirin's, ibuprofen's, and naproxen's anti-inflammatory effects. Paracetamol can relieve pain in mild arthritis but has no effect on the underlying inflammation, redness, and swelling of the joints. Its mechanism of action is less well understood than the one of the NSAIDs but it is believed that it targets COX-2 much more than COX-1. Since COX-1 is normally present in a variety of areas of the body, including in particular the stomach, paracetamol's relative indifference to COX-1 make it a safe drug with good gastrointestinal (GI) properties. However, high doses of paracetamol can cause liver damage, in particular in conjunction with alcohol.

To conclude this small excursion into the world of widely used painkillers, it should also be mentioned that subsequent attempts to selectively target only COX-2 have produced very mixed results, illustrated by the Vioxx story: Merck launched Vioxx in 1999 as a painkiller targeting, in particular, patients suffering from arthritis pain. Unfortunately, the drug had to be withdrawn in 2004 because of negative side effects such as heart attacks and stroke. However, another COX-2 inhibitor, Pfizer's drug Celebrex, is still on the market and prescribed for arthritis pain.

4.3 Stage 2: Preclinical Research

During the preclinical testing phase, the lead compounds (or drug candidates) found to be effective in modulating a biochemical pathway governed by the selected target, are tested in laboratories and in animals to determine whether they have a chance to be safe and effective when used on humans.

ADME is a frequently used acronym to describe the properties required to successfully move a drug candidate through the preclinical phase.

"A" stands for *absorption*. For a typical orally active compound to reach a tissue, it usually must be absorbed by mucous surfaces like the digestive tract before reaching the bloodstream and being taken up by the target cells. Factors such as poor compound solubility, gastric emptying time, intestinal transit time, chemical instability in the stomach, and inability to permeate the intestinal wall can all reduce the extent to which a drug is absorbed after oral administration. Absorption critically determines the compound's *bioavailability,* defined as the fraction of an administered drug dose that reaches the systemic circulation. Drugs that absorb poorly when taken orally must be administered in some less desirable way (intravenously, by inhalation, etc.).

"D" stands for *distribution* and is defined as transfer of a drug to its effector site, most often via the bloodstream. The compound may be distributed into muscles and organs. Factors affecting drug distribution include regional blood flow rates and molecular size. The compound may also bind to blood serum proteins, forming a complex. For drugs supposed to reach the brain, Distribution can be a serious problem as the blood–brain barrier (BBB) must be overcome. The BBB is a highly selective permeability barrier that separates the circulating blood from the brain extracellular fluid (BECF) in the central nervous system (CNS). It allows the passage of water, some gases, and lipid soluble molecules, as well as the selective transport of molecules such as glucose and amino acids that are crucial to neural function. On the other hand, it prevents the entry of lipophilic, potential neurotoxins. Historically, the concept of the BBB was first proposed in 1900. It was not until the introduction of the scanning electron microscope to the medical research fields in the 1960s that the actual membrane could be observed and proved to exist.

"M" stands for *metabolism*. Compounds begin to break down as soon as they enter the body. For small-molecule drugs, the liver is playing a crucial role: drug metabolism is carried out in the liver by redox enzymes, termed cytochrome P450 enzymes. The initial (parent) compound is usually converted to new compounds called metabolites. When metabolites are pharmacologically inert, metabolism deactivates the administered dose of parent drug, thereby reducing the effects on the body. However, metabolites may also be pharmacologically active, sometimes even more so than the parent drug.

"E" stands for *excretion*. Drug molecules and their metabolites are removed from the body via excretion, usually through the kidneys (urine) or through the liver and gut (feces). It is important that excretion is as complete as possible, to avoid the accumulation of foreign substances. In addition to the kidney (excretion through urine) and the liver and gut (biliary or fecal excretion), there can also be a role played by the lungs.

Whenever conducting preclinical studies, the effect a candidate drug has on the kidney and the liver of the patient must be well researched and understood, to assess important toxicity risks.

Pharmacokinetics (PK) is the pharmacological discipline concerned with what happens to substances administered externally to a living organism. PK attempts to discover the fate of a drug from the moment that it is absorbed up to the point at which it is completely eliminated from the body.

Pharmacodynamics (PD) is the study of the biochemical and physiological effects of drugs on the body.

PK may be therefore be simply defined as what the body does to the drug, whereas PD may be defined as what the drug does to the body.

During the preclinical phase, in vitro laboratory and animal studies are used to determine the viability of the drug candidate, to make sure it is worth moving into clinical development. If the R&D organization decides to move forward it needs to make sure that the drug's synthesis can be scaled up to ensure manufacturability and that a plan for clinical development of the drug candidate is put in place.

The first two stages of drug discovery are typically taking six to seven years to complete. Most compounds fail during the two stages: According to PhRMA, only 5 of 10,000 compounds, on average, will move from the preclinical phase into clinical development.

Before doing so, the drug company has to submit to the FDA an *investigational new drug (IND)* application that summarizes the data collected so far, demonstrates manufacturability, and outlines plans for the clinical trials. Clinical trials may begin 30 days after the IND application submission, unless the FDA orders a delay. The FDA will not issue a formal approval of an IND application but it can prohibit clinical trials if it determines that human volunteers would be exposed to an unreasonable and significant risk of illness or injury.

The FDA's role in the development of a new drug begins when the drug's sponsor, having screened the new molecule for pharmacological activity and acute toxicity potential in animals, wants to test its therapeutic potential in humans. At that point, the molecule changes in legal status and becomes a new drug subject to specific requirements of the drug regulatory system.

The IND application must contain information in three areas:

I. Animal pharmacology and toxicology studies—preclinical data to permit an assessment as to whether the product is reasonably safe for initial testing in humans.

II. Manufacturing information—information to ensure that the IND sponsor can adequately produce and supply consistent batches of the drug.

III. Clinical protocols and investigator information—detailed protocols for proposed clinical studies to assess whether the initial-phase trials will expose subjects to unnecessary risks. The IND sponsor must further commit to obtain informed consent from the clinical research subjects and to obtain review of the study by an institutional review board (IRB).

4.4 Stages 3–4: Clinical Trials and FDA Review

As shown in Fig. 4.1, the clinical development stage consists of three phases, known as phase 1, phase 2, and phase 3 clinical trials.

In phase 1, sponsors conduct safety studies on about 20 to 100 healthy volunteers in order to identify potential side effects and determine dosage levels.

In phase 2, the drug is tested on 100 to 500 volunteers who are suffering from the disease targeted by the drug. The goal of phase 2 is to determine the drug's effectiveness.

Frequently, phase 2 is divided into phase 2A and phase 2B, where phase 2A trials are pilot studies to evaluate efficacy (and safety) of selected patient populations with the disease or condition to be treated, and phase 2B clinical trials are well-controlled studies representing a most rigorous evaluation of a medicine's efficacy. Phase 2B clinical trials are sometimes referred to as pivotal trials.

In phase 3 clinical trials, the drug is tested on about 1000 to 5000 volunteers to gather enough data on the drug's safety and effectiveness to convince the FDA about approval. The formal review and approval process is initiated by the sponsor via a new drug application (NDA). On average, the three phases of the clinical trial stage require a total of seven years until successful completion.

The fourth and final stage is the FDA review stage, covering the FDA's review and final approval of the submitted NDA.

The NDA contains scientific and clinical data submitted by the sponsor and should convincingly demonstrate that the drug is safe and effective and satisfying an unmet medical need. It takes about 1.5 years to complete the review process and obtain the FDA's approval.

The FDA classifies NDAs by chemical type and therapeutic potential. Seven chemical types are considered, namely:

1. NMEs that have never been approved before and are therefore considered innovative
2. New salts of previously approved drugs
3. New formulations of previously approved drugs
4. New combinations of two or more previously approved drugs
5. Duplication (new manufacturer) of already marketed drug products
6. New indications (claims) for already marketed drugs (including switch from a prescription to over-the-counter drug)
7. Already marketed drug products without approved NDAs (e.g., drugs such as aspirin that were marketed prior to the creation of the FDA in 1906[165])

When classifying an NDA by its therapeutic potential, the FDA compares the new drug to existing products already on the market. Priority is given to those NDAs that appear to have significant therapeutic benefits in the treatment, diagnosis, or prevention of disease. All others are classified as standard.

The US government has set certain performance goals for the FDA, including a goal to complete its initial review and to act on 90% of all priority NDAs within 6 months and to act on 90% of standard NDAs within 10 months.

Sometimes, phase 3B trials are conducted after submission of the NDA to generate more data needed for final FDA approval.

As shown in Fig. 4.1, typically only one out of 10,000 compounds initially identified as potential drug candidates will be approved by the FDA.

After approval and launch, there may be more clinical studies called phase 4 trials, to provide additional details about the medicine's safety and efficacy profile. Different formulations, dosages, durations of treatment, medicine interactions, and other medicine comparisons may be evaluated and new age groups, races, and other types of patients can be studied. In particular, the detection of previously unknown adverse reactions and related risk factors are an important aspect of many phase 4 studies.

Since 1971, the World Health Organization (WHO) via its Collaborating Centre for International Drug Monitoring (or Uppsala Monitoring Centre, located in Uppsala, Sweden) is collecting data about adverse reactions to approved drugs and stores the results in a publicly available database, VIGIBASE.[166] *Pharmacovigilance* is the field of risk management practiced by the biopharmaceutical industry and by national governments to protect patients and take steps whenever there is emerging evidence that a drug (or a combination of interacting drugs) may have serious side effects.

Biopharmaceutical drug discovery and development is a capital-intensive high-risk undertaking and requires sponsors with experience, financial strength, and staying power. It takes on average more than $1 B to bring a new NME to market. As a consequence, the industry has seen consolidation toward a small number of strong global companies.

However, despite a wave of mergers and acquisitions since the 1980s, even the largest biopharmaceutical companies are struggling with in-house translation of the new biomedical knowledge into successful identification and validation of new drug targets, thereby slowing progress toward greater numbers of approved new innovative NMEs.

The above-mentioned GAO report is pointing to a shortage of skills with strong enough scientific background to accelerate the current pace of drug discovery and is closing with a recommendation to create PPPs between government/academic research and the biopharmaceutical industry and to encourage industry to engage in pre-competitive collaboration.

4.5 The Emerging Importance of "Biologics"

Biological medical products, or "biologics,"[167] are medicinal products such as:

- vaccines
- blood and blood products for transfusion and/or manufacturing into other products

- allergenic extracts, which are used for both diagnosis and treatment (e.g., allergy shots)
- human cells and tissues used for transplantation (e.g., tendons, ligaments, bone)
- gene therapies
- cellular therapies
- tests to screen potential blood donors for infectious agents such as human immunodeficiency virus (HIV), etc.

Biologics are created by biological processes rather than being chemically synthesized like small-molecule drugs.

Technically, biologics are a subset of biopharmaceuticals. However, the term "biopharmaceuticals" usually refers to macromolecular products like protein-based and nucleic-acid-based drugs, while the term "biologic" is used more often when the medical product is composed of cellular- or tissue-based products (e.g., stored-packed red blood cell units). Biologics are usually produced by biotechnology (BT) methods and do require advanced (and sometimes difficult to control) processes for development and manufacturing that differ from small-molecule-based drug manufacturing. Biologics are typically resulting from leading edge biomedical research and may be used to treat a variety of medical conditions for which no other treatments are available.

Historically, biologics were usually extracted from the bodies of animals (including other humans). Important traditional biologics include whole blood and other blood components and organs and tissue transplants. More recently, since the emergence of biotech industry in the 1970s, the primary ways to develop biologics is to use *recombinant DNA (rDNA) technology* and *monoclonal antibodies (mabs)*.

rDNA molecules are formed by laboratory methods of genetic recombination (such as molecular cloning) to bring together genetic material from multiple sources, creating sequences that would not otherwise be found in biological organisms.

The DNA sequences used in the construction of rDNA molecules can originate from plants, bacteria, fungi, etc., and may include human DNA. Even DNA sequences that do not occur anywhere in nature may be created by chemical synthesis of DNA and incorporated into recombinant molecules. Using rDNA technology, any DNA sequence may be created and introduced into any of a very wide range of living organisms.

Proteins that can result from the expression of rDNA within living cells are termed "recombinant proteins."

rDNA is created by two main methods, polymerase chain reaction (PCR) and cloning. While PCR replicates DNA in the test tube, molecular cloning involves replication of the DNA within a living cell.

Both concepts—PCR and cloning—have been introduced already in Chapter 3 of this book.

In most cases, organisms containing rDNA have apparently normal phenotypes, that is, their appearance, behavior, and metabolism are usually unchanged. However, phenotypic changes can occur when a recombinant gene has been chosen to generate biological activity in the host organism. For instance, toxicity to the host organism can be induced by the recombinant gene product if it is overexpressed or expressed within inappropriate cells or tissues. rDNA can also have deleterious effects: insertional inactivation can be used to knock out genes in the host cell to determine their biological function and importance.

rDNA is widely used in BT, medicine, and research. For example:

- *Recombinant human insulin* almost completely replaced insulin obtained from animal sources (e.g., pigs and cattle) for the treatment of diabetes. Recombinant insulin is typically synthesized by inserting the human insulin gene into *Escherichia coli*, which then produces insulin for human use.
- *Recombinant human growth hormone* (HGH, somatotropin) is administered to patients whose pituitary glands generate insufficient quantities to support normal growth and development. Before recombinant HGH became available, HGH for therapeutic use was obtained from pituitary glands of cadavers, an unsafe practice that carried the risk of Creutzfeldt–Jacob disease transfer to patients requiring HGH.
- *Recombinant hepatitis B vaccine*: Hepatitis B infection is controlled through the use of a recombinant hepatitis B vaccine, which contains a form of the hepatitis B virus

(HBV) surface antigen that is produced in yeast cells. The development of the recombinant subunit vaccine was an important and necessary development because HBV, unlike other common viruses such as polio virus, cannot be grown in vitro.

- *Epoetin alfa* (Amgen's drug Epogen, FDA approval 1989) is human erythropoietin produced in cell culture using rDNA technology. Erythropoietin stimulates erythropoiesis (increases red blood cell levels) and is used to treat anemia, commonly associated with chronic renal failure and cancer chemotherapy.

rDNA technology sparked the Biotech revolution. The first US patent on rDNA was filed by Stanford University in 1974, listing as the inventors Stanley N. Cohen and Herbert W. Boyer; this patent was awarded in 1980. While Dr. Cohen pursued an academic career, Dr. Boyer went on to cofound Genentech in South Francisco. The first licensed drug generated using rDNA technology was human insulin, developed by Genentech. In 1982, synthetic "human" insulin was approved by the US FDA, thanks largely to Genentech's partnership with insulin manufacturer Eli Lilly and Company (Lilly), which shepherded the product through the FDA approval process. The product (Humulin) was then licensed to and manufactured by Lilly and became the first-ever approved *genetically engineered* human therapeutic.

Another big wave of biotech breakthroughs is based on the concept of mabs. To explain the concept, let's first go back to the work by Paul Ehrlich, who shared the 1908 Nobel Prize for physiology or medicine[168] for his research into immunity. Ehrlich not only developed chemotherapy but is most famous for introducing the idea of "magic bullets," substances that could selectively target organisms that caused diseases such as sleeping sickness, typhoid, and syphilis. We know that our body's immune system is making antibodies, proteins that recognize invading microbes and other attacking organisms and then try to destroy the dangerous targets.

The mabs are defined as "monospecific antibodies that are made by identical immune cells that are all clones of a unique parent cell."

Following Ehrlich's magic-bullet idea, the goal is to produce mabs that specifically bind to a given substance and can then serve to detect or purify that substance.

In 1882, Robert Koch, the discoverer of the tubercle bacillus, assembled a team to tackle the newly forming field of immunotherapeutics. Team members in Berlin's Institute of Infectious Diseases were Ehrlich, Emil von Behring, Erich Wernicke, and Shibasaburo Kitasato. Led by Behring, the team developed the first working antisera against diphtheria, tetanus, meningitis and pneumonia.

In 1984, Georges Köhler, César Milstein, and Niels Kaj Jerne shared the Nobel Prize for physiology or medicine for theories concerning the specificity in development and control of the immune system and the discovery of the principle for production of mabs.[169]

Milstein and Köhler were able to turn the normally evil feature of tumor cells, the capacity to proliferate forever, into a very beneficial property: During a hectic two year period, 1975–1976, they developed a technique allowing them at will to fish up exactly those rare antibody-producing cells that they wanted from a sea of cells. These cells were then fused with tumor cells from another species (e.g., mouse) creating hybrid cells with eternal life and capacity to produce the very same antibody in high quantity. Köhler and Milstein called these hybrid cells *hybridomas*, and as all cells in a given hybridoma come from one single hybrid cell, the antibodies made are monoclonal. The first application of hybridoma technology was to use a line of myeloma cancer cells that had lost their ability to secrete antibodies and to fuse them with healthy antibody-producing B cells. The mab technology allowed scientists to grow huge quantities of pure antibodies aimed at specific selected targets, leading to the design of new diagnostic tests and therapeutics. By injecting a payload of mabs into the bloodstream, the antibodies were headed straight to their disease target.

Drugs based on mab technology are all named *xxxx*mab. The first drug that received FDA approval was *Rituxan* (rituximab) for the treatment of non-Hodgkin's lymphoma, in 1998. It was developed by IDEC, since merged with Biogen. Rituximab is a mab against the protein CD20, which is primarily found on the surface of immune system B cells. Rituximab destroys B cells and is therefore used to treat diseases that are characterized by excessive numbers of B cells, overactive B cells, or dysfunctional B cells. This includes many lymphomas, leukemias, transplant rejection, and autoimmune disorders. IDEC teamed up with Genentech to get FDA approval and to comarket Rituxan. In parallel, Genentech developed *Herceptin* (trastuzumab), another mab therapeutic, for the treatment of breast cancer associated with the HER2/neu receptor. Herceptin was approved by the FDA in 1998 and is the first important example of a drug requiring a diagnostic test to ensure efficacy, an example of *personalized medicine*: Only patients who express the *Her2* gene in their breast cancer tissue have a chance to benefit from treatment with Herceptin.

Remicade (infliximab) was approved by the FDA for the treatment of psoriasis, Crohn's disease, ankylosing spondylitis (an inflammatory disease that can cause vertebrae in the spine to fuse together and causing a hunched-forward posture), psoriatic arthritis, rheumatoid arthritis, and ulcerative colitis. Infliximab was developed by Junming Le and Jan Vilcek at New York University School of Medicine and developed by Centocor (now J&J).

Humira (adalimumab) has been approved for rheumatoid arthritis, psoriatic arthritis, ankylosing spondylitis, Crohn's disease, ulcerative colitis, moderate to severe chronic psoriasis, and juvenile idiopathic arthritis. Adalimumab was the first fully human mab drug approved by the FDA and was discovered through a collaboration between BASF Bioresearch Corporation (Worcester, MA) and Cambridge Antibody Technology. Humira is now owned by AbbVie.

Avastin (bevacizumab) is a recombinant humanized mab that inhibits angiogenesis by targeting vascular endothelial growth factor A (VEGF-A). Angiogenesis, the physiological process through which new blood vessels form from pre-existing vessels, is a normal and vital process in growth and development, as well as in wound healing and in the formation of granulation tissue. However, it is also a fundamental step in the transition of tumors from a benign state to a malignant one. VEGF-A stimulates angiogenesis in a variety of diseases, especially in cancer. Bevacizumab was the first clinically available, angiogenesis inhibitor in the United States and is used against a number of cancers (in combination with chemotherapy). It also has ophthalmological use for treatment of age-related macular degeneration (AMD) and diabetic retinopathy, where blood vessels around the retina grow abnormally and leak fluid, causing the layers of the retina to separate. Bevacizumab was based on research into angiogenesis by Judah Folkman (Harvard), who developed a drug that could cure cancer in mice. Genentech scientist Napoleone Ferrara, who was hired to work on a childbirth drug that was unsuccessful, also worked in his spare time on angiogenesis proteins that led to bevacizumab.

Although it took more than 20 years from Nobel Prize–winning research to first therapeutic and commercial success, mabs have now been generated and approved to treat cancer, cardiovascular disease (CVD), inflammatory diseases, macular degeneration, transplant rejection, multiple sclerosis (MS), and viral infection. In August 2006 the Pharmaceutical Research and Manufacturers of America reported that US companies had 160 different mabs in clinical trials or awaiting approval by the FDA.

It is clear biologics require process and manufacturing technologies very different from traditional small-molecule drugs. Those technologies are not easy to fully master and control. Therefore, generic versions of biologics—*biosimilars*—are much more difficult to produce.

Biologics generally may be quite sensitive to changes in manufacturing processes. Follow-on manufacturers do not have access to the originator's molecular clone and original cell bank, nor to the exact fermentation and purification process, nor to the active drug substance. They do have access to the commercialized innovator products. Differences in impurities and/or breakdown products can have serious health implications. This has created a concern that copies of biologics might perform differently than the original branded version of the product.

A major disadvantage of biologics is the need to administer such drugs by parenteral routes, that is, routes other than the digestive tract (oral) or topical application. There are several parenteral routes, usually requiring injection. Therefore, most protein drugs are limited to diseases where an injectable drug is acceptable. In most cases, biologics are focused on serious medical conditions; they frequently have a lifesaving quality.

4.6 Biomarkers and Stratified/Personalized Medicine

A "biomarker," or biological marker, generally refers to a measured characteristic that may be used as an indicator of some biological state or condition. The term occasionally also refers to a substance whose presence indicates the existence of living organisms.[170]

Biomarkers are often measured and evaluated to examine normal biological processes, pathogenic processes, or pharmacologic responses to a therapeutic intervention.

Biological markers (biomarkers) have been defined by the US FDA[171] as a "characteristic that is objectively measured and evaluated as an indicator of normal biological processes, pathogenic processes, or pharmacological responses to a therapeutic intervention."

The FDA's commitment to biomarkers is reflected in its critical path initiative,[172] launched already in 2004 and aimed at the incorporation of recent scientific advances such as genomics and advanced imaging technologies into the drug development process, as explained by Woodcock and Woosley[173] and illustrated in Fig. 4.3.

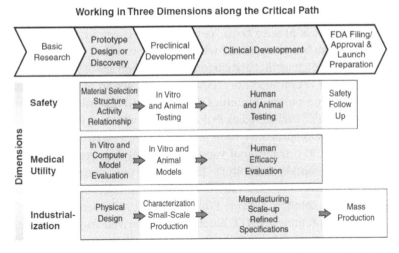

Figure 4.3 The critical path of drug development according to Woodcock and Woosley.

Biomarkers related to measurements that provide information about the safety and efficacy (or medical utility) of drug candidates are believed to hold the promise of increasing the productivity of biopharmaceutical R&D. It is expected that biomarker-based drug development will enable better and earlier decision making and that (genomic) biomarkers will pave the way toward targeted therapeutics and, ultimately, stratified and personalized medicine.

Let us take a closer look at the impact of biomarker based drug discovery and development on the business processes of a biopharmaceutical R&D organization, following the excellent review by the MIT Center for Biomedical Innovation (CBI).[174] The third author, Frank L. Douglas, joined the CBI after a long career in the pharmaceutical industry, including the position of global head of Aventis R&D. The authors point out that there is a continuum of patient therapy with empirical traditional medicines on the one end and individualized medicine on the other. In between lies the field of stratified medicine, where specific patient populations can be characterized by clinical biomarkers. Clinical biomarkers link patient populations to treatments.

While "stratified" is a suitable term to use for the stratification of patient populations by means of clinical biomarkers, it has not been widely accepted and will therefore be replaced below by the somewhat less precise term "personalized." However, we will still use the term "individualized," whenever applicable.

Most medicines that are currently prescribed are of the *empirical* type. Good examples are NSAIDs (see Section 4.1) for pain relief; proton-pump inhibitors (PPIs) for GI disorders, such as omeprazole[175] (marketed as Prilosec by AstraZeneca; FDA approval 1989) or esomeprazole (trade name Nexium,[176] AstraZeneca; FDA approval 2001); traditional vaccines; and antidepressant selective serotonin-reuptake inhibitors (SSRIs) such as Prozac[177] (Lilly; FDA approval 1987), Paxil[178] (GlaxoSmithKline; FDA approval 1996), and Zoloft[179] (Pfizer; FDA approval 1997). Serotonin, or 5-hydroxytryptamine (5-HT), biochemically derived from the amino acid tryptophan, is primarily found in the GI tract, platelets, and the CNS of animals, including humans. It is popularly thought to be a contributor to feelings of well-being and happiness. In the CNS, serotonin has various functions, including the regulation of mood, appetite, and sleep, but also cognitive functions, including memory and learning. Modulation of serotonin at synapses (structures that permit a nerve cell, a neuron, to pass an electrical or chemical signal to another cell) is thought to be a major action of several classes of pharmacological antidepressants.

Individualized medicines are not widely used today but will gain in importance as part of regenerative medicine (see Section 5.6 below) and as we learn more and more about gene-disease relationships and individual responses of the immune system. For instance, tumor cells extracted from a cancer patient can be used to develop a vaccine to be administered to this patient. Another way to design an individualized treatment program for a particular cancer patient would be to perform a genomic analysis along with testing of antibodies and then pick a combination of drugs selected by a panel of experts or recommended by sophisticated decision support algorithms. Whereas the practice of individualized medicine is still rare and may—for financial reasons—never be a realistic choice for the majority of patients, a strong case can be made for *personalized medicine*. If clinical biomarkers can be found that have been validated and accepted by the FDA, the associated treatments should generate better patient outcomes with less side effects. Drug makers should then co-develop diagnostic tests along with specific drugs target those drugs to patient populations identified as responders to the test. Examples of such drugs are already in use and include the breast cancer drug trastuzumab[180] (marketed by Genentech as Herceptin; FDA approval 1998) and the cancer drug imatinib[181] (marketed by Novartis as Gleevec; FDA approval 2001) that is used for the treatment of multiple cancers, in particular chronic myelogenous (or myeloid or myelocytic) leukemia (CML), a cancer of the white blood cells.

The Herceptin development timeline has been documented in detail by Genentech.[182] What's important for our current discussion of personalized medicine is the fact that Herceptin is a mab that interferes with the HER2/neu receptor. Hence it will only be effective as a breast cancer drug for ~25% of patients belonging to a subpopulation that is defined by overexpression of HER2/neu in cancer tissue. The HER receptors are proteins that are embedded in the cell membrane and communicate molecular signals from outside the cell (molecules called epidermal growth factors, or EGFs) to inside the cell and turn genes on and off. In certain types of breast cancer, HER2 is overexpressed and causes cancer cells to reproduce uncontrollably. In 1996, researchers at Memorial Sloan Kettering Cancer Center (MSKCC) in New York published a clinical study that showed that the targeting of growth factor receptors caused regression of human cancer. After initiating a partnership with diagnostics company DAKO, Genentech conducted successful phase 3 trials that showed that Herceptin, in combination with chemotherapy, increased time to disease progression and response rates for the stratified breast cancer patient population identified by the HER2/neu test. HER2/neu is the clinical biomarker associated with Herceptin.

Likewise, the breakpoint cluster region (BCR)-Abelson (ABL)-positive tyrosine kinase genotype is the clinical biomarker associated with patient populations suffering from CML that benefit from Gleevec, an inhibitor of this kinase. CML was the first malignancy to be linked to a clear genetic abnormality, the chromosomal translocation known as the Philadelphia chromosome, so named because it was first discovered and described in 1960 by two scientists from Philadelphia, Peter Nowell and David Hungerford.[183] They identified a translocation in which parts of two chromosomes, #9 and #22, swap places, resulting in a fusion gene that is created by juxtapositioning the *ABL1* (Abelson, the name of a leukemia virus) gene on chromosome 9 (region q34) to a part of the *BCR* gene on chromosome 22 (region q11).

As a result, patients afflicted by this genetic anomaly carry an elongated chromosome #9 and a truncated chromosome #22, as shown in Fig. 4.4.

The BCR-ABL fusion gene encodes a kinase that may generate uninterrupted phosphorylation of tyrosine residues in proteins, a nonstop functional state often responsible for the initiation or progression of cancer. In the presence of tyrosine kinase inhibitors (TKIs) such as Gleevec, the binding of ATP is blocked, phosphorylation is prevented, and BCR-ABL-expressing cells either have a selective growth disadvantage or undergo cell death.

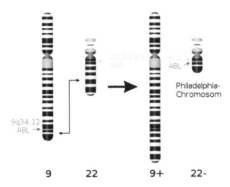

9   22   9+   22-

Figure 4.4 Philadelphia chromosome. Source: Wikipedia.

Gleevec was found to inhibit the progression of CML in a majority of patients (65%–75%), sufficiently to achieve regrowth of their normal bone marrow stem cell population. However, because some leukemic cells persist in nearly all patients, the treatment has to be continued indefinitely. Still, because of Gleevec and other follow-on TKI drugs, CML has become the first cancer in which a standard medical treatment may result in a normal life expectancy.

On the basis of the apparent success stories of Herceptin and Gleevec, the biopharmaceutical industry has made serious attempts to incorporate personalized medicine ideas into their R&D strategies. As demonstrated by Herceptin and Gleevec, drugs that were targeted to patient populations could generate similar return of investment (ROI) as one-size-fits-all blockbuster drugs such as Prilosec.

However, traditional clinical practice has to be modified to accommodate personalized medicine: In addition to traditional differential diagnosis, the provider must add a clinical biomarker assessment step and the health insurance payer must agree to pay for the extra test. In return, the new treatment will not be wasted on patients that won't benefit and it should generate better outcomes, a higher probability for responders, patients who belong to the specific subpopulation identified by the clinical biomarker, to benefit from the treatment. Everybody wins: The patient's chance for survival increases, the provider's success rate increases, the payer's efficiency of providing funds will be higher, and the drug developer's ability to charge a premium for the stratified medicine is improving as well.

In our examples we have focused on cancer as the primary disease area for personalized medicine. However, the more we learn about genetics and disease mechanisms, the more reasons we can find to introduce personalized medicine.

Already, antibiotics are matched to specific resistant infections. Another disease area is human immunodeficiency virus (HIV)/acquired immunodeficiency syndrome (AIDS), a condition caused by the human immunodeficiency virus that leads to AIDS. AIDS causes progressive failure of the human immune system to protect against life-threatening infections and cancers. Today's HIV/AIDS therapies are tailored to the viral variant present for a specific patient.

Unfortunately, there may be situations where clinical biomarkers are not yet accompanied by effective treatments designed to go hand in hand with the diagnostic test. Alzheimer's disease (AD) is such a disease area: Beta-amyloid is a sticky protein that is implicated in the condition and it can be traced by medical imaging techniques. Other tests look for proteins (like tau[184]) known to indicate cognitive dysfunction in the spinal fluid, in the eyes, and even noses of Alzheimer's patients, years ahead of actual memory loss problems. It can only be hoped that treatments—at least preventive measures—will soon catch up with the existing and forthcoming diagnostic tests.

What may be expected in the future is a transition from empirical to personalized medicine, even in disease areas that so far have not been touched by the clinical biomarker concept. The use of cholesterol-lowering drugs such as *statins* could move in this direction, although the case for doing so is still not strong enough. Statins (or 3-hydroxy-3-methyl-glutaryl–coenzyme A [HMG-CoA] reductase inhibitors) are a class of drugs used to lower cholesterol levels by inhibiting the enzyme HMG-CoA reductase (or HMGCR), the rate-controlling enzyme of the metabolic pathway that produces cholesterol. HMGCR plays a central role in the production of cholesterol in the liver, which produces about 70% of total cholesterol in the body. High cholesterol levels have been associated with CVD and statins have been found to prevent CVD in those who are at high risk.

Patients do actually react differently to statins and a genetic test could help in the selection of the optimal statin for a given patient. However, the added inconvenience and cost of finding the optimal statin have so far outweighed the possible incremental benefit over a randomly chosen one.

Necessary conditions for personalized medicine include therefore (i) a biological characteristic with the potential to induce differential patient responses to a therapy, (ii) multiple therapeutic options that have heterogeneous responses, and (iii) a clinical biomarker that can link therapies to a subset of patients likely to show that different response.

From a business point of view, three criteria must be satisfied for the biopharmaceutical industry to engage in the development of a personalized medicine, namely:

1. Technical feasibility of identifying a patient population prospectively and already included in the planning of clinical trials. Note that the number of patients enrolled can be reduced by including only a given patient population pre-selected by means of the clinical biomarker. FDA validation of the biomarker is an important factor.
2. Attractive economics: As a rule of thumb, a drug must show potential for $500 M peak annual sales to justify the costs of R&D and marketing. More about this criterion is given further below.
2. A sustainable franchise, characterized by significantly improved patient outcomes, strong patent protection, and regulatory market protection due to small population size ("orphan drug status"). Orphan drugs are developed specifically to treat a rare medical condition, the condition itself being referred to as an "orphan disease." Governments in the United States and in Europe provide financial incentives, such as extended exclusivity periods, to encourage the development of such drugs.

What is the business case for development of a drug targeted to a subpopulation, a personalized drug?

Figure 4.5 shows the investment and ROI phases. Their durations are dictated by the clinical development phases and FDA approval time and by the patent life of the drug, respectively.

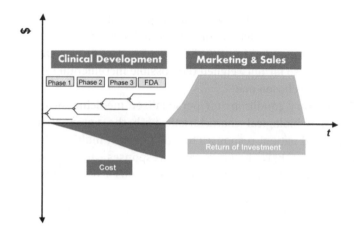

Figure 4.5 Investment (red) and ROI (green) cycles of a drug.

During the ROI phase it is assumed that it takes time to introduce the drug, convince doctors and patients about its benefit, and eventually reach peak sales until expiration of the patent. Obviously, pricing power disappears as generic copies of the off-patent drug reach the market.

Now let's investigate how a stratified/personalized medicine could modify the pattern shown in Fig. 4.5.

In their *Nature Reviews/Drug Discovery* paper, Trusheim, Berndt, and Douglas establish a classic reference business case by assuming that a one-size-fits-all empirical drug requires $1 B to develop, reaches sales of $500 M per year, and enjoys a 10-year patent life. With a typical profit margin of 80%, the lifetime gross profit for the drug would be $4 B, four times greater than the cost.

However, current trends are eroding this 4.0 payback ratio. Assuming that development costs of an empirical drug are doubling to $2 B, that due to increased development times and regulatory delays the patent life is shrinking to six years, and that—caused by government actions—gross margins are declining from 80% to 50%, the lifetime gross profit for a $500 M drug becomes $1.5 B, leading to a payback ratio of <1. Even if the drug generates revenues of $2 B per year, resulting in a lifetime gross profit of $6 B, the payback ratio will fall from 4.0 to 3.0.

Now let's assume the stratified/personalized case and assume that all healthcare stakeholders are embracing the new concepts, as advocated in FDA's critical path vision. With well-chosen patient populations selected by means of a validated clinical biomarker, the number of patients to be recruited for clinical trials could be reduced and development times perhaps cut in half, from 10 years (reference case above) to 5 years. Due to smaller clinical trial sizes, the development cost for this "bright future" scenario could perhaps be cut to only $250 M. Furthermore, drug performance will be so much superior to the current trend scenario that governments may agree to reimbursement rates that preserve the 80% profit margin the industry has enjoyed in the past. Even with peak sales of only $200 M, the lifetime gross profit will be as high as 80% $\times$ 15 $\times$ $200 M = $2.4 B, compared to a cost of only $250 M. That's a payback ratio of 9.6!

Even in a more pessimistic scenario where development times still remain at 10 years resulting in a $500 M cost, gross margins decline to 65% due to ever-increasing government and managed care buying power, the business case would still close: Lifetime gross profit will be 65% $\times$ 10 $\times$ $200 M = $1.3 B, 2.6 times more than the cost.

This financial analysis suggests that the industry would be well advised to move toward a biomarker driven future, with drugs targeted at patient populations that will enjoy better outcomes and less side effects than in the past.

However, progress along the critical path toward personalized medicine has been slow: There are not enough validated biomarkers, that is, biomarkers that are recognized by the scientific community, and some major pharma companies are still struggling with full incorporation of critical path concepts into their everyday R&D operations.

Almost 10 years after FDA's launch of the critical path initiative, the status in the industry has been summarized by the Tufts Center for the Study of Drug Development,[185] as follows:

- 94% of all pharma companies are investing in personalized medicine research.
- 81% of pharmas have entered into personalized medicine partnerships.

- 50% of all clinical trials are involved in the collection of DNA samples of participating patients.
- 30% of all pharmas require a biomarker for all compounds in development.
- 10% of all compounds currently in phases 2B–4 (postapproval) have an associated companion diagnostic.

Finally, it should be stated that biomarker-based drug development is critically dependent on collaboration between the private and public sectors. Several organizations such as the Critical Path Institute (Tucson, AZ) and the NIH are actively promoting such collaborations.

More detailed information about the role of biomarkers in drug development can be found in a recent handbook edited by Bleavins, Carini, Jurima-Romet, and Rahbari.[186]

4.7 The Past and Future of Biopharmaceutical R&D

After several decades of highly profitable operations and ever increasing R&D budgets, the industry is seriously rethinking how to improve R&D productivity and address the unacceptable failure rates leading to the well-documented stagnation of FDA approvals.

Starting in the 1980s, the industry tried to solve its productivity challenges by investing in advanced new technologies and by consolidation based on a seemingly never-ending series of mergers and acquisitions.

Examples of new technologies and paradigm shifts that were embraced by the biopharmaceutical industry during the past 50 years are listed below:

- 1960s Mechanism-based drug discovery: rational biochemical screening
- 1970s Structural biology and molecular modeling: rational drug design
- Early 1980s New era of biologics—rDNA and mabs
- Late 1980s Knockout mouse: genetically engineered mice with inactivated existing genes[187]
- Early 1990s Combinatorial chemistry and HTS

- Late 1990s uHTS and miniaturization/automation
- 2000s Genetics, genomics, proteomics, stem cells, etc.
- 2010s Nanomedicine for targeted drug delivery, regenerative medicine
- 2015s Immunotherapy, gene therapy

Unfortunately, the impact of all those new technologies on both drug discovery failure rates and the number of FDA approvals has been disappointing. Despite serious attempts by the industry to "fail early" instead of investing resources into late-stage drug developments that have to be abolished due to insufficient efficacy or drug safety (due to problems with bioavailability, metabolic properties, toxicity, etc.), overall R&D productivity remains poor, in particular when compared to other industries.

Despite such gloom, the industry has thrived and achieved strong financial results due to a "blockbuster business model" that—at least in the past—could support high failure rates: Whenever a drug made it to the market and satisfied an important unmet medical need, the healthcare sectors of the Western developed countries were willing to richly compensate the industry for its efforts. As long as patent protection lasted for FDA-approved NMEs, the owner of the patent could charge a premium and therefore afford continued R&D investments.

In addition, the industry consolidated from a large number of midsize companies into a small number of global giants. Let us consider an example, SANOFI, to illustrate this wave of industry consolidation:

- In 1990, the French chemical/pharmaceutical company Rhone-Poulenc (Lyon-Paris) formed Rhone-Poulenc-Rorer (RPR) with R&D labs located in a suburb of Paris.
- In 1997, German chemical/pharma giant Hoechst (Frankfurt) formed the life sciences company Hoechst-Marion-Roussel by merging its life sciences operations with US pharma Marion Merrell Dow (formed 1995) and French pharma Roussel Uclaf.
- In 1999, RPR and HMR agreed to merge and form Aventis, with major research labs in France, Germany, and the United

States. Aventis corporate headquarters were placed near Strasbourg, Alsace, France.

- In 2003, Aventis had 75,000 employees, a revenue of >$20 B, and a profit of >$3 B.
- In 1973, SANOFI (French for Health & Finance) was formed in Paris as a subsidiary of the French oil company Elf-Acquitaine, later merged into TOTAL.
- In 1999, Sanofi merged with Synthelabo, a former subsidiary of cosmetics giant L'Oreal.
- In 2003, Sanofi-Synthelabo had 33,000 employees, a revenue of $10.5 B, and a profit of $2.7 B.
- In 2004, involving highest levels of government in France (President Chiraq) and Germany (Chancellor Schroeder), Sanofi-Synthelabo acquired Aventis and created Sanofi-Aventis (S-A)[188] with corporate headquarters in Paris.
- In 2011, S-A acquired Genzyme, a top US biotech company (11,000 employees) with headquarters in Boston, MA.
- Later in 2011, Sanofi-Aventis dropped the name Aventis and became Sanofi. Sanofi Pasteur is part of Sanofi and entirely focused on vaccines.

At the end of 2014, Sanofi achieved global sales of ~$39B with ~110,000 employees. Profit was slightly down compared to the combined profits 10 years ago, but R&D investments remained high at ~$6B.

Similar stories can be told about Pfizer (Pharmacia including Upjohn and Farmitalia, Warner-Lambert Parke-Davis [WLPD], Wyeth/ American Home Products, including Ayerst, American Cyanamid—Lederle Labs); Astra-Zeneca (created through the merger of Swedish Astra with UK's Zeneca, going back to Imperial Chemical Industries, a chemical company founded in 1870 by dynamite inventor Alfred Nobel); British GSK (Glaxo, Wellcome, SmithKline, Beecham); US-based Merck (now including Schering-Plough); Swiss Novartis (merger of Basel-based Ciba-Geigy and Sandoz) and Roche (including Boehringer-Mannheim Diagnostics, Syntex, Chugai, and Genentech); German Bayer (including Schering), etc.

While in 1985 the total market share of the top 10 global pharmaceutical companies was about 20%, by 2002 it had grown to 48%[189] and is now above 60%.

More mergers and acquisitions are expected to occur by 2020, probably driven by a desire to dominate certain market segments or to be among the top three global leaders by market share. To achieve that goal, we can see, for example, Novartis to sell its vaccines business and acquire the cancer business of a competitor that has a stronger position in vaccines but is not a dominant player in oncology. We may see more specialization by disease area as global pharmaceutical companies can no longer afford to support a broad R&D portfolio across all therapeutic areas.

Some mergers are even contemplated to minimize corporate taxes. Prompted by the high US corporate tax rate, Pfizer generated media attention in 2014 by announcing its intention to acquire AstraZeneca for $117 B and to move the corporate headquarters from New York to London. However, AstraZeneca rejected the offer because of fears of both British and Swedish governments of the loss of research jobs. Those fears seem to be justified, given the long history of Pfizer acquisitions and recent cuts in R&D budgets:

- 1999/2000 acquisition by Pfizer of WLPD for $116 B
- 2002/2003 acquisition by Pfizer of Pharmacia for $60 B
- 2009 acquisition by Pfizer of Wyeth for $68 B
- 2014 (so far) unsuccessful attempt by Pfizer to acquire AstraZeneca for $117 B

As Pfizer made those three major acquisitions, spending a total of $244 B, its market capitalization only moved by ~$60 B between 1999 (~$125 B) and 2014 (~$185 B). Pfizer's market cap reached $290 B in 2000 after the WLPD acquisition but dropped down to ~$188 B in 2002. After peaking again at ~$269 B in 2003 after the Pharmacia acquisition, it fell all the way down to ~$119 B in 2008, due to patent expirations and disappointing clinical trial results (2006) of a much anticipated Lipitor successor. Lipitor, the cholesterol-lowering statin that came to Pfizer via WLPD, has been the most successful blockbuster ever, beating out the previous leader, the GI drug Prilosec/Nexium, developed by Astra in Sweden.

After the Wyeth acquisition in 2009, Pfizer had to cut R&D budgets and streamline operations to improve its shareholder value. Pfizer employees paid a heavy price: Since 2011, Pfizer has closed research labs in Ann Arbor, MI (where Lipitor was developed);

Groton, CT; and Sandwich, UK. As an interesting side note it can be mentioned that both Groton and Sandwich were originally chosen as manufacturing sites for the production of antibiotics, a big part of Pfizer's business during and after World War II. Both sites were former navy shipyards. Unfortunately, the locations turned out be less attractive during the recent "Big Pharma" trend to locate drug discovery labs very close to leading medical research universities.

As mentioned above, Pfizer's so far unsuccessful attempt to acquire AstraZeneca was motivated by the high US corporate tax rate of 35%. Another rather complex merger designed to avoid the high US tax rate is planned between AbbVie—the Chicago-based pharmaceutical company that was created in 2013 by demerging Abbott Labs[190]—and the Irish drug maker Shire.

It's a deal that will be further complicated by involvement of Mylan, a Pennsylvania-based drug maker.

Unless the US government intervenes and objects to the "inversion" trend by closing a current tax loophole, AbbVie will pay $53 B to buy Shire and move its headquarters to Europe. Then Mylan will acquire AbbVie's international generic drug business from AbbVie for $5.3 B and reincorporate in the Netherlands.

However, just continuing to execute more mergers and acquisitions—for whatever reason—will not secure the future of this industry.

With increasing pressure on drug prices by the world's health-care systems, the biopharmaceutical industry has to rethink its business model and focus on R&D productivity improvements. Here is a list of important trends that may be implemented by the year 2020:

- Use biomarker-based drug development: For each new drug there should be a (*companion*) *diagnostic* test that identifies the patient population that will benefit.
- Increase investment in translational medicine for the validation of targets and small, speedy clinical studies designed using sensitive endpoint biomarkers.[191]
- Take advantage of the FDA's preferential (fast track) treatment of new drug applications for "orphan diseases" where only a small percentage of patients is suffering from the condition.

- Close unproductive isolated research labs and move to research campus locations closer *to* academic medical research centers in hot spots such as Boston (Harvard, MIT, etc.), New York, San Francisco, San Diego, and London.
- Form PPPs with academic and government research centers.
- Explore collaborations (even with competitors!) in precompetitive areas such as disease understanding and target identification.
- Decrease emphasis on animal models, as they frequently are inaccurate predictors of drug efficacy in humans.
- More critically evaluate medical research literature, in particular in therapeutic areas such as cancer. Use cognitive computing and big data analytics to assist scientists in digesting literature, extracting relevant facts, and forming hypotheses.
- Exploit the emerging highly predictive power of modeling and simulation, taking advantage of improved accuracy of algorithms and vastly improved price–performance characteristics of supercomputers.
- Make the most of genetics and genomics as more and more gene–disease relationships are uncovered. Translate breakthroughs in DNA sequencing technologies and genomic data analysis into benefit for patients (genomic medicine).
- Place increased emphasis on prevention, that is, focus on vaccines.
- Incorporate nanomedicine (for targeted drug delivery, diagnostic tests, regenerative medicine, etc.) into the research and process development environment.
- Insist on best practices in portfolio management, hopefully leading to a reduction of attrition rates as a candidate drug makes its way through the R&D process pipeline. We may also expect therapeutic portfolios to be trimmed by strategic top management decisions to focus on a few therapeutic segments instead of trying to cover all disease areas.

Table 4.1 shows a list of the top 31 global pharma companies, sorted by market capitalization (not sales revenues) as of September 2014.

Table 4.1 Top global pharmaceutical companies

#	Company name	Country (corp HQs)	Sep 2014 Market cap ($B)	2013 Sales rev ($B)
1	J&J (*) - Ph $28B	USA	291.39	73.54
2	Roche (incl Genentech) (*) - Ph $39B	Switzerland	250.30	51.00
3	Novartis	Switzerland	228.84	59.33
4	Pfizer (incl Wyeth)	USA	185.69	50.33
5	Merck (incl Schering Plough)	USA	174.66	43.55
6	Gilead	USA	165.42	17.44
7	Sanofi (incl Genzyme)	France	146.59	43.50
8	NovoNordisk	Denmark	120.06	14.83
9	GSK	UK	116.92	40.79
10	Bayer (incl Schering) (*) - Ph $15B	Germany	113.17	53.22
11	Amgen	USA	105.37	19.46
12	AstraZeneca	UK	95.37	25.96
13	AbbVie	USA	88.08	19.26
14	BMS	USA	83.67	16.21
15	Biogen Idec	USA	81.26	8.34
16	Celgene	USA	73.70	7.03
17	Lilly	USA	68.59	21.20
18	Boehringer-Ingelheim	Germany	NA	18.28
19	Actavis	Ireland	70.52	11.78
20	Shire	Ireland	48.22	5.39
21	TEVA	Israel	44.94	20.54
22	Merck KGaA (*) - Ph. $8.5B	Germany	37.09	14.47
23	Takeda (incl Millennium)	Japan	35.86	16.25
24	Astellas	Japan	31.97	11.13
25	Alexion	USA	31.60	1.92
26	Otsuka	Japan	19.13	14.55
27	UCB	Belgium	18.26	4.61
28	My lan	USA	17.88	7.13
29	Daiichi-Sankyo	Japan	12.46	10.75
30	Eisai	Japan	11.70	5.56
31	Dr. Reddy's Labs	India	8.39	2.28
	Top Global Pharma Total		2777	710

*not pure Pharma (Ph) - revenue incl. medical devices, diagnostics, chemicals.

Note that a few companies in Table 4.1—marked with *—include other business areas in addition to pharmaceuticals. In terms of pure pharmaceutical (Ph) revenue, Novartis tops the list ahead of Pfizer, Merck, Sanofi, GSK, and Roche. The high valuations of "newcomers" Gilead, Celgene, Actavis, Shire, Alexion, and Mylan are remarkable.

It will also be interesting to watch India and China as emerging pharmaceutical powers.

India already has a few strong companies that started out as manufacturers of generic drugs and may now move into the space of research-based pharmaceutical companies. In April 2014, Sun Pharma acquired Ranbaxy for $3.2 B[192] and moved into fifth place globally as a generic pharmaceutical company. Teva is the global leader.

China is currently behind India but several global pharmaceutical companies have invested in China and established R&D labs. The largest domestic Chinese pharmaceutical company is Luye Pharma, with a market capitalization of $3B (at the Hongkong exchange). SIMCERE was taken private in 2013 for $503 M.[193] A lot can be expected from Chinese pharmaceutical and biotech start-ups during the next decade.

There will always be a need for the translation of medical research breakthroughs into new diagnostic tests and treatments of patients. Biopharmaceutical companies will continue to prosper and thrive as long as they focus on good science, keep an open mind, and understand that tomorrow's world will be more open and that the organization best positioned to collaborate may win future competitive battles.

In the following section we will explore if the life sciences industry could perhaps learn from the semiconductor industry.

4.8 Semiconductors: Progress through Collaboration

Successful introduction of new technologies now depends on a broader and deeper industrial collaboration, aligning key stakeholders across the industry ecosystem.[194]

SEMATECH (from **Sem**iconductor **M**anufacturing **Tech**nology), a not-for-profit consortium, was conceived in 1986, and has broad engagement with various sectors of the semiconductor R&D community, including chipmakers, equipment and material suppliers, universities, research institutes, and government partners. The group is funded by member dues and was set up 1988 as a partnership between the US government and 14 U.S.-based semiconductor manufacturers to solve common manufacturing problems.

Without going too much into details, it is a fair statement that the semiconductor industry has demonstrated for over 25 years how technical progress can be accelerated by collaboration, by sharing/cross-licensing of intellectual property (IP)/patents and by an unrelenting focus on quality control and yield improvement.

The basic IP philosophy of the semiconductor industry can be summarized as "freedom of action": Rather than jealously guarding new inventions and exclusively licensing an important patent protecting a novel idea, the scientists are encouraged to communicate and share. Patent portfolios of one specific company are routinely cross-licensed to the community of both partners and competitors. Even the concept of strict confidentiality is redefined to serve the common goal of progress through collaboration: IBM's IP attorneys have added a "residuals clause" to all confidentiality agreements involving their scientists. While confidentiality is respected and honored, residual knowledge stored in the brains of exposed researchers cannot be erased. In other words, if exposure to confidential information triggers new ideas and concepts in the heads of scientists, those new concepts will not be covered by the agreement and will not belong to the partner with whom the agreement was signed. The scientist who was exposed to confidential information and uses a concept he/she picked up in another context will have full freedom of action to pursue this new idea.

For the semiconductor industry, exclusive relationships with a chosen partner based on the exclusive licensing of a key technology are not desirable. They may be attractive for the chosen licensee, but they would most certainly slow down progress for the whole industry and restrict freedom of action for the licensing organization.

Returning to SEMATECH and its 25-year history, the progress-through-collaboration strategy has enabled major transitions in advanced lithography, new materials, and processes for sub-20 nm manufacturing and 450 mm wafer integration.

Projecting forward, as the industry pushes into the sub-14 nm realm, major challenges lie ahead: how will industry collaborations change as they address heterogeneous packaging, 3D device structures, nanodefectivity, and 450 mm wafer technology? During the next decade, the opportunities and challenges of heterogeneous integration, nanodefects, and the post-450 mm era will require new innovations—not only in technology and manufacturing, but maybe also in the business of collaboration.

4.9 Life Sciences Industry: Exclusive IP Deals and Limited Collaboration

The life sciences industry, on the other hand, has developed an industry culture that is very different from the semiconductor model described above. Aside from a few areas labeled "precompetitive," the research-based drug companies have been competing fiercely with each other.

Instead of freedom of action, the IP strategy is mostly based on exclusive licensing deals. The industry's "blockbuster business model" requires jealously guarded ownership of IP associated with the chemical structure of the drug molecule that—until patent expiration 20 years later—guarantees billion-dollar sales. Rather than working together to solve common industry problems, life sciences companies are trying to seek a competitive advantage, wherever possible. Confidentiality is supposed to be airtight and any violations are argued in court.

It is difficult to argue with success and the life sciences industry has indeed been very successful. Progress in getting a better understanding of disease mechanisms and associated targets for drugs has also been good, but it must be said that the industry has benefited greatly from investments by the public sector, by academic and government research. The question is about the future of the industry: Is it time to rethink the role of IP, and could the life sciences industry learn from the semiconductor industry?

Is it time to increase the precompetitive areas of R&D and to take PPPs seriously as ways to enhance our understanding of disease mechanisms and focus on collaborative solutions to a number of challenging health problems?

4.10 IMI and AMP

Recently, there have been encouraging signs that the industry is indeed starting to move into the direction indicated above. Two examples are demonstrating the new trend, *innovative medicines initiative (IMI)* and *accelerating medicines partnership (AMP)*.

4.10.1 *Innovative Medicines Initiative*

The IMI[195] is Europe's largest public–private initiative aiming to speed up the development of better and safer medicines for patients.

The IMI supports collaborative research projects and builds networks of industrial and academic experts in order to boost pharmaceutical innovation in Europe. The IMI is a joint undertaking between the European Union and the European Federation of Pharmaceutical Industries and Associations (EFPIA).[196] The EFPIA brings together 33 European national pharmaceutical industry associations as well as 40 leading companies and claims to be the voice on the European Union scene of 1900 companies committed to researching, developing, and bringing patients new medicines that will improve health and the quality of life around the world.

The IMI now has 40 projects that are up and running as a result of the successful launches of its first six calls for proposals. There is strong emphasis on vaccines, based on the fact that 80% of all vaccines are produced in Europe. Other focus areas include:

- antibacterial resistance and "new drugs for bad bugs"
- neurodegenerative diseases, autism, pain
- diabetes research focused on beta cell function
- IT projects focused on health records, medical information, knowledge management, and expert systems
- pharmacovigilance and epidemiology
- biomarkers for oncology, systemic autoimmune diseases, respiratory disease
- quantitative imaging for cancer research
- stem cell research

IMI funding is provided both by the European Commission's Seventh Framework Programme (FP7) at a level of €1 B and by matching of FP7 funding by mainly in kind contributions worth at least another €1 B from the pharmaceutical companies that are members of the EFPIA.

The IMI governing board is composed of 10 board members representing equally the 2 founding members of the IMI: 5 from the European Commission, representing the European Union, and 5 from the EFPIA, representing the research-based pharmaceutical industry in Europe.

It may be too early to judge the IMI by its results. It will be important to keep European politics out of funding decisions and to provide enough sustained funding for meaningful research initiatives.

4.10.2 *Accelerating Medicines Partnership*

The AMP[197] is a bold new venture between the NIH, 10 bio-pharmaceutical companies, and several nonprofit organizations to transform the current model for developing new diagnostics and treatments by jointly identifying and validating promising biological targets of disease.

The AMP will begin with three- to five-year pilot projects in three disease areas:

- AD
- Type 2 diabetes
- Autoimmune disorders of rheumatoid arthritis and systemic lupus erythematosus (lupus)

For each pilot, scientists from the NIH and industry have developed research plans aimed at characterizing effective molecular indicators of disease called biomarkers and distinguishing biological targets most likely to respond to new therapies.

The ultimate goal is to increase the number of new diagnostics and therapies for patients and reduce the time and cost of developing them.

AMP partners include the US government (NIH, FDA), non-profit organizations (Alliance for Lupus Research, Alzheimer's Association, American Diabetes Association, Lupus Foundation of America, Lupus Research Institute, Foundation for the NIH, Geoffrey

Beene Foundation, PhRMA, Rheumatology Research Foundation, the United States against Alzheimer's), and the following 10 pharmaceutical companies:

- AbbVie
- Biogen Idec
- Bristol-Myers Squibb
- GlaxoSmithKline
- Johnson & Johnson
- Lilly
- Merck
- Pfizer
- Sanofi
- Takeda

The goal of the AMP is to expedite translation of scientific knowledge into next-generation treatments in a unified way that maximizes the efforts of the NIH and industry. Over the next three to five years the data and analyses generated from the partnership will be made public for all scientists to use. The partners have developed research plans and are sharing costs, expertise, and resources in an integrated governance structure to enable the best contributions to science from all members.

The AMP initiative alters the traditional approach to R&D activities. It will be interesting to watch how far the member companies will go in sharing data and building new precompetitive knowledge.

Chapter 5

Nanomedicine

He who would learn to fly one day must first learn to stand and walk and run and climb and dance; one cannot fly into flying.

—Friedrich Nietzsche (1844–1900)

Before diving into the heart of the matter of nanomedicine, let's briefly discuss nanotechnology and—since there is such a close connection between nano- and information technology (IT), let's also take a look at the important role of computers in nanomedicine. We will then focus on materials used in nanomedicine, and discuss the impact of those materials on medical diagnostics (including medical imaging), on drug delivery, and regenerative medicine.

5.1 Computers in Nanomedicine

Done properly, computer simulation represents a kind of "telescope for the mind," multiplying human powers of analysis and insight just as a telescope does our powers of vision.

—Mark Buchanan, physicist and author

Since the 1980s, nanotechnology has played an important role in the advancement of computer technology, and likewise, computer

Nanomedicine: Science, Business, and Impact
Michael Hehenberger
Copyright © 2015 Pan Stanford Publishing Pte. Ltd.
ISBN 978-981-4613-76-7 (Hardcover), 978-981-4613-77-4 (eBook)
www.panstanford.com

technology has been feeding advances in nanotechnology. It is widely acknowledged that IBM's research lab in Zuerich-Rueschlikon (ZRL), Switzerland, is the birthplace of nanotechnology. In 1986, IBM ZRL scientists Gerd Binnig and Heinrich Rohrer earned the Nobel Prize for physics for their invention of the scanning tunneling microscope (STM).[198] They shared the prize with Ernst Ruska, who in 1933 first designed an electron microscope. In the late 1920s, Ruska found that a magnetic coil could act as a lens for electrons and that such an electron lens could be used to obtain an image of an object irradiated with electrons. In the same way that optical glass lenses can be combined to form a microscope, he used coils to build an electron microscope. Since electrons have wavelengths that are about 100,000 times shorter than those of visible light photons, electron microscopes can achieve better than 0.05 nm resolution and magnifications of more than a million times. In comparison, ordinary light microscopes are limited by diffraction to about 200 nm resolution and about 1000 times smaller magnifications (below 2000$\times$). (However, as we will see in Section 5.5, new ways have been found to cheat this "Abbe[199] limit"!)

Whereas the electronic microscope can be considered an extension of the optical light microscope, the STM is based on a completely different concept, namely, the principle of "feeling" instead of "seeing." Imagine a mechanical finger, which may be a very fine needle that is moved across the surface of the structure to be investigated. Binnig and Rohrer were able to create a tip of the needle consisting of only a few atoms. When such a fine tip is moved across a surface at a height of a few atomic diameters, the finest atomic details in the surface structure can be registered. By registering the tip's movements in the vertical direction as it traverses the surface, they could obtain a sort of topographical map, equivalent to the image obtained in an electron microscope. To get the desired results, the tip of the needle must be very small and direct contact with the surface must be avoided to take advantage of a phenomenon called *tunneling*, which enables the flow of a small electric current between the needle tip and the surface.[200] Note that tunneling is a quantum effect that cannot be explained with concepts of classical physics (Newtonian mechanics). Tunneling plays an essential role in several physical phenomena, such as radioactivity and the Stark effect.[201] The splitting of spectral lines first observed by Stark for the hydrogen atom in an electric field is really caused by a small probability that electrons may "tunnel," that is, escape from the atomic nuclei.[202]

Binnig and Rohrer were able to measure the magnitude of the tunneling current through the tip of needle kept at a distance of, typically, 2–3 atomic diameters above the surface. The movement of the needle is regulated by a piezoelectric servo mechanism, which in turn is controlled by the tunneling current. The piezoelectric effect is defined as the internal generation of electrical charge resulting from an applied mechanical force in crystalline materials. It is a reversible process, that is, piezoelectric materials also exhibit the reverse piezoelectric effect: generation of a mechanical strain resulting from an applied electrical field. Binnig and Rohrer were able to utilize both tunneling and piezoelectricity to achieve their breakthrough. In doing so, they took advantage of related expertise that was readily available in IBM Research's environment: For instance, disturbing vibrations from the environment were eliminated by building the STM upon a heavy permanent magnet floating freely in a dish of superconducting lead. By controlling the horizontal movement of the stylus via piezoelectric elements that scanned the surface in two perpendicular directions, they achieved a horizontal resolution of approximately 2 Å[203] (0.2 nm) and vertical resolution of approximately 0.1 Å (0.01 nm). This makes it possible to depict individual atoms, that is, to study in the greatest possible detail the atomic structure of the surface being examined.

In the press release issued by the Royal Swedish Academy of Sciences, the expected impact of this invention was described as follows.[204]

It is evident that this technique is one of exceptional promise, and that we have so far seen only the beginning of its development. Many research groups in different areas of science are now using the STM. The study of surfaces is an important part of physics, with particular applications in semiconductor physics and microelectronics. In chemistry, surface reactions play an important part, for example, in connection with catalysis. It is also possible to fixate organic molecules on a surface and study their structures. Among other applications, this technique has been used in the study of DNA molecules.

The press release was prophetic: The invention of the STM would launch a whole new field of microscopy, scanning probe microscopy (SPM), a branch of microscopy that creates images of surfaces using a physical probe that scans the specimen. SPM turned out to be an important enabler of nanotechnology because it provided the means to measure and test nanodevices in new and more accurate ways, thereby giving a huge boost to this new area of science and technology. By mentioning organic molecules including DNA, the press release further anticipated the first beginnings of nanomedicine.

SPM resolution varies somewhat from technique to technique, but some probe techniques reach a rather impressive atomic resolution. They owe this largely to the ability of piezoelectric actuators to execute motions with a precision and accuracy at the atomic level or better on electronic command. Piezoelectric techniques are key to the successful use of SPM.

But let's return to the interplay of nanotechnology and computers. Clearly, computers are required to analyze the data generated by SPM instruments: There is no way to see what the instrument is measuring—the image has to be reconstructed and visualized by a computer application, a new type of imaging software. Figure 5.1 shows that visualization is an integral part of the STM instrument.

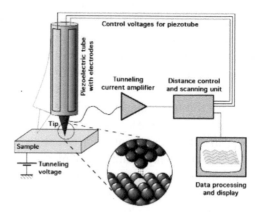

Figure 5.1 Schematic view of a scanning tunneling microscope. Source: Wikipedia.

In addition, faster and smaller microprocessors can be used to push the limits of nanotechnology by means of modeling and simulation of new materials and devices.

In a paper published in 1965 and titled "Cramming More Components onto Integrated Circuits,"[205] Gordon E. Moore, a cofounder of Intel, made the observation that the number of transistors on integrated circuits is doubling approximately every two years. It turned out that this Moore's law has held true for more than four decades.

In 1974, Robert Dennard and coworkers at the IBM T.J. Watson Research Center wrote a seminal paper describing metal–oxide–semiconductor field-effect transistor (MOSFET) scaling rules for obtaining simultaneous improvements in transistor density, switching speed, and power dissipation.[206] The scaling principles described by Dennard and his team were quickly adopted by the semiconductor industry as a roadmap for providing systematic and predictable transistor improvements.

Moore and Dennard received the most prestigious Institute of Electrical and Electronics Engineers (IEEE) Award, the "Medal of Honor," in 2008 and 2009, respectively.[207] Their contributions helped propel the semiconductor industry to revenues of more than $200 B per year, feeding into a trillion-dollar-a-year electronics industry.

An emerging challenge will be limitations caused by the physics of atomic dimensions: As semiconductor technology is pushing the limits of miniaturization, the industry will encounter new technical hurdles due to quantum effects. By switching to other materials it will be possible to delay the inevitable, but it is highly likely that Moore's law will reach an endpoint dictated by the laws of physics sometime around the year 2030 (if not earlier).

Let's go back to the role computers play in nanotechnology: During the 1980s, semiconductor feature size went below 1 μm (1000 nm) and the transistor count surpassed 1 million (Fig. 5.2). Clearly, it became necessary to introduce more precise instrumentation for failure analysis, and SPM technologies began to complement optical and electron microscopes to do the job.

In addition to ever-improving manufacturing technologies and quality control techniques, the semiconductor industry is using

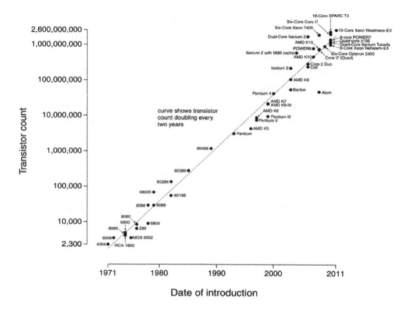

Figure 5.2 Plot of CPU transistor counts against dates of introduction. Note the logarithmic vertical scale; the line corresponds to exponential growth with transistor count doubling every two years. Source: Wikipedia.

circuit simulation on high-performance computers to simulate and replicate the behavior of an actual electronic device or circuit. Simulation software allows for modeling of circuit operation and is an invaluable analysis tool.

Simulating a circuit's behavior before actually building it can greatly improve design efficiency by making faulty designs known as such and by providing insight into the behavior of electronics circuit designs. In particular, for integrated circuits the tooling is very expensive and probing the behavior of internal signals is extremely difficult. Therefore, almost all circuit design relies heavily on computer simulation.

Another important area where computers pave the way for new advances in nanotechnology is the design of new materials. Whenever an understanding of molecular mechanisms is required to move forward, *modeling and simulation* will have to complement experimental laboratory work. Methods of quantum chemistry will be applied whenever quantum effects are responsible for desired properties of materials, and semiempirical methods and classical mechanics approximations are used whenever full ab initio quantum mechanical treatments are too complex and time consuming. The 2013 Nobel Prize for chemistry was awarded to Martin Karplus, Michael Levitt, and Arieh Warshel for the development of multiscale models for complex chemical systems. In the press release of the Nobel Committee, the combined use of classical and quantum mechanics and the interplay between experimental laboratory activities and computer simulations are aptly described as follows:

> The strength of classical physics was that calculations were simple and could be used to model really large molecules. Its weakness, it offered no way to simulate chemical reactions. For that purpose, chemists instead had to use quantum physics. But such calculations required enormous computing power and could therefore only be carried out for small molecules.
>
> This year's Nobel Laureates in chemistry took the best from both worlds and devised methods that use both classical and quantum physics. For instance, in simulations of how a drug couples to its target protein in the body, the computer performs quantum theoretical calculations on those atoms in the target protein that interact with the drug. The rest of the large protein is simulated using less demanding classical physics.
>
> Today the computer is just as important a tool for chemists as the test tube. Simulations are so realistic that they predict the outcome of traditional experiments.[208]

But there is another important use of computers in nanotechnology, in particular when being applied to the biomedical domain as nanomedicine, namely, *text analytics*: It's very difficult if not impossible for scientists to keep up with the ever increasing number of scientific publications. It's not only a question of volume but also quality. What reports of breakthrough results can be trusted? Text analytics (also referred to as computational linguistics or natural-language understanding) has been around for a long time but had to develop and mature before entering into the mainstream of life sciences research.[209] The universal use of search engines such as Google has prepared scientists to go beyond keyword searches and ask for tools to help with *hypothesis generation* enabled by *feature extraction*—the creation of a reduced representation of data by extraction of relevant information from large volumes of input data according to specific criteria chosen to perform the desired task—using this reduced representation instead of the full size input, and *machine learning*. In simple terms, we refer to *machine learning* when we give computers the ability to learn, that is, improve the performance at given tasks with experience. Typically, the computer is fed a training set of data and then asked to apply the acquired knowledge to another, related data set.

By improving algorithm development and data processing the field of text analytics has entered the era of cognitive computing. By the end of the year 2010, a major milestone of cognitive computing was reached by IBM Research when the Watson computer was able to beat the Jeopardy champions Ken Jennings and Brad Rutter.[210,211]

Clearly, the Jeopardy playing computer system Watson was smarter than a search engine. More than 40 years after the publication of Simmons' paper[212] on "Natural Language Question-Answering Systems," the Watson–Jeopardy team was able to beat the human brain at tasks previously thought as too difficult to do for even the most powerful supercomputer. Watson could not only understand complicated sentences in natural language, but also perform temporal and geospatial reasoning and statistical paraphrasing, a way to bridge the gap between, for example, specialized medical terminology and everyday language. To perform at the highest level in the game of Jeopardy, Watson also had to develop confidence measures to pick the best answer among possible options. A detailed description of all aspects of the Watson–Jeopardy system has been published by D. A. Ferrucci et al. in the *IBM Journal of R&D*.[213]

To illustrate the limitations of keyword searches and Watson's ability to perform cognitive computing—defined as computing systems that learn and interact naturally with people to extend what either humans or machine could do on their own, and helping human experts make better decisions by penetrating the complexity of big data[214]—let us take a look at hidden links among terms across collections of documents. The term *hidden links* refers to work by Swanson starting in the mid-1980s that uncovered relationships between disease syndromes and dietary or other chemical substances that could treat diseases.[215] For example, Swanson found a potential connection between Raynaud's disease— a condition that causes fingers, toes, nose, and ear tips to feel numb and cool in response to cold temperatures or stress—and acid substances in fish oil. The connections were hidden, not actually expressed in specific sentences or documents: the disease syndrome was connected to an intermediate cluster of concepts, having to do with blood aggregation and viscosity. In an entirely different document context, these concepts were in turn connected to dietary or chemical substances. Swanson used manual methods and clever inference to deduce his hidden links but later developed specialized computer tools for supporting these analyses. The Watson system went a step further and developed an ability to automatically detect the hidden links needed to resolve difficult Jeopardy puzzles such as the following: "On hearing of the discovery of George Mallory's body, he told reporters he still thinks he was first."

The Watson computer system has to generate a set of connections based on the first part of the sentence and then make sense of the second part and find the missing link. For students of the history of mountaineering, the "discovery of George Mallory's body" is clearly pointing to Mount Everest, the mountain that Mallory tried to climb in the 1920s. On the other hand, Hillary and Tenzing were the first to reach the top of Mount Everest, and since Sir Edmund Hillary was probably approached by reporters more frequently than Tenzing Norgay, it would make sense to respond to this Jeopardy puzzle with "Hillary" as the answer.

Our conclusion from this example is that text analytics has come a long way and that cognitive computing promises to assist scientists in not only digesting and analyzing textbooks and publications but also improving their ability to form associations, detect hidden links, and perhaps even generate new hypotheses to be validated by experiments.

Let us also mention the possible role of *social computing*, in particular *crowdsourcing*, in nanomedicine. As explained above, text analytics can be very helpful and valuable when large volumes of scientific literature have to be digested and analyzed. What may still be a remaining issue is the assessment of quality and reliability of published results. In its October 19–25, 2013, issue, *The Economist* addressed this problem by putting the words "HOW SCIENCE GOES WRONG" on its cover.[216] There is the frequently quoted paper by Ioannides that states that "most published research findings are probably false"[217] and there are studies by major life sciences companies according to which the results of only 6 of 53 cancer research studies (AMGEN[218]) and only a quarter of 67 seminal biomedical studies (BAYER[219]) could be reproduced. In addition to creating a new scientific discipline called metaresearch (research about research),[220] a possible answer to this problem may be "Verification of scientific claims by means of crowdsourcing." By inviting experts in a particular field of biomedical research to validate a given scientific result, the quality of scientific results can be enhanced and "self-correction" of scientific results ensured. By inviting a motivated community of scientists to compete on the solution of a specific problem and providing incentives for winners of such a competition, progress in a chosen field can be accelerated. The DREAM project in systems biology provides an excellent example.[221,222]

Similarly, the EteRNA project demonstrates how online video games—used to engage a large number of scientists playing the game—can simulate a biochemistry lab and design RNA sequences that will fold into a target structure. EteRNA grew out of an online video game called Foldit, a protein-folding game created in 2008 by molecular biologists (Baker Lab)[223] and computer scientists (Zoran Popovic team)[224] at the University of Washington, Seattle.[225] Members of the Foldit team then conceived of EteRNA, a "version of Foldit for RNA." Now managed by scientists at Carnegie Mellon University and Stanford University, EteRNA engages users to solve puzzles related to the folding of RNA molecules. By January 2014, 133,000 players were participating, 4000 of them actually doing research. The goal is to generate new rules for RNA folding that can be captured in algorithms.[226]

Next, to show the role of computers in nanomedicine for an integrated medical device and drug delivery system, let us discuss the Artificial Pancreas project. Aside from playing a major role in the digestion of food, the pancreas regulates glucose levels by producing two hormones, glucagon and insulin.

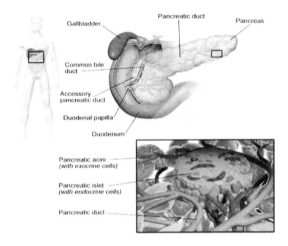

Pancreatic Tissue

Figure 5.3 Endocrine function of human pancreas. Source: Wikipedia.

- The alpha cells of the pancreas produce glucagon. Glucagon raises blood glucose levels by stimulating the liver to metabolize glycogen into glucose molecules and to release glucose into the blood.
- The beta cells of the pancreas produce insulin. Insulin lowers blood glucose levels by stimulating the absorption of glucose by liver, muscle, and fat (adipose tissue).

A healthy pancreas (see Fig. 5.3) provides insulin, when needed, to remove excess glucose from the blood. When blood glucose levels fall below a certain level, glucagon production kicks in and makes sure that enough glucose will be available for energy production in all tissues.

Note that the third hormone, somatostatin, as produced by delta cells, is not yet considered for inclusion in the Artificial Pancreas project. After release, somatostatin travels to the heart and then to systemic circulation, where it will exert its digestive system effects (reducing secretion in the stomach).

Energy (via adenosine-5′-triphosphate, ATP) is needed for all intracellular metabolic events, including normal functioning of the central nervous system (CNS).

Type 1 diabetes (T1D), juvenile diabetes, or insulin-dependent diabetes mellitus (IDDM) is an autoimmune disorder where the beta cells of the patient's pancreas have lost their ability to produce insulin. T1D is fatal without external supply of insulin. Insulin injections turn IDDM into a chronic disease that (except for pancreas or beta cell transplantation) currently has no cure, although promising clinical studies point to a future possibility of prevention.[227]

The Artificial Pancreas project, orchestrated by the Juvenile Diabetes Research Foundation (JDRF)[228] in collaboration with industry (i.e., Medtronic, Dexcom, and Johnson & Johnson), attempts to recreate the metabolic function of the pancreas by combining continuous glucose monitoring (CGM) with an insulin pump controlled by software that processes CGM signals and other patient-specific information to administer just the right amount of insulin to keep the glucose level of the diabetic patient at a normal level. There are several important challenges associated with this ambitious project:

- There is still no way to perform CGM at the required accuracy without implanting a device designed to transmit signals to the outside. Even so, there is an unavoidable delay of automatic insulin disposal into the bloodstream as administered by the insulin pump, caused by the time it takes for blood glucose levels to increase after a meal.
- Even with fast-acting insulin, there is a delay in bringing glucose levels down to normal levels after a meal.
- Computer algorithms to control the insulin pump lack the sophistication needed to perform the functions of a working pancreas that can process signals associated with food ingestion, pace of insulin absorption, the amount of insulin already in circulation, and exercise.

As recently reported in *Science*,[229] researchers are working on two new types of control algorithms, (1) model predictive control (MPC), which is capable of tuning the model to the particular patient's physiology, and (2) a "fuzzy logic/truth variable" approach, which makes decisions based on replication of physicians' best practices in treating diabetic patients.

Still, because of the high risks involved in the administration of too high doses of insulin (leading to hypoglycemia or so-called glucose "hypos," potentially followed by unconsciousness and even death) and too low doses of insulin (leading to long-term complications, including vascular and heart disease, blindness, kidney failure, and the risk of amputated limbs), the Food and Drug Administration (FDA) will be very reluctant to approve artificial pancreas devices without successful long-term clinical studies.

What has been approved so far is a device that contains an insulin pump capable of receiving signals from a CGM transmitter that shuts off basal insulin administration whenever glucose levels are getting too low.[230] Another way to reduce risk could be to use a two-hormone pump that dispenses both insulin and glucagon, just like a healthy human pancreas (shown above in Fig. 5.3).

Translated into requirements for the computer algorithm controlling such a pump, it would certainly present new challenges but could also create significant benefits for type 1 diabetics.

In conclusion, the Artificial Pancreas project exemplifies the opportunities created by combining nanodevices with computer algorithms in order to emulate the functions of a complicated human organ.

5.2 Biocompatible Nanoparticles and Targeted Drug Delivery

We define biomaterials as materials that are used in contact with biological systems. Biocompatible nanoparticles play an important role in nanomedicine, in particular for drug delivery.

5.2.1 Dendrimers

Dendrimers are star-shaped synthetic polymers that are repetitively branched. A dendrimer is typically symmetric around the core and often adopts a spherical 3D morphology. A dendron, on the other hand, usually contains a single focal point and may lack spherical symmetry, as shown in Fig. 5.4.

However, the terms "dendrimer" and "dendron" are frequently mixed up and used interchangeably. The first dendrimers were

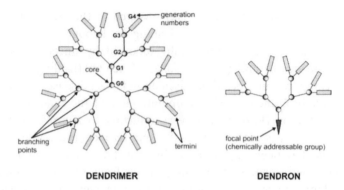

DENDRIMER   DENDRON

Figure 5.4   Synthesis of dendrimers and dendrons. Source: Wikipedia.

synthesized by Fritz Vögtle, University of Bonn, Germany, in 1978,[231] followed by researchers at Allied Corporation (R.G. Denkewalter, 1981) and Dow Chemical (Donald Tomalia, 1983).

Major contributions were also made by George Newkome and by Jean Fréchet (convergent synthesis—see below) during the late 1980s. Dendrimer popularity then greatly increased, resulting in more than 5000 scientific papers and patents by the year 2005. Dendrimers and dendrons are usually highly symmetric, spherical compounds characterized by structural perfection. Their properties tend to be dominated by the functional groups on the molecular surface. Dendritic encapsulation of functional molecules allows for the isolation of the active site, a structure that mimics that of active sites in biomaterials. This active site is always required for repeated reactions to build the tree-like chains.

Dendrimers can be considered to have three major portions: a core, an inner shell, and an outer shell. Ideally, a dendrimer can be synthesized to have different functionality in each of these portions to control properties such as solubility, thermal stability, and attachment of compounds for particular applications. It is possible to precisely control the size and number of branches on the dendrimer. Unlike most polymers, it is possible to make dendrimers water soluble by functionalizing their outer shell with hydrophilic groups.

There are two defined methods of dendrimer synthesis, *divergent* synthesis—from the core outward (see Fig. 5.5)—and *convergent* synthesis (Fig. 5.6).

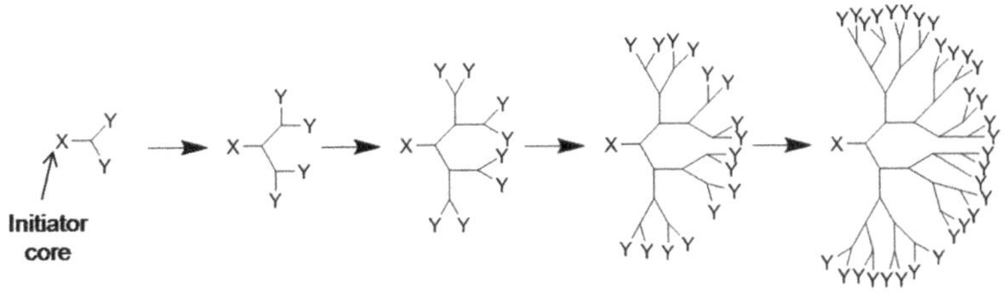

Figure 5.5 Divergent synthesis of dendrimers. Source: Wikipedia.

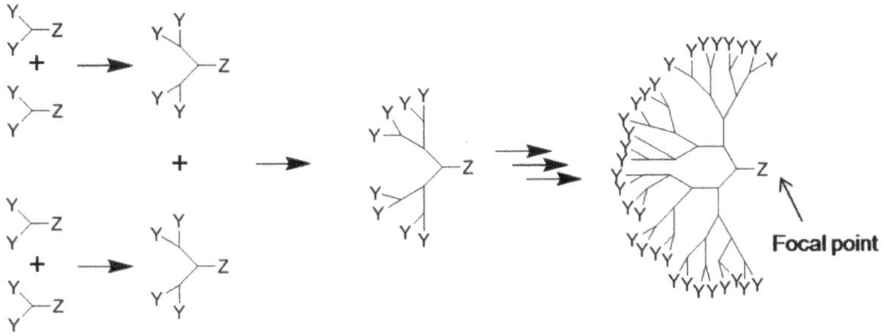

Figure 5.6 Convergent synthesis of dendrimers. Source: Wikipedia.

In convergent synthesis, dendrimers are built from small molecules that end up at the surface of the sphere. Reactions are then proceeding inward. The last step is to attach the core. This method makes it much easier to correct imperfections along the way, so the final dendrimer is characterized by particles of uniform size in a dispersed phase, that is, monodisperse. However dendrimers made this way cannot be made as large as those made by divergent methods.

Because the actual reactions consist of many steps needed to protect the active site, it is difficult to synthesize dendrimers using either method, even by applying principles of *click chemistry*, a way to chemically join small molecular units and to quickly and reliably produce dendrimers.[232]

Dendrimers have very strong potential for applications where other chemicals are conjugated to the dendrimer surface: one dendrimer molecule may have hundreds of possible sites to couple to another active molecule. Dendrimers can therefore act as detecting agents (e.g., dye molecules), imaging agents (see below), or pharmaceutically active compounds.

On the other hand, dendrimer architecture and selective functionalization of core and surface can make them *solubilizing agents.* For instance, dendrimers with hydrophobic core and hydrophilic periphery have shown to exhibit *micelle-like* behavior.

To understand those terms and concepts, we need to briefly review colloid chemistry.

A *colloid* is a substance in which microscopically dispersed insoluble particles are suspended throughout another substance. The term "colloidal suspension" refers unambiguously to the overall mixture. Unlike a solution, whose *solute* and *solvent* constitute only one phase, a colloid has a dispersed phase (the suspended particles) and a continuous phase (the medium of suspension). To qualify as a colloid, the mixture must be one that does not settle or would take a very long time to settle appreciably. The dispersed-phase particles have a diameter of between approximately 1 and 1000 nm.

A *micelle* is an aggregate of surfactant molecules dispersed in a liquid colloid. *Surfactants* are compounds that lower the surface tension between two liquids (or between a liquid and a solid). Surfactants may act as detergents, wetting agents, emulsifiers, foaming agents, and dispersants.

Figure 5.7 shows a micelle of oil in aqueous suspension, such as might occur in an emulsion of oil in water. In this example the surfactant molecules' oil-soluble tails project into the oil, while the water-soluble ends remain in contact with the water phase.

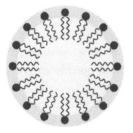 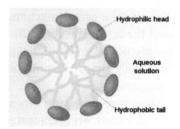

Figure 5.7 Micelle of oil in aqueous suspension. Source: Wikipedia.

The lipophilic tails of the surfactant ions remain inside the oil because they interact more strongly with oil than with water. The polar heads of the surfactant molecules coating the micelle interact more strongly with water, so they form a hydrophilic outer layer that forms a barrier between micelles. This inhibits the oil droplets, the hydrophobic cores of micelles, from merging into fewer, larger droplets (*emulsion breaking*) of the micelle.

A typical micelle in aqueous solution forms an aggregate with the hydrophilic head regions in contact with surrounding solvent, sequestering the hydrophobic single-tail regions in the micelle center. This phase is caused by the packing behavior of single-tail lipids in a bilayer. The difficulty filling all the volume of the interior of a bilayer, while accommodating the area per head group forced on the molecule by the hydration of the lipid head group, leads to the formation of the micelle. This type of micelle is known as a normal-phase micelle (oil-in-water micelle). Inverse micelles have the head groups at the center with the tails extending out (water-in-oil micelle). Micelles are approximately spherical in shape.

Solubilization is a short form for micellar solubilization, a term used in colloidal and surface chemistry. Solubilization may occur in a system consisting of a solvent, an association colloid (i.e., a colloid that forms micelles), and at least one other component called the solubilizate. The solubilizate is the component that undergoes solubilization. *Solubilization is the process of incorporation of the solubilizate into or onto the micelles.*

Examples of micellar solubilization are:

- laundry washing using detergents;
- for formulations of poorly soluble drugs in solution form in the pharmaceutical industry; and
- dispersants for the cleanup of oil spills.

Solubilization is illustrated in Fig. 5.8.

Note that solubilization is different from *dissolution*, which is defined as the process by which a solute forms a solution in a solvent. The solute, in the case of solids, has its crystalline structure disintegrated as separate ions, atoms, and molecules form. An example of dissolution is the making of a saline water solution by

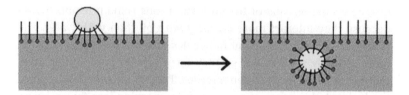

Figure 5.8 Schematic of micellar solubilization of fatty substance in water using a dispersant. Source: Wikipedia.

dissolving table salt (NaCl) in water. The salt is the solute and water is the solvent.

As mentioned above, thanks to hydrophilic outer side chains, dendrimers can be made solubilizing agents, able to encapsulate hydrophobic molecules such as drugs. Many pharmaceutical drugs are hydrophobic in nature, causing drug formulation problems. However, dendrimeric scaffolding can be used to encapsulate as well as to solubilize the drugs, a particularly promising approach for the delivery of anticancer drugs.

There are currently three methods for using dendrimers in drug delivery:

(1) The drug is covalently attached to the periphery of the dendrimer to form dendrimer *pro-drugs*, medications that are administered in an inactive or less than fully active form and then converted to their active form through a normal metabolic process

(2) The drug is coordinated to the outer functional groups via ionic interactions.

(3) The dendrimer acts as a unimolecular micelle by encapsulating a pharmaceutical through the formation of a dendrimer-drug supramolecular assembly.

The use of dendrimers as drug carriers by encapsulating hydrophobic drugs is a potential method for delivering highly active pharmaceutical compounds that may not be in clinical use due to their limited water solubility and resulting suboptimal pharmacokinetics.

In addition to drug delivery, there is ongoing research on using dendrimers for gene transfection, the process of deliberately introducing nucleic acids into cells. The challenge is to move DNA into cells without damaging or inactivating the DNA.

Another possible use of dendrimers is in diagnostics, for example, to measure pH, defined as the negative decimal logarithm of the hydrogen ion (or proton) H^+ concentration. Solutions with a pH less than 7 are said to be acidic and solutions with a pH greater than 7 are basic or alkaline. Pure water has a pH very close to 7.

Finally, dendrimers are investigated for use as blood substitutes.

5.2.2 *Liposomes*

Another important class of biocompatible nanoparticles are the liposomes.

Liposomes are made of phospholipids (see Section 2.3) and are named after the Greek words *lipo* (*fat*) and *soma* (body). We already encountered liposomes in Section 3.4 when we discussed the eukaryotic cell.

A liposome is an artificially prepared spherical vesicle composed of a lamellar phase lipid bilayer.

Liposomes were first described by British hematologist Alec Bangham in 1961. They were discovered when Bangham added negative stain to dry phospholipids. His microscope pictures served as the first real evidence for the cell membrane being a bilayer lipid structure.

It is often speculated that, millions of years ago, eukaryotic cells formed as a lipid bilayer structure, to protect the cytoplasmic hydrophilic environment from the external aqueous environment.

Similarly, a liposome encapsulates a region of aqueous solution inside a hydrophobic membrane. Whereas hydrophilic solutes cannot readily pass through the lipids, hydrophobic chemicals can be dissolved into the membrane, as shown in Fig. 5.9.

To deliver the molecules to sites of action, the lipid bilayer can fuse with other bilayers such as the cell membrane, thus delivering the liposome contents.

Some liposome research is focused on ways to avoid detection by the body's immune system, specifically, the cells of the mononuclear phagocyte system (MPS). The MPS is a part of the immune system that consists of cells that accumulate in lymph nodes, the spleen, and

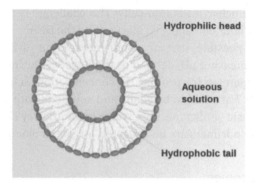

Figure 5.9 Liposome formed by phospholipids in an aqueous solution. Source: Wikipedia.

the liver. Such drug delivery vehicles are sometimes called "stealth" liposomes.

By coating liposomes with polyethylene glycol (PEG), which is inert in the body, liposomes gain a longer circulatory life—typically up to five times—for the drug delivery mechanism. However, too much PEG coating may introduce adverse effects, as discussed further below, with Doxil as the example.

In addition to a PEG coating, most stealth liposomes also have some sort of biological species attached as a ligand to the liposome in order to enable binding via a specific expression on the targeted drug delivery site. These targeting ligands could be monoclonal antibodies (mabs) (making an immunoliposome), vitamins, or specific antigens. Targeted liposomes can target nearly any cell type in the body and deliver drugs that would normally be delivered in a nontargeted, systemic way. The toxicity of certain drugs can be reduced significantly if delivered only to diseased tissues.

Polymersomes, morphologically related to liposomes, can also be used this way.

The use of liposomes for transformation or transfection of DNA into a host cell is known as lipofection.

In addition to gene and drug delivery applications, liposomes can be used as carriers for the delivery of dyes to textiles, pesticides to plants, enzymes and nutritional supplements to foods, and cosmetics to the skin.

5.2.3 *Targeted Drug Delivery*

In traditional drug delivery systems such as oral ingestion or intravascular injection, the medication is normally distributed in a systemic way through the patient's blood circulation. Targeted drug delivery seeks to concentrate the medication in the tissues of interest, while reducing the relative concentration of the medication in the remaining tissues. The goal of a targeted drug delivery system is to prolong drug circulation, localize drug delivery, target specific tissues or organs, and introduce a controlled drug interaction with the diseased tissue. The advantages to the targeted release system are reduction in the frequency of the dosages taken by the patient, a more uniform effect of the drug, reduction of drug side effects, and reduced fluctuation in circulating drug levels. The disadvantage of a nanomedical targeted drug delivery system may be higher cost and the need for additional clinical research associated with regulatory hurdles that must be overcome before health providers can administer such treatments to patients.

As discussed above, there are different types of drug delivery vehicles, such as liposomes, polymeric micelles, dendrimers, and other nontoxic polymers. An ideal drug delivery vehicle must be nontoxic, biocompatible, and nonimmunogenic, that is, avoid recognition by the host's defense mechanisms.

There have even been attempts to use DNA as a structural material, to construct a "DNA box" with a controllable lid, synthesized using the DNA origami method. This structure could encapsulate a drug in its closed state, and open to release it only in response to a desired stimulus.[233]

A relatively simple and straightforward way to deliver drug candidates that are hydrophobic, that is, poorly soluble in water, is to reduce drug crystal size to the submicron (or nano) range, sometimes referred to as *drug nanonization*, or the use of drug *nanocrystals*. A decrease of drug particle size will increase the effective surface area and therefore the drug dissolution rate and oral bioavailability. In some cases, nanozitation removes restrictions such as the need to take the drug only along with food. Nanocrystal preparation methods include media milling, high-pressure homogenization, and nanoprecipitation.[234] Since the

Table 5.1 Orally delivered drugs formulated with nanocrystal/nano-emulsion technology

Nanomedicine name	Drug name	Pharma	(FDA) approval	Indication
Rapamune	Sirolimus	Pfizer	1999	Kidney Transplant - Immunosuppression
Tricor	Fenofibrate	AbbVie	2001	Hypercholesterolemia
Emend	Aprepitant	Merck & Co.	2001	Postop. Cancer (Nausea, Vomiting)
Neoral (capsule)	Cyclosporine	Novartis	1995	post-organ transplant
Norvir (capsule)	Ritonavir	AbbVie	1996	HIV
Megace ES (enhanced stability)	Megestrol	PAR Pharma	2005	Breast Cancer

beginning of the 1990s, Elan Drug Technologies (acquired in 2011[235] by Alkermes, Ireland)[236] has pioneered the use of nanocrystals for oral bioavailability and the use nanocrystals suspended in water (nanosuspensions) for intravenous or pulmonary drug delivery.[237]

Table 5.1 shows a list of drugs delivered orally by means of nanocrystal and nanoemulsion (formulated in gelatin capsules) technology.

Figure 5.10 illustrates the various ways targeted drug delivery can be based on liposomes as delivery vehicles.

Targeted drug delivery can be applied to many therapeutic areas including cardiovascular and metabolic disease. However, the most important application of targeted drug delivery is to treat cancerous tumors.

The earliest example of a targeted drug delivered by a nanomedical device is Doxil.

Doxil is a PEGylated (PEG-coated) liposome-encapsulated form of doxorubicin, invented by Yechezkel Barenholz (Hebrew University, Israel) and developed into an FDA-approved anticancer drug by Barenholz and Alberto Gabizon, sponsored by Sequus Pharmaceuticals, Menlo Park, USA. Sequus was founded in 1981 as Liposome Technology, Inc., changed its name to Sequus in 1995, and became part of ALZA, which in 2001 was acquired by Johnson & Johnson.

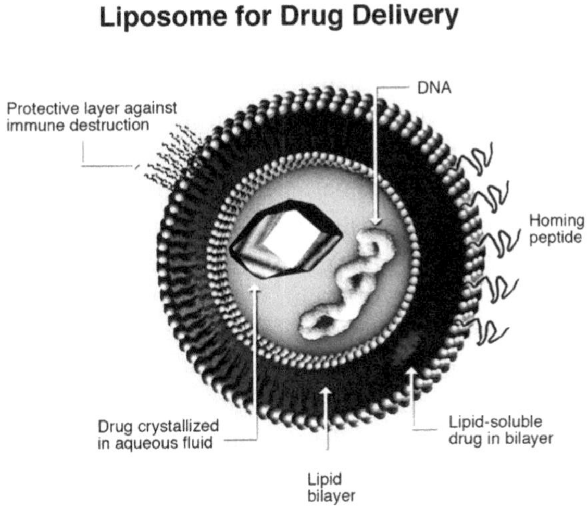

Liposome for Drug Delivery

Figure 5.10 Ways to use liposomes for drug delivery. Source: Wikipedia.

Approved in 1995, Doxil was the first FDA-approved liposomal nanodrug and is now owned and marketed by Johnson & Johnson.

There is now also a generic version of Doxil, named Lipodox, made by Sun Pharma, Mumbai, India.

Myocet is a non-PEGylated liposomal doxorubicin made by Enzon Pharmaceuticals for Cephalon in Europe and for Sopherion Therapeutics in the United States and Canada. Myocet is approved in Europe and Canada for treatment of metastatic breast cancer but is not yet approved by the FDA for use in the United States.

Doxorubicin is a drug used in cancer chemotherapy and derived by chemical semisynthesis from a bacterial species. It is an anthracycline antibiotic (but not meant to treat bacterial infections) and closely related to the natural product daunomycin or daunorubicin, codiscovered in the 1950s by Farmitalia (Milan, Italy) and by French researchers. Daunorubicin was found to slow or stop the growth of cancer cells in the body but could produce fatal cardiac toxicity. Still, it is used as the starting material for semisynthetic manufacturing of doxorubicin and other chemotherapy agents. Doxorubicin is commonly used to treat blood cancers (leukemias, Hodgkin's lymphoma) as well as cancers of the bladder, breast, stomach, lung, ovaries, and thyroid, soft tissue sarcoma, multiple myeloma, and others. It is often used in combination chemotherapy as a component of various chemotherapy regimens.

The key benefit of Doxil's liposomal encapsulation of doxorubicin is reduction of cardiotoxicity. By limiting the cardiotoxicity of doxorubicin through liposomal encapsulation, Doxil can be used safely in concurrent combination with other cancer drugs.

Unfortunately, Doxil's PEG coating is leading to preferential concentration of Doxil in the skin, resulting in a side effect called palmar-plantar erythrodysesthesia (PPE), more commonly known as hand-foot syndrome. Following administration of Doxil, small amounts of the drug can leak from capillaries in the palms of the hands and soles of the feet. The result of this leakage is redness, tenderness, and peeling of the skin that can be uncomfortable and even painful. This side effect limits Doxil dosage.

Myocet avoids the PEG coating and therefore the associated side effect. However, doxyrubicin release rates in the tumor environment may be negatively impacted by the lack of PEGylation. Time will tell which liposomal doxorubicin will gain the upper hand for clinical use.

Since 1995, the year of first FDA approval of a nanomedicine, many other drugs have been encapsulated by liposomes and micelles for improved targeting and reduction of certain side effects.

A list of drugs (in alphabetical order) delivered by means of liposome technology is given in Table 5.2.

Promising new research is focused on the development of polymer-coated liposomes as an oral delivery system for degradation-sensitive and poorly absorbed drugs, such as proteins and peptides. Without special protection, such drugs cannot efficiently transported through the intestinal membrane in the gastrointestinal (GI) tract. By coating liposomes with thiolated (chitosan-thyoglycolic acid) polymers and creating "thiomers," mucoadhesion, permeation, and efflux-pump inhibitory properties can be improved. Preliminary in vivo studies in rats have shown promising results.[238]

Due to the great flexibility in designing polymers with desired shape, size and other properties, we can expect a lot of innovation in the field of polymer drugs or polymer therapeutics. As we will see below in our discussion of infectious diseases, it may even

Table 5.2 Liposomal nanodrugs

Nanomedicine name	Drug name	Bio-pharma company	(FDA) approval	Indication
Abelcet	Amphotericin B	Enzon	1995	Fungal infections
Ambisome	Amphotericin B	Gilead, Astellas	2000	Fungal infections
Curosurf	Poractant alfa	Chiesi Farmaceutici	1998	Pediatric Respiratory Distress Syndrome
DaunoXome	Daunorubicin	Gilead	1996	HIV related Saucoma
DepoCyt	Cytarabine	Pacira (Skye Pharma)	2007	Meningitis
DepoDur	Morphine	Skye Pharma, Endo	2004	Postsurgical analgesia
Diprivan	Propofol	AstraZeneca	2001	Pediatric Anaesthetic
Doxil/Caelix	Doxorubicin	J&J - Ortho	1995	Various cancers
Epaxal	Inactive Hepat A Virus	Crucell (Berna Biotech)	2006	Hepatitis A
InflexalV	Flu Vaccine	Crucell (Berna Biotech)	1997	Influenza
Marquibo	Vincristine	Spectrum Pharma	2012	Leukemia (ALL), Melanoma
Mepact	Mifarmurtide	Takeda	Europe	Osteosarcoma
Myocet	Doxorubicin	Zeneus, GP Pharm	Europe, Canada	Breast cancer
VisuDyne	Verteporfin	QLT, Novartis	2001	Age-related Macular Degeneration

be possible to develop antimicrobial polymers that could rival antibiotics in the prevention and treatment of multidrug-resistant infections.

A special category of polymer therapeutics includes PEG–protein conjugates that are intended for parenteral administration of viral infections and tumors[239] and are designed to overcome some of the limitations of biologics including poor stability, short plasma half-life, and immunogenicity. For instance, the therapy of liver diseases (hepatitis) was significantly improved by PEGylation of interferon α-2a (Pegasys, Roche Pharma) and interferon α-2b (PegIntron, Schering-Plough/Merck).

The only polymeric drugs approved so far for oral administration are Renagel and Renvela by Genzyme, now part of Sanofi. The polymeric nanodrug Renvela was approved in 2007 by the FDA and in 2009 by the European authorities for the control of serum phosphorus in patients with chronic kidney disease (CKD). Renvela (sevelamer carbonate) and Renagel (sevelamer hydrochloride) both contain the active ingredient Sevelamer, contraindicated in patients with hyperphosphatemia. When taken with meals, it binds to dietary phosphate and prevents its absorption.

Another active area of nanomedical research is based on the desire to improve the administration of cancer chemotherapy by means of the drug *paclitaxel*. Paclitaxel was discovered in 1967 as a result of a US National Cancer Institute (NCI)-funded screening program; the drug was isolated from the bark of the Pacific yew, *Taxus brevifolia*, a conifer native to the Pacific Northwest of North America. The drug was named taxol and developed commercially by Bristol-Myers Squibb (BMS). Later, the generic name has changed to paclitaxel, with BMS continuing to sell the compound under trademark as Taxol. Paclitaxel's mechanism of action involves its stabilization of cellular microtubules; as a result, it interferes with the normal breakdown of microtubules during cell division. Paclitaxel is used to treat patients with lung, ovarian, breast, bladder, prostate, head, and neck cancers, melanomas, etc. It is widely used worldwide. Paclitaxel's standard formulation with polyethoxylated castor oil (Cremophor EL by BASF) and ethanol is causing serious side effects such as nausea and vomiting, loss of appetite, change in taste, thinned or brittle hair, pain in the joints of the arms or legs lasting two to three days, changes in the color of the nails, and tingling in the hands or toes.

Nanomedical research is aimed at reduction of those side effects. Abraxis BioScience developed *Abraxane*, in which paclitaxel is bonded to albumin, the most abundant protein in human blood plasma.

Albumin transports hormones, fatty acids, and other compounds, buffers pH, and maintains osmotic pressure. Abraxane uses albumin as an alternative delivery agent to the toxic solvent delivery method based on castor oil. Abraxane was approved by the US FDA in January 2005 for the treatment of breast cancer. By reducing acute toxicity in patients, Abraxane allows paclitaxel administration at higher doses with reduced side effects.

Another nanodrug based on paclitaxel is Genexol-PM developed by the South Korean biopharmaceutical company Samyang. Genexol-PM is a polymeric micelle-based formulation of paclitaxel, showing superior efficacy against cancer cells and less adverse reactions. It

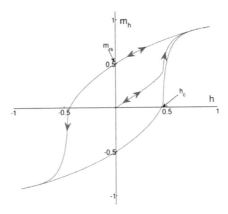

Figure 5.11   Hysteresis loop—magnetization **m** against a magnetic field **h**.
Source: Wikipedia.

has been approved in South Korea for breast and lung cancers and is
undergoing phase 3 trials in the United States.

Magnetic nanoparticles produce heat when subjected to an alternating magnetic field. As a consequence, if magnetic nanoparticles
are put inside a tumor and the patient is placed in an alternating
magnetic field of well-chosen amplitude and frequency, the tumor
temperature will rise. *Magnetic fluid hyperthermia* (MFH) is the
name of an experimental cancer treatment based on biocompatible
ferromagnetic iron oxide nanoparticles. Whenever a ferromagnetic
material is placed under a magnetic field, the magnetic moments
become aligned to the direction of the field. Even after removal
and then reversal of the magnetic field, residual magnetism is
retained and a coercive force is required to overcome this residual
magnetism. Hence, application of an alternating magnetic field
results in a "hysteresis loop" and the work necessary to close this
loop (Fig. 5.11) will result in the release of energy, dissipated as heat
from the magnetic material.

The resulting heat—if generating temperatures above the physiological level, that is, 40–45°C, can be used to destroy tumor cells. As
opposed to application of microwaves, MFH can be localized to the
positions of the magnetic nanoparticles, at frequencies as low as 10^5
Hz, not harmful to healthy living tissue. On the other hand, typical
microwave frequencies fall in the range of 10^{10} to 10^{12} Hz and would
seriously harm healthy surrounding tissues.

By choosing biocompatible magnetic nanoparticles of optimal size and coercivity, it is now possible to achieve selective interstitial heating of tumors. By conjugating them with tumor specific ligands, the nanoparticles can be delivered directly into the tumor tissue. Although still in its infancy, MFH holds great promise as a nanomedical way to treat tumors using hyperthermia and to do so with minimal side effects.[240]

Other nanomedical technologies for targeted drug delivery include the following:

- *Buckyballs* (already introduced in Section 2.2 and rewarded with the 1966 Nobel Prize for chemistry[241]) and other C_{60} derivatives have been tested as vehicles for the insertion of DNA or RNA into a cell (transfection). There is also ongoing research with paclitaxel-embedded buckysomes,[242] spherical nanostructures with a hydrophilic surface.
- *Carbon nanotubes* (CNTs), allotropes of carbon with a cylindrical nanostructure, are long, hollow structures with the walls formed by one-atom-thick sheets of carbon, called graphene. Among other CNT-based research projects pioneered by Stanford University, biocompatible CNTs are investigated as delivery vehicles for paclitaxel to cancer cells.[243]

Nanomedical research, being new, is also creating new health and safety concerns. *Nanotoxicity* is an important new field of research as nanoparticles are introduced into biomedical research.

5.2.4 *The Business of Nanotherapeutics*

The translation of nanomedical technologies into the clinical environment is still driven mostly by start-ups and small/medium companies. Big pharmaceutical companies have not yet fully embraced nanomedicine although they are watching the space with great interest. However, as seen from the information included in Table 5.1, a number of global biopharmaceutical companies have nanomedical products on the market or in late-stage clinical trials.

Here is a list of major biopharmaceutical players by nanomedical technology area:

- Nanocrystals: Pfizer, AbbVie, Merck & Co, Lilly, Johnson & Johnson
- Nanoemulsions: Novartis, AbbVie
- Polymeric: Sanofi/Genzyme, Pfizer, Merck & Co, Roche, Amgen, UCB Pharma, Teva
- Liposomes: Johnson & Johnson, Teva, Gilead, Takeda, AstraZeneca, Novartis
- Nanoparticles: Bayer
- Nanocomplex: Sanofi, Takeda

The main goals of nanotherapeutics are to improve drug solubility, to guide drugs to the desired location of action with increased precision, and to enhance transport across biological barriers. There must be significant improvements in bioavailability, pharmacokinetics (PK), efficacy, and safety due to nanomedicine both for global biopharmaceuticals to invest resources and for regulatory authorities to approve nanomedicines.

In its Draft Guidance from August 2002, the FDA wrote the following:

> A drug substance in a liposome formulation is intended to exhibit a different pharmacokinetic and/or tissue distribution (PK/TD) profile from the same drug substance (or active moiety) in a nonliposomal formulation given by the same route of administration. The complete characterization of the PK/TD profile of a new liposome drug product is essential to establish the safe and effective dosing regimen of the product . . .
>
> A. Bioanalytical Methods . . . for liposomal drug products the bioanalytical method should also be capable of measuring encapsulated and unencapsulated drug substance.
>
> B. In Vivo Integrity (Stability) Considerations . . . if the bioanalytical method can distinguish between encapsulated and unencapsulated drug substance, the in vivo stability of the liposome should be determined. . . .[244]

Similar instructions are provided by the European Medicines Agency (EMA) in a reflection paper issued on February 21, 2013.[245]

While academic research publications are excited about the many opportunities to improve efficacy and reduce side effects with nanomedicine, publications such as the *Journal of Controlled Release* are much less bullish.

Here are two recent examples of statements published in 2012 and mentioned by Novartis in an industry session of CLINAM 7/2014,[246] a leading global nanomedicine conference, organized by the European Foundation for Clinical Nanomedicine and held annually in Basel, Switzerland:

> Almost the whole decade of 2000s has been consumed by developing various nanoparticles for targeted drug delivery to tumors, and the results are not, on the whole, encouraging.[247]
>
> These insights strongly suggest that besides making ever more nanomedicine formulations, future efforts should also address some of the conceptual drawbacks of drug targeting to tumors, and that strategies should be developed to overcome these shortcomings.[248]

Still, Novartis, a company known for its excellent research, is pursuing an ambitious nanomedicine program that includes work on nanoparticles, albumin, micelles, liposomes and polymeric drugs.

With Doxil generating revenues close to $0.5 B, there may well be a nanomedical (>$1 B) blockbuster drug reaching the market in the near future. It could be both personalized, as discussed in Section 4.6, and targeted to an organ or a tumor site.

5.3 Biomedical and Molecular Imaging

There are no facts, only interpretations.

—Friedrich Nietzsche (1844–1900)

Nanomedicine is a highly interdisciplinary field in which the physical sciences play an important role. This is particularly true for biomedical imaging.

The practice of medicine and study of biology have always relied on visualization to study anatomy, to get an understanding of biologic function and to detect and treat disease.

For a long time the microscope served as the principal tool but modern medical imaging started with the discovery of X-rays by Wilhelm Conrad Roentgen, recipient in 1901 of the first Nobel Prize for physics in recognition of the extraordinary services he has rendered by the discovery of the remarkable rays subsequently named after him.[249] Doctors instantly realized this new photographic technique could help them look inside the human body without surgery, and within weeks were using X-rays to diagnose bone fractures, locate embedded bullets, and identify causes of paralysis.

Today, Roentgen's X-rays are just one modality of many that are used in the field of biomedical imaging.

The processing of medical images by means of sophisticated computer software has enabled 3D imaging and sometimes even added a fourth dimension—time—to the interpretation of structures and processes captured by imaging modalities such as computed tomography (CT), magnetic resonance imaging (MRI), positron emission tomography (PET), ultrasound (US), and electroencephalography (EEG).

Imaging modalities span a wide spectrum of technologies and applications, from structural imaging focused on anatomy/morphology to functional imaging, which adds an element of time and is used to investigate physiological processes such as metabolism, to molecular imaging, mostly based on the tracing of radioactive isotopes of common elements such as phosphorus, sodium, potassium, magnesium, sulfur, calcium, manganese, iron, copper, and zinc.

Molecules containing such radioactive isotopes have first been thoroughly studied by George de Hevesy who won the Nobel Prize for chemistry 1943 for for his work on the use of isotopes as tracers in the study of chemical processes.[250] Imaging techniques and medical treatments based on George de Hevesy's radiotracers are often referred to as *nuclear medicine.* They incorporate a radioactive tracer atom into a larger pharmaceutically-active molecule, a *radiopharmaceutical*, which is designed to localize in the body. An example is fludeoxyglucose (Fig. 5.12), in which fluorine-18 is incorporated into deoxyglucose.

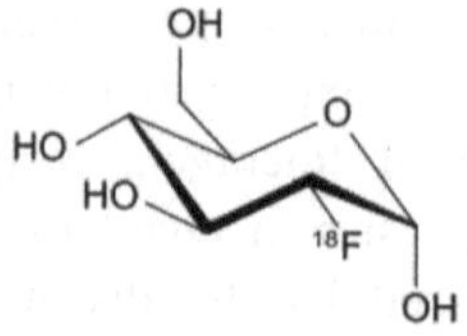

Figure 5.12   Fluorodeoxyglucose (^{18}F). Source: Wikipedia.

The imaging modality based on X-rays is often called *radiography* and is still the most widely used technique in medical practice, for example, for detection of bone fractures and pathological changes in the lung. CT scans are used to obtain a 3D image. Although predominantly a structural imaging technique, CT scans can be turned into a functional tool by means of contrast agents and by repeated application and subsequent analysis of changes, such as, for example, functional analysis of tumor size. Diagnostic *angiography* is a medical imaging modality used for the visualization of the inside or lumen of blood vessels, particularly the arteries, veins and the heart chambers. Contrast media is injected into the blood vessels for the study.

Mammography is an imaging modality that uses low energy X-rays specifically for imaging of breast tissue

Medical ultrasonography (or US) uses high-frequency broadband sound waves in the megahertz range that are reflected by tissue to varying degrees to produce images. Important uses include imaging of the fetus in pregnant women, the abdominal organs, heart, breast, muscles, tendons, arteries, and veins. While it may provide less anatomical detail than techniques such as CT, the US technique has several advantages, including low cost, ease of use/portability, and real-time/dynamic capability. It is very safe to use and does not appear to cause any adverse effects.

MRI has already been introduced at the end of Section 2.4 when we discussed nuclear magnetic resonance (NMR). MRI instruments are impressive technological achievements and are powerful tools in exploring both morphological features (including soft tissues like muscles, ligaments, and even the brain) and functional changes. Like X-ray-based CT, MRI in its basic form creates a 2D image of a thin slice of the body. Modern MRI instruments are capable of producing

3D images as a generalization of the single-slice, tomographic, concept. Unlike CT, MRI does not involve the use of ionizing radiation and is therefore not creating similar health hazards. There appears to be no limit to the number of scans to which an individual can be subjected, in contrast with X-ray and CT. However, there are potential health risks associated with tissue heating from exposure to the magnetic radiofrequency field on top of the static magnetic field.

Functional magnetic resonance imaging (fMRI) is used to understand how different parts of the brain respond to external stimuli. Assuming a linear coupling between neural activity and metabolic activity, it is possible to use blood oxygenation levels—a measure of metabolic activity—to measure neural activity. Since blood flow increases when neurons require glucose and oxygen, the ratio of oxygenated to deoxygenated blood will be higher than normal in areas of increased metabolism. Since oxygenated hemoglobin is diamagnetic (i.e., essentially nonmagnetic) whereas deoxygenated hemoglobin is paramagnetic, deoxyhemoglobin will disturb the applied magnetic field more than oxyhemoglobin, causing an increase of the magnetic resonance signal. This blood-oxygen-level-dependent (BOLD) effect is the major source of contrast in fMRI images of the brain.

The sensitivity of MRI can be enhanced either by increasing the magnetic field or by adding contrast agents (signal molecules). Basic MRI magnetic field strengths are typically 20,000 to 60,000 times stronger than the earth's magnetic field, making MRI instruments large, heavy, and expensive.

However, there are also recent research efforts aimed at significant reduction of the size and cost of MRI instruments. The ability of MRI to take images at higher resolution than the submillimeter range has been limited by the fundamental sensitivity limitations of coil-based inductive detection. Conventional MRI techniques yield a resolution of about 3 μm per side of a volume element (voxel). To overcome this resolution limit, John Sidles at University of Washington proposed an alternative detection method, called magnetic resonance force microscopy, or MRFM.[251] MRFM overcomes the sensitivity limitations of inductive detection by using the ultrasensitive detection of magnetic forces. IBM Research scientists, the previous inventors of scanning tunneling and atomic force microscopy, have led the effort in developing MRFM techniques, starting with the earliest demonstrations.[252] A number of academic groups, including Harvard's Amir Jacoby,[253] are now working on the realization of MRI systems that can produce nanoscale images and may one day allow researchers to peer into the atomic structure of individual molecules.

Back to conventional MRI, two widely used MRI contrast agents are the nanoparticles gadolinium (Gd) and iron oxide. Gd particles are used in *DCE-MRI*, where DCE stands for dynamic contrast enhanced. DCE-MRI is the acquisition of serial magnetic resonance images before, during, and after the intravenous administration of a Gd contrast agent. The low-molecular-weight Gd complex is highly diffusible outside of the brain and can easily pass the capillary walls. The signal intensities in the contrast-enhanced images reflect therefore a composite of vessel enhancement and tissue uptake. DCE-MRI has proven to be a reliable biomarker for imaging changes introduced by antivascular or chemotherapeutic drugs before volumetric changes of the tumor occur.

Iron oxide nanoparticles can be classified as superparamagnetic iron oxides (SPIOs), ultrasmall superparamagnetic iron oxides (USPIOs), and very small paramagnetic iron oxides (VSPIOs) of size 4–8 nm.

As opposed to Gd particles, use of MRI with Fe oxide particles is still more in the clinical research stage, for example, for early detection of vulnerable plaques in the cardiovascular diagnostics, not yet part of standard clinical care.

However, outside the field of MRI, there is a new emerging imaging modality, magnetic particle imaging, based on magnetic nanoparticles and codeveloped by Philips and Bruker[254] and aimed at providing new insights into cardiovascular disease (CVD), cancer, and stem cell therapies.

PET is a molecular imaging technique that produces a 3D image of functional processes in the body. The system detects pairs of gamma rays emitted indirectly by a positron-emitting radiotracer that is introduced into the body on a biologically active molecule. 3D images of tracer concentration within the body are then constructed by computer analysis. If the biologically active molecule chosen for PET is fluorodeoxyglucose (FDG), an analogue of glucose—see

Fig. 5.4-1, the concentrations of tracer imaged will indicate tissue metabolic activity associated with glucose uptake. The uptake of ^{18}F-FDG by tissues is a marker for the tissue uptake of glucose, which in turn is closely correlated with certain types of tissue metabolism. After ^{18}F-FDG is injected into a patient, a PET scanner can form 2D or 3D images of the distribution of 18F-FDG within the body.

Use of this tracer to explore the possibility of cancer metastasis (i.e., spreading to other sites) is the most common type of PET scan in standard medical care (90% of current scans).

The FDG radiotracer is also used for brain research and early detection of cognitive dysfunction such as Alzheimer's disease (AD)/dementia. Since PET scans of the brain measure energy consumption associated with glucose uptake, they monitor neural activity in the most direct way. PET brain scan spatial resolution can be reduced to 1.5–2 mm by means of the industry-leading high-resolution research tomography (HRRT) (Siemens) instrument along with highly sophisticated and computationally intensive image reconstruction software.

PET scans are increasingly read alongside CT or MRI scans, combining two modalities to achieve coregistration, resulting in both anatomic and metabolic information.

Figures 5.13 and 5.14 show examples of fused PET-CT and PET-MRI images, respectively.

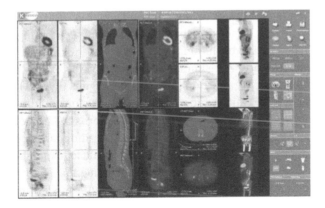

Figure 5.13   Fused PET-CT body images. Source: Wikipedia.

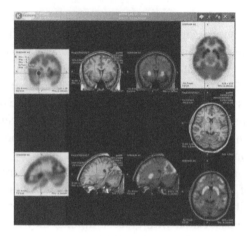

Figure 5.14   Fused PET-MRI brain images. Source: Wikipedia.

Another nuclear medicine modality is *single-photon emission computed tomography* (SPECT).

SPECT is similar to PET in its use of radioactive tracer material, but it's based on gamma rays, acquired by moving a gamma camera around the patient. The tracer (frequently technetium isotope Tc-99m) used in SPECT is first injected into the body of the patient and then emits gamma radiation that is measured directly. Technetium is the chemical element with atomic number 43 and the symbol Tc, the first element in the periodic table without any stable isotopes. SPECT spatial resolution is only about 1 cm, but SPECT scans are significantly less expensive than PET scans. By combining SPECT with CT, the spatial resolution can be improved. SPECT can be used to complement any 2D gamma imaging study, for example, blood flow, tumor imaging, infection (leukocyte) imaging, thyroid imaging, detection of bone cancer, and renal scans (perfusion and drainage of the kidneys). SPECT-CT can be used to detect prostate cancer.

Both PET and SPECT rely upon the use of radiopharmaceuticals, thus enabling the above-mentioned field of nuclear medicine that extends beyond diagnostics into surgical treatments.

In addition to the above-mentioned imaging modalities, there are other special techniques that typically are confined to particular parts of the body.

Electrocardiography (EKG or ECG) is used to measure the heart's electrical conduction system. It picks up electrical impulses generated by cardiac tissue and translates them into a waveform. The waveform is then used to measure the rate and regularity of heartbeats and the presence of any damage to the heart. *EEG* is the recording of electrical activity along the scalp. EEG refers to the recording of the brain's spontaneous electrical activity over a short period of time, usually 20–40 minutes, as recorded from multiple electrodes placed on the scalp. If placed directly on the brain (after partial removal of the skull), the (invasive!) technique is referred to as electrocorticography (ECoG).

ECoG may be performed either in the operating room during surgery (intraoperative ECoG) or outside of surgery (extraoperative ECoG). Because a craniotomy (a surgical incision into the skull) is required to implant the electrode grid, ECoG is an invasive procedure.

Magnetoencephalography (MEG) is an imaging technique used to measure the magnetic fields produced by electrical activity in the brain. MEG offers a very direct measurement of neural electrical activity with very high temporal resolution but relatively low spatial resolution.

Optical microscopy remains an important field, in particular in the field of cytopathology, defined as the branch of pathology that studies and diagnoses diseases on the cellular level. It is used in the diagnosis of cancer, but also helps in the diagnosis of certain infectious diseases and other inflammatory conditions. Cytopathology is generally used on samples of free cells or tissue fragments, in contrast to histopathology, which studies whole tissues. Cytopathologic tests are sometimes called smear tests because the samples may be smeared across a glass microscope slide for subsequent staining and microscopic examination. Pathologists are highly trained to identify cell and tissue abnormalities. It has been widely accepted that human vision can detect and distinguish between shapes better than computers. However, recent advances in the use of pattern recognition and machine learning have created a situation where image recognition software like the one developed by Definiens[255] is beginning to rival the leading experts in cancer pathology.

Superresolution microscopy is a form of light microscopy. Due to the diffraction of light, the resolution of conventional light microscopy is limited as stated by Ernst Abbe in 1873. Abbe was the cofounder and scientific mind behind the Jena-based Carl Zeiss[256] company (Germany). Abbe worked out that the attainable resolution of a precise wide-field microscope working with visible light is limited to ~250 nm. However, in 2014, the Nobel Prize for chemistry[257] was awarded to Eric Betzig, William Moerner, and Stefan Hell for the development of super-resolved fluorescence microscopy, which brings optical microscopy into the nanodimension.

What they did was bypassing Abbe's seemingly unmovable limitation, in two different ways:

- Stimulated emission depletion (STED) microscopy was developed by Stefan Hell in 2000. Two light amplification by stimulated emission of radiation (laser) beams are utilized; one stimulates fluorescent molecules to glow, while another cancels out all fluorescence except for that in a nanometer-sized volume. Scanning over the sample, nanometer for nanometer, yields an image with a resolution better than Abbe's stipulated limit.
- Betzig and Moerner, working separately, laid the foundation for the second method, single-molecule microscopy. The method relies upon the possibility to turn the fluorescence of individual molecules on and off. Scientists image the same area multiple times, letting just a few interspersed molecules glow each time. Superimposing these images yields a dense super-image resolved at the nanolevel. In 2006 Betzig utilized this method for the first time.

In what has become known as nanoscopy, scientists can now visualize the pathways of individual molecules inside living cells. They can see how molecules create synapses between nerve cells in the brain; they can track proteins involved in Parkinson's, Alzheimer's, and Huntington's diseases as they aggregate; they follow individual proteins in fertilized eggs as these divide into embryos.

Regarding the business of biomedical imaging, it is a multibillion-dollar market with global players such as General Electric (GE), Siemens, Philips, and Toshiba. Their financials will be included below in Table 5.3. In addition, there are several smaller companies specializing on niche areas.

Among them are the optical and electronic microscope leaders Zeiss,[258] Leica,[259] Nikon,[260] Olympus,[261] JEOL,[262] and Meiji.[263]

5.4 Nanodiagnostics

A good decision is based on knowledge and not on numbers.

—Plato (428/427 or 424/423 BC to 348/347 BC)

Medical diagnosis (often abbreviated Dx) is the determination of which disease or condition is causing a patient's signs and symptoms. A diagnostic test is any kind of medical test performed to aid in the diagnosis (detection of disease).

Diagnostic tests in vivo (Latin for *within the living*) are performed on whole living organisms, usually animals, including humans and plants. In Section 5.4 we discussed biomedical imaging as an example of in vivo diagnostics.

Diagnostic tests in vitro (Latin for *within the glass*) are performed on a partial (or dead) organism, in a laboratory environment using test tubes, petri dishes (a petri dish being a shallow cylindrical glass or plastic lidded dish that biologists use to culture cells), etc.

According to the industry leader, Roche, in vitro diagnostics (IVD) provides the medical information doctors need to guide 60% of all clinical decisions.[264]

Nanotechnology is playing an increasingly important role in diagnostics. As microprocessors are becoming nanoprocessors and consuming less and less power, it is possible to use implants to monitor the in vivo variation of important biological parameters, such as, for example, blood glucose levels by means of CGM, as discussed at the end of Section 5.1. It is also possible to miniaturize test equipment for in vitro use, potentially enabling hand-held devices for use in remote areas without easy access to hospitals or testing labs.

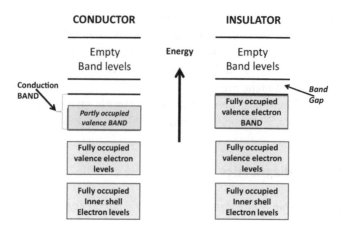

Figure 5.15 Valence band, conduction band, and band gap.

Let's discuss a few nanotechnologies that have already been applied successfully in electronics and materials science and are currently tested in a nanodiagnostic context.

A good example is *quantum dot* technology, a hot new technology area with existing applications in consumer electronics. According to Wikipedia, a quantum dots is a "nanocrystal made of semiconductor materials that are small enough to exhibit quantum mechanical properties."

To understand what that means we need to go back to our discussion of semiconductors and metals in Section 2.3: A quantum dot is a semiconductor nanostructure that confines the motion of conduction band electrons, valence band holes, or excitons (bound pairs of conduction band electrons and valence band holes) in all three spatial directions (Fig. 5.15).

We remember that a semiconductor is nothing but an insulator with a very small band gap.

The confinement can be due to electrostatic potentials (generated by external electrodes, doping, strain, impurities), the presence of an interface between different semiconductor materials (e.g., in core–shell nanocrystal systems), the presence of the semiconductor surface (e.g., semiconductor nanocrystal), or a combination of these. Quantum dots are grown by advanced epitaxial techniques in nanocrystals produced by chemical methods or by ion implantation, or in nanodevices made by state-of-the-art lithographic techniques. *Epitaxy* refers to the deposition of a crystalline overlayer on a crystalline substrate. *Photolithography* (also termed optical lithography or ultraviolet [UV] lithography) is a process used in semiconductor industry (in particular complementary metal–oxide–semiconductor [CMOS] technology) to pattern parts of a thin film or the bulk of a substrate. It uses light to transfer a geometric pattern from a photomask to a light-sensitive chemical photoresist on the substrate. A series of chemical treatments then either engraves the exposure pattern into or enables deposition of a new material in the desired pattern upon the material underneath the photoresist.

A quantum dot has a discrete quantized energy spectrum. The corresponding Schrödinger wave functions are spatially localized within the quantum dot, but extend over many periods of the crystal lattice.

Small quantum dots, such as colloidal semiconductor nanocrystals, can be in the range of 2 to 10 nm, corresponding to 10 to 50 atoms in diameter and a total of 100 to 100,000 atoms within the quantum dot volume. If the size of the quantum dot is small enough that the quantum confinement effects dominate (typically less than 10 nm), the electronic and optical properties are highly tunable.

Applications of quantum dots can be found in transistors, solar cells, light-emitting diodes (LEDs), and diode lasers. Most recently, quantum dots have been inserted into television sets to reduce power requirements and to enhance the dynamic range of colors encoded in the picture (e.g., IQ technology).[265]

In biology, quantum dots are replacing traditional dyes due to increased brightness and increased stability. Cellular imaging has seen major advances over the past decade. The improved photostability of quantum dots allows the optical acquisition of many consecutive focal-plane images that can be reconstructed into a high-resolution 3D image.

Another application is the real-time tracking of molecules and cells over extended periods of time. For instance, antibodies, peptides, DNA, or small-molecule ligands can be used to target quantum dots to specific proteins on cells, for example, a four-month observation of lymph nodes in mice.[266]

Quantum dots have also been employed for in vitro imaging of single-cell migration in real time, with uses in embryogenesis, cancer metastasis, stem cell therapeutics, and lymphocyte immunology.

Microfluidics is a multidisciplinary field with practical applications to the design of systems in which small volumes of fluids will be handled. Microfluidics emerged in the beginning of the 1980s and is used in the development of inkjet print heads, DNA and protein chips (see below), and lab-on-a-chip (LOC) technology. An LOC is a device that integrates one or several laboratory functions on a single chip of only millimeters to a few square centimeters in size.

Microfluidics deals with the behavior, precise control and manipulation of fluids that are geometrically constrained to a small, typically submillimeter, scale. Also called microfluidics, volumes can get as small as nanoliters (nL, or 10^{-9} L), picoliters (pL, or 10^{-12} L), and even femtoliters (fL, or 10^{-15} L).

The behavior of fluids at the microscale can differ from macrofluidic behavior in that factors such as surface tension, energy dissipation, and fluidic resistance start to dominate the system. Microfluidics studies how these behaviors change, and how they can be worked around, or exploited for new uses.

A *DNA microarray* (also called DNA chip) is a collection of microscopic DNA spots attached to a solid surface. DNA microarrays are used to measure the expression levels of large numbers of genes simultaneously or to genotype multiple regions of a genome. Each DNA spot contains picomoles (10^{-12} moles) of a specific DNA sequence, known as probes (or reporters or "oligos").

Similarly, a *protein microarray* (or protein chip) is a high-throughput method used to track the interactions and activities of proteins, and to determine their function. Its main advantage lies in the fact that large numbers of proteins can be tracked in parallel. The chip consists of a support surface such as a glass slide, nitrocellulose membrane, bead, or microtitre plate, to which an array of capture proteins is bound. Probe molecules, typically labeled with a fluorescent dye, are added to the array. Any reaction between the probe and the immobilized protein emits a fluorescent signal that is read by a laser scanner.

Protein chips have emerged as a promising approach for a wide variety of applications, including the identification of protein–

protein interactions, protein–phospholipid interactions, and small-molecule targets. They can also be used for clinical diagnostics and monitoring disease states. However, a protein chip requires a lot more steps in its creation than does a DNA chip, thus inhibiting commercialization. A long list of technical challenges include:

- finding a surface and a method of attachment that allows the proteins to maintain their secondary or tertiary structure and thus their biological activity and their interactions with other molecules;
- producing an array with a long shelf life so that the proteins on the chip do not denature over a short time;
- identifying and isolating antibodies or other capture molecules against every protein in the human genome;
- quantifying the levels of bound protein while assuring sensitivity and avoiding background noise;
- extracting the detected protein from the chip in order to further analyze it; and
- reducing non-specific binding by the capture agents.

Finally, the capacity of the chip must be sufficient to allow as complete a representation of the proteome to be visualized as possible. Abundant proteins overwhelm the detection of less abundant proteins such as signaling molecules and receptors, which are generally of more therapeutic interest.

Microfluidics is used extensively in DNA sequencing, as already discussed in Section 3.7. For instance, ion-sensitive field-effect transistors (ISFETs), used for measuring ion (e.g., proton) concentration in solution, combine microfluidics with transistors. The challenge is to expose the sensors to fluids, while protecting the electronics.[267]

Another interesting area of nanotechnology with potential relevance for nanodiagnostics is *silicon photonics*, the study and application of photonic systems that use silicon as an optical medium. Silicon photonics is actively researched by many electronics manufacturers including IBM and Intel, as well as by academic research groups such as Michal Lipson's team at Cornell University.[268] It is seen a means for using optical interconnects to provide faster data transfer both between and within microchips. Microphotonic

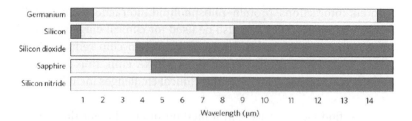

Figure 5.16 Optical transparency of silicon, germanium, and Si compounds used as substrates. Source: Wikipedia.

components operate in the infrared, most commonly at the 1.55 μm wavelength used by most fiber-optic telecommunication systems.

As seen from Fig. 5.16, silicon is transparent up to a wavelength close to 9 μm.

Medical applications (not yet attempted commercially but feasible as low cost applications[269]) could take advantage of silicon's midinfrared transparency to develop diagnostic devices for noninvasive glucose monitoring (absorption wavelength 2–2.5 μm) and the monitoring of exhaled breath to generate, process, and detect signals associates with molecules such as:

- nitric oxide (NO)
- ammonia (NH_3)
- carbonyl sulfide (OCS)
- alkanes (C_nH_{2n+2})
- formaldehyde (CH_2O)
- acetaldehyde (C_2H_4O)
- acetone (C_3H_6O)
- carbon disulfide (CS_2)
- ethane (C_2H_6)

that have their absorption bands in the range of 3 to 7 μm and are associated with medical conditions such as asthma, ulcers, renal and liver function, breast cancer, lung cancer, and organ rejection.

A hot area of cancer diagnostics is the detection of *circulating tumor cells (CTCs)*, cells that have entered the bloodstream from a primary tumor. CTCs are assumed to constitute seeds for the subsequent growth (or metastasis) of additional tumors in vital distant organs, triggering a mechanism that is responsible for the vast majority of cancer-related deaths.

Modern cancer research has demonstrated that CTCs derive from clones in the primary tumor, validating Ashworth's observation that "cells identical with those of the cancer itself being seen in the blood may tend to throw some light upon the mode of origin of multiple tumors existing in the same person."[270] The significant efforts put into understanding the CTCs' biological properties have demonstrated the critical role CTCs play in the metastatic spread of carcinoma and that these CTCs reflect molecular features of cells within tumor masses. CTCs thus could be considered a liquid biopsy which reveals metastasis in action, providing live information about the patient's disease status. The major challenge for CTC researchers is the prevailing difficulty of CTC purification that allows the molecular characterization of CTCs. Several methods have been developed to isolate CTCs in the peripheral blood and essentially fall into two categories, biological methods and physical methods.

Biological methods are separation based on antigen–antibody bindings. Antibodies against tumor-specific biomarkers are used. The most common technique is magnetic nanoparticle-based separation (immunomagnetic assay) as used in CellSearch or magnetic-activated cell sorting (MACS), a method for separation of various cell populations depending on their surface antigens (CD molecules) invented by Miltenyi Biotec. Other techniques under research include microfluidic separation.

Physical methods are often filter based, enabling the capture of CTCs by size. ScreenCell[271] is a filtration-based device that allows sensitive and specific isolation of CTCs from human whole blood in a few minutes.

Finally, we should also mention DNA and RNA sequencing as another example of nanodiagnostics, positioned to gain important clinical relevance as the "$1000 genome" is becoming reality and as more and more gene–disease relationships are uncovered.

Nanodiagnostics is a fertile ground for applications of new nanotechnologies that may already have been applied in other fields such as electronics and materials science. What's slowing down their adoption are regulatory and safety issues as well as the reluctance by established industry leaders to displace existing products that have

Table 5.3 Device and diagnostics companies (including medical imaging and sequencing)

Company name	Country	Sep-14 Market cap ($B)	TEV ($B)(1)	2013 Sales ($B)
J&J (*) Healthcare	USA	291.39	278.05	73.54
GE (*) Conglomerate	USA	259.60	629.16	142.31
Roche (*) Pharma + Diagnostics	Switzerland	250.30	NA	51.00
Siemens (*) Conglomerate	Germany	114.20	NA	101.37
Abbott Labs	USA	63.37	63.82	21.85
Medtronic	USA	62.99	61.37	17.08
Danaher (incl Beckman Coulter) (*)	USA	53.55	53.43	18.00
Thermo Fisher (incl. Life Tech) (*)	USA	48.82	62.98	14.88
Baxter	USA	40.07	47.59	16.36
Philips (*) Conglomerate	Netherlands	29.06	NA	29.00
Illumina	USA	25.00	25.17	1.09
Becton Dickinson	USA	22.40	23.81	7.92
Agilent	USA	19.31	18.86	6.89
Toshiba (*) Conglomerate	Japan	18.45	NA	24.00
Boston Scientific	USA	16.49	20.7	7.22
Qiagen	Netherlands	5.62	6.14	1.30
Bio-Rad	USA	3.48	3.26	2.05
Bruker	USA	3.40	3.26	1.87
PacBio (note deal with Roche)	USA	0.41	0.32	0.04
BGI (incl Complete Genomics)	China	0.8 (2)	NA	NA

*Diagnostics Revenue is only part of larger corporate sales
(1) TEV (Total Enterprise Value) is the value of a company, incorporating equity, debt, and cash. It is essentially a way of measuring what it would cost to buy the company.
(2) Forbes estimated value of BGI: $800M

not yet reached the end of their useful life. When is the right time to move from a microtechnology to a nanotechnology platform if there is no compelling business argument?

In Table 5.3 we summarize the financials of the leading diagnostics and medical device companies.

Compared to the leading pharmaceutical companies discussed in Section 4.7, diagnostic companies are less profitable and therefore have to live with more limited research budgets. Even if a companion diagnostic is required for a prescription drug, pharmaceutical companies are not willing to share the rich compensation received for a patent protected drug that meets a medical need. For this reason, diagnostic research business models usually do not rely on deep basic in-house research activities. Typically, new diagnostic technologies are developed by academic research groups and then tested and commercialized by start-up companies. The established global players will then form risk sharing partnerships with negotiated milestone payments, perhaps followed by an acquisition based upon careful evaluation of market potential.

It is therefore difficult for nanomedicine to achieve quick penetration of the diagnostics market. However, even if the adoption may be slow, it will be inevitable.

5.5 Regenerative Medicine: Stem Cells, Gene Therapy, and Immunotherapy

The human mind will not be confined to any limits.

—Johann Wolfgang von Goethe (1749–1832)

According to the National Institutes of Health (NIH), *regenerative medicine*[272] is defined as "the process of creating living, functional tissues to repair or replace tissue or organ function lost due to age, disease, damage or congenital effects."

A congenital disorder, a condition existing at birth and often before birth, may be the result of genetic abnormalities, the intrauterine (uterus) environment, errors of morphogenesis, infection, or a chromosomal abnormality.

Mentioned first in the early 1990s, the term "regenerative medicine" refers to clinical therapies that may involve the use of stem cells, undifferentiated biological cells that can differentiate into specialized cells and can divide (through mitosis) to produce more stem cells. In mammals, there are two broad types of stem cells, *embryonic* and *adult stem cells.*

To understand the ongoing research activities based on stem cells, we need to review some basic developmental cell biology concepts, not yet fully introduced in Sections 3.3 and 3.4.

Let's start with the *zygote*, the initial cell and earliest stage of the embryo generated when two gamete cells are joined by means of sexual reproduction. After a series of cleavage divisions, once the early embryo has divided into 16 cells, it begins to resemble a

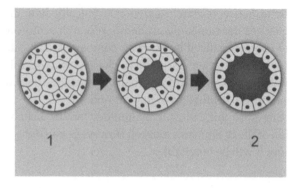

Figure 5.17 During blastulation, the morula (1) turns into the blastula (2) (or blastocyst). Source: Wikipedia.

mulberry and is called *morula*. Within a few days after fertilization, cells on the outer part of the morula become bound tightly together and a cavity is formed inside the morula. As indicated in Fig. 5.17, this process, called *blastulation*, results in a hollow ball of cells known as the *blastula* or (in mammals) the *blastocyst*.

The blastocyst's outer cells will become the first embryonic epithelium. The outer layer of the blastocyst is also called the *trophoblast*. The trophoblast gives rise to the placenta.

Some cells, however, will remain trapped in the interior and will become the inner cell mass (ICM), shown in black in Fig. 5.17. *Embryonic stem cells* (ESCs) are derived from the ICM.

A stem cell must possess two properties:

- Self-renewal: The ability to go through numerous cycles of cell division while maintaining the undifferentiated state. Two differentiation mechanisms ensure that a stem cell population is maintained:
  - Obligatory asymmetric replication: A stem cell divides into one mother cell that is identical to the original stem cell, and another daughter cell that is differentiated.
  - Stochastic differentiation: One stem cell develops into two differentiated daughter cells, and another stem cell undergoes mitosis and produces two stem cells identical to the original.

- Potency: The capacity to differentiate into specialized cell types. There are five different types of potency:

  - Totipotent (aka. omnipotent) stem cells can differentiate into embryonic and extraembryonic cell types. Such cells can construct a complete, viable organism. These cells are produced from zygotes and by their first few divisions during formation of the morula in Fig. 5.17.

  - Pluripotent stem cells are the descendants of totipotent cells and can differentiate into nearly all cells, that is, cells derived from any of the three germ layers: endoderm (interior stomach lining, GI tract, the lungs), mesoderm (muscle, bone, blood, urogenital), or ectoderm (epidermal tissues and nervous system). ESCs are defined as pluripotent cells derived from the ICM of the blastocyst, as shown in Fig. 5.17. In humans, ESCs no longer exist after about five days of development.

  - Multipotent stem cells can differentiate into a number of cell types but only those of a closely related family of cells.

  - Oligopotent stem cells can differentiate into only a few cell types, such as lymphoid or myeloid stem cells.

  - Unipotent cells can produce only one cell type, their own, but have the property of self-renewal, which distinguishes them from nonstem cells (e.g., progenitor cells, muscle stem cells).

The next phase in the embryonic development is *gastrulation*. The single-layered blastula is reorganized into a trilaminar (three-layered) structure known as the *gastrula*. These three germ layers are known as the ectoderm, mesoderm, and endoderm, as shown in Fig. 5.18.

The *archenteron* (or digestive tube) is the primitive gut that forms during gastrulation in the developing embryo. It develops into the digestive tract of an animal.

Gastrulation is then followed by *organogenesis*, when individual organs develop within the newly formed germ layers. Each layer gives rise to specific tissues and organs in the developing embryo. Following gastrulation, cells in the body are either organized into sheets of connected cells (as in epithelia), or as a mesh of

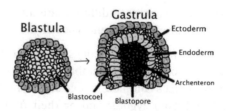

Figure 5.18 During gastrulation, the blastula develops into the gastrula. Source: Wikipedia.

isolated cells, such as mesenchyme. *Epithelial* and *mesenchymal* cells differ in phenotype as well as function. Epithelial cells are closely connected to each other and are bound by a basal lamina at their basal surface. Mesenchymal cells have a spindle-shaped morphology and interact with each other only through focal points. The epithelial–mesenchymal transition (EMT) is essential for numerous developmental processes including mesoderm formation and neural tube formation. EMT has also been shown to occur in wound healing, in organ fibrosis and in the initiation of metastasis for cancer progression.

Figure 5.19 illustrates the three germ layers: endoderm (internal), mesoderm (middle), and ectoderm (external).

- The endoderm gives rise to the epithelium of the digestive system and respiratory system and organs associated with the digestive system, such as the liver and pancreas.
- The mesoderm is found between the ectoderm and the endoderm and gives rise to somites, which form muscle; the cartilage of the ribs and vertebrae; the dermis; the notochord, blood and blood vessels; bone; and connective tissue.
- The ectoderm gives rise to the epidermis and to the neural crest and other tissues that will later form the nervous system via the process of neurulation. The neural tube and neural crest cells will become the CNS, along with melanocytes, facial cartilage, and the dentin of teeth, and the epidermal cell region will give rise to the epidermis, hair, nails, sebaceous glands, olfactory and mouth epithelium, and eyes.

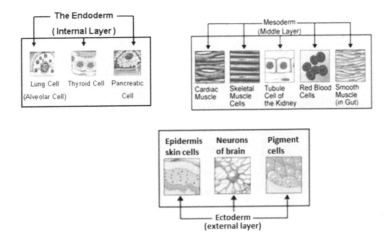

Figure 5.19 The three germ layers: endoderm, mesoderm, and ectoderm. Source: Wikipedia.

After gastrulation, all pluripotent ESCs have differentiated and become multipotent.

Differentiation is an example of epigenetic change, as discussed above in Section 3.10.

Tissue-specific stem cells are also called adult stem cells. They exist in most of the body's tissues such as the blood, brain, liver, intestine, and skin. These cells are committed to becoming a cell from their tissue of origin, but they still have the broad ability to become any one of these cells. Stem cells of the bone marrow, for example, can give rise to any of the red or white cells of the blood system. Stem cells in the brain can form all the neurons and support cells of the brain, but can't form nonbrain tissues. Unlike ESCs, scientists have not been able to grow adult stem cells indefinitely in the lab.

In adult organisms, stem cells act as a repair system for the body, replenishing adult tissues. For instance, such adult stem cells have been found in the placenta and in the umbilical cord of newborn infants. The cord blood cells that some people bank after the birth of a child are a form of adult blood-forming stem cells. Adult stem cell treatments have been successfully used for many years to treat leukemia and related bone/blood cancers through bone marrow

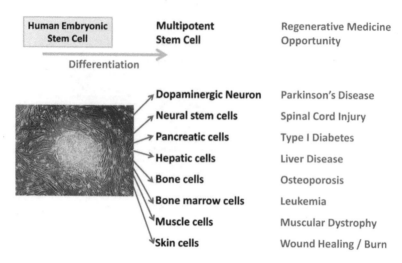

Figure 5.20 Regenerative medicine opportunity based on embryonic stem cells.

transplants. Adult stem cells are also used in veterinary medicine to treat tendon and ligament injuries in horses.

The promise of ESCs for regenerative medicine is illustrated in Fig. 5.20.

However, there is still a long way from promise to successful clinical care of patients. In addition, since the generation of ESCs involves destruction (or at least manipulation) of the preimplantation stage embryo, there has been much controversy surrounding their use. In 2001, US President George W. Bush and Pope John Paul II made clear that the destruction of human embryos for the purpose of generation of new ESC lines violated their ethical principles. Further, because ESCs can only be derived from embryos, it has so far not been feasible to create patient-matched ESC lines.

Fortunately, there may be another option.

An *induced pluripotent stem cell* (iPSC) is a cell taken from any tissue from a child or adult that has been genetically modified to behave like an ESC. Shinya Yamanaka, Kyoto University, Japan, created the first iPSC from a mouse in 2006 and shared the Nobel Prize 2012[273] with John Gurdon, who—already in 1962—demonstrated that mature (frog) cells contained the genetic information needed to form all types of cells. By means of successful reversal of stem cell differentiation, Yamanaka[274] was able to prove that embryonic development is not a "one-way-street."

Since iPSCs can be derived directly from adult tissues, they not only bypass the need for embryos but can be matched with patients. It is the promise of iPSCs that each individual could have their own pluripotent stem cell line. These unlimited supplies of autologous cells could be used to generate transplants without the risk of immune rejection. While iPSC technology has not yet advanced to a stage where therapeutic transplants have been deemed safe, iPSCs are readily being used in personalized drug discovery efforts and understanding the patient-specific basis of disease.

Here is a list of historic milestones (1908–2007) in stem cell research:[275]

- 1908: The term "stem cell" was proposed by the Russian histologist Alexander Maksimov (1874–1928) at a meeting of the hematologic society in Berlin.
- 1960s: Joseph Altman and Gopal Das present scientific evidence of adult neurogenesis, ongoing stem cell activity in the brain; their reports contradict Nobel laureate Cajal's "no new neurons" dogma.
- 1963: Becker, McCulloch, and Till illustrate the presence of self-renewing cells in mouse bone marrow.
- 1968: Bone marrow transplant is done between two siblings to treat severe combined immunodeficiency.
- 1978: Hematopoietic stem cells (HSCs) are discovered in human cord blood.
- 1981: Mouse ESCs are derived from the ICM by scientists Martin Evans, Matthew Kaufman, and Gail R. Martin. Martin coins the term "embryonic stem cell."
- 1992: Neural stem cells are cultured in vitro as neurospheres.
- 1997: Leukemia is shown to originate from an HSC, the first direct evidence for cancer stem cells.
- 1998: James Thomson and coworkers derive the first human ESC line at the University of Wisconsin–Madison.[276]
- 1998: John Gearhart (Johns Hopkins University) extracts germ cells from fetal gonadal tissue (primordial germ cells)

before developing pluripotent stem cell lines from the original extract.

- 2000s: Several reports of adult stem cell plasticity are published.
- 2006: Mouse induced pluripotent stem cells (iPSCs) are published by K. Takahashi and S. Yamanaka.
- 2007: Mario Capecchi, Martin Evans, and Oliver Smithies win the 2007 Nobel Prize for Physiology or Medicine[277] for their discoveries of principles for introducing specific gene modifications in mice by the use of ESCs.

Stem cell therapy it is the use of stem cells to treat or prevent a disease or condition. Bone marrow transplant is a crude form of stem cell therapy that has been used clinically for many years without controversy. No stem cell therapies other than bone marrow transplant are widely used, but animal experiments are conducted in many research labs, aimed in particular at stem cell treatments for neurodegenerative diseases and conditions, diabetes, heart disease, and other diseases listed in Fig. 5.20.

Another possible use of stem cells is rejuvenation. The aging process is difficult to understand because of the complexity of system-wide changes that take place as we grow older. Different parts of our bodies are affected in various ways and changes can be slow and steady or feel like a rapid decline. What is known is that aging is related to a decrease in our ability to regenerate new tissue, causing joints, blood vessels, and other parts of our body to function differently than in young persons.

Animal experiments can be helpful although the results cannot be assumed to translated directly to humans. However, a promising example may be the recent finding by Dr. Amy Wagers and coworkers at the Harvard Stem Cell Institute[278] that a protein in the blood of young mice, called GDF11, was able to quickly reverse symptoms of heart failure in older mice. Growth differentiation factor 11 (GDF11), also known as bone morphogenetic protein 11 (BMP-11) is a protein that in humans is encoded by the *GDF11* gene. Once injected, the protein caused the older hearts in mice to reduce in size and thickness so they resembled the healthy hearts of younger mice. This suggests that there may be some common signal, traveling in the bloodstream, that drives the body's response to getting older. The Harvard team could then show that GDF11 is not only heart specific but is able to restore skeletal muscle stem cell function and enhance muscle repair after injury. They also explored what the protein is doing at the cellular level, making it a potential "fountain of youth" based on improved repair mechanisms of DNA damage in muscle stem cells. GDF11 seems to reverse the inability of aged cells to properly differentiate and make mature muscle cells, a mechanism that is needed for adequate muscle repair.

GDF11 also reverses aging in the brain. Older mice injected with the protein experienced an increase in neural stem cells and renewed development of blood vessels. The older mice treated with the protein even recovered function in their ability to smell odors, like mint, typically only detected by younger mice.

As Wagers and collaborators work to learn more about GDF11, aging, and stem cells, they are hopeful that the protein, or a drug developed from the protein, can be brought to clinical trials in the near future.

Regenerative medicine is a new way of treating injuries and diseases that uses specially-grown tissues and cells, laboratory-made compounds, and artificial organs. Combinations of these approaches can amplify our natural healing process in the places it's needed the most or take over the function of a permanently damaged organ.

An early example of cell therapy is the treatment of urinary incontinence by means of muscle stem cells, as already practiced at the McGowan Institute of Regenerative Medicine at the University of Pittsburgh Medical Center (UPMC).[279] Instead of injecting collagen to add bulk to the bladder sphincter, a patient's own stem cells are isolated from a small biopsy of the patient's thigh, replicated in the laboratory over several weeks, and then injected into the bladder sphincter. The stem cells persist in the bladder for up to six months and grow to become the same sort of muscle that the bladder sphincter is made of, improving its strength.

Even more ambitious projects are focused on the development of artificial organs such as artificial hearts, livers and lungs. They may first serve as a "bridge to transplant" devices while patients are waiting for organ transplants. However, the long-term vision for, for example, ventricular heart assist devices would be to use them to give a heart a rest while it repairs itself with the help of cell therapy, such as a patient's own stem cells injected into damaged areas of their heart.

A promising technology applied to regenerative medicine is 3D bioprinting, defined as "the process of generating spatially-controlled cell patterns using 3D printing technologies, where cell function and viability are preserved within the printed construct." The goal is to use modified inkjet printer technology to print small pieces of tissues as varied as skin, cartilage, blood vessels, liver, lung, and heart.

Scaffold-free printing has been pioneered by Gabor Forgacs, who cofounded Orgonova (San Diego) in 2007 and has developed an extrusion-based printing process that can deposit cell aggregates in spheroid or cylindrical form.

Aside from the creation of artificial organs, a very interesting application of bioprinting is the generation of slivers of human liver tissue for biopharmaceutical laboratory use. Those liver tissues can be used to test the toxicity of new drugs before embarking on expensive clinical trials with patients.

The term "tissue engineering" refers to methods that promote the regrowth of cells lost to trauma or disease. The manipulation of artificial and natural materials can provide structure and biochemical instructions to young cells as they grow into specific kinds of tissue. These materials are called *scaffolds* because they provide support and materials for tissue regrowth in the same way that a scaffold supports a building under construction. During bioprinting, the scaffolds act like a prosthesis until the cells are able to take over and lay down their own matrix. The scaffold has then to be removed at the appropriate moment. Flat structures like skin and cartilage are easier to print than tubular structures, such as blood vessels or windpipes. The most complex organs to create will be the heart, liver or kidneys. Solid organs will be next frontier.

Already, it is possible to repair a patient's damaged *esophagus* (~20 cm long fibromuscular food tube) using natural scaffolds seeded with the patient's own cells to encourage the growth of healthy tissue instead of scar tissue.

The Wake Forest Institute for Regenerative Medicine (WFIRM)[280] is engineering laboratory-grown organs and their successful implantation into humans. Already in 2006, the WFIRM reported the successful engineering of bladders that were grown from the patients' own cells, thereby avoiding the risk of rejection. The study involved young patients who had poor bladder function because of a congenital birth defect. Their bladders were not pliable and the high pressures could be transmitted to their kidneys, possibly leading to kidney damage. They had urinary leakage, as frequently as every 30 minutes. The WFIRM's director, Dr. Anthony Atala, reported that the bladders showed improved function over time—with some patients being followed for more than seven years.

Dr. Atala also directs the National Regenerative Medicine Foundation, which recently received funding from the federal government to create a Soldier Treatment and Regeneration Consortium to research how to treat burns and grow limbs for wounded soldiers.

In September 2014, the WFIRM reported a potential major regenerative medicine milestone, successful endothelial cell coating of the vessels in a human-sized pig kidney that kept the vessels open during a four-hour testing period. This research is part of a long-term project to use pig kidneys to make scaffolds that could potentially be used to build replacement kidneys for human patients with end-stage renal disease. After removing all animal cells from the organ—leaving only the organ structure or "skeleton"—a patient's own cells would then be placed in the scaffold, making an organ that the patient theoretically would not reject.

If proven successful, the new method could potentially be applied to other complex organs that scientists are working to engineer, including the liver and pancreas.

Given the global backlog in organ transplants and the high cost of transplantation medicine, the promise of regenerative medicine would provide significant patient and health economic benefits.

In the United States, kidney transplants cost about $150,000, liver transplants $250,000, and heart transplants $860,000.

In addition, after having received a transplant, patients have to take life-long daily doses of immunosuppressant drugs that frequently cause serious side effects.

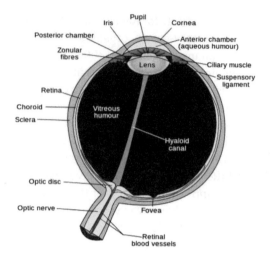

Figure 5.21 The human eye and its connection to the brain via the optic nerve. Source: Wikipedia.

Restoring vision, creating a bionic eye, is another important challenge addressed by several regenerative medicine research teams. Let's take a look at the human eye—Fig. 5.21—and see where nanomedicine can be used to restore lost function.

The human retina (Latin for *net*) is a light-sensitive layer of tissue lining the inner surface of the eye. Through the cornea and the lens, the retina receives an image of the visual world. Light striking the retina initiates a cascade of chemical and electrical events that ultimately trigger nerve impulses. These are sent to various visual centers of the brain through the fibers of the optic nerve. In human (and all vertebrate) embryonic development, the retina and the optic nerve originate as outgrowths of the developing brain.

Hence the retina is considered part of the CNS and is actually brain tissue.

The retina is a layered structure with several layers of neurons interconnected by synapses. There are two types of photoreceptor cells, rods and cones. Rods function mainly in dim light and provide black-and-white vision, while cones support daytime vision and the perception of color.

The eye's visual perception proceeds in four stages, namely, photoreception, transmission to bipolar cells, transmission to ganglion cells (which also contain photoreceptors, the photosensitive ganglion cells), and transmission along the optic nerve. At each synaptic stage there are also laterally connecting horizontal and amacrine (relay neuron) cells. The optic nerve is a central tract of many axons of ganglion cells connecting primarily to the lateral geniculate body, a visual relay station in the rear of the forebrain, and other parts of the brain. In adult humans, the entire retina is about 22 mm in diameter and contains about 7 million cones and 75 to 150 million rods.

Bionic eyes aim at restoring basic visual capabilities to people suffering from retinal defects such as retinitis pigmentosa, a genetic eye condition leading to irreversible destruction of retinal light receptors from the outside inward, from the center outward, or in sporadic patches, progressively reducing the ability of the eye to detect light. Another reason for reduction or loss of visual capabilities can be damage to the optic nerve as a result of glaucoma or head trauma. Blindness affects ~40 million persons worldwide.

Prosthetic devices[281] work at different levels downstream from the initial reception and biochemical conversion of incoming light photons at the back of the retina. Implants can stimulate the bipolar cells directly downstream of the photoreceptors, for example, or the ganglion cells that form the optic nerve. Alternatively, for pathologies such as glaucoma or head trauma that compromise the optic nerve's ability to link the retina to the visual centers of the brain, prostheses have been designed to stimulate the visual system at the level of the brain itself.

While brain prostheses have yet to be tested in people, clinical results with retinal prostheses have already been obtained.

One bionic eye strategy developed by Bionic Vision Australia (BVA), coordinated by the Bionic Institute in Melbourne,[282] is to use a retinal implant connected to a video camera, fitted to a pair of eyeglasses, to convert images into electrical impulses that activate remaining retinal cells, which then carry the signals back to the brain. BVA ran a series of preclinical animal studies between 2009 and 2012, demonstrating the safety and efficacy of a prototype suprachoroidal (see Fig. 5.21) implant, made up of a silicone carrier with 33 platinum disc-shaped electrodes that can be activated in various combinations to elicit the perception of rudimentary patterns, much like pixels on a screen. In 2012, BVA commenced a pilot trial with three end-stage retinitis pigmentosa patients who were barely able to perceive light. The electrode array was joined to a titanium connector affixed to the skull behind the ear, permitting neurostimulation and electrode monitoring without the need for any implanted electronics. In all three patients, the device proved stable and effective, providing enough visual perception to better localize light, recognize basic shapes, orient in a room, and walk through mobility mazes with reduced collisions. Future clinical trials are planned for the testing of an improved fully implantable device with twice the number of electrodes.

Similar projects are conducted at Retina Implant AG[283] in Reutlingen, Germany, a start-up with a system—Alpha IMS—approved in Europe for experimental clinical use, by the French company Pixium Vision[284] collaborating with l'Université Pierre et Marie Curie UPMC (Paris), Stanford University, and the University of Ulm, Germany; the Italian Institute of Technology in Genoa and Guglielmo Lanzani at the Institute's Center for Nanoscience and Technology in Milan; the Boston Retinal Implant project;[285] and the Nano Bioelectronics and Systems Research Center of Seoul National University in South Korea.

Osaka University, Japan, working with the Japanese company NIDEK, has been developing an intrascleral prosthetic device, which, unlike the Australian BVA devices, is implanted in between the layers of the sclera rather than in the suprachoroidal space. In a clinical trial of this device, often referred to as suprachoroidal-transretinal stimulation (STS), two patients with advanced retinitis pigmentosa showed improvement in spatial resolution and visual acuity over a four-week period following implantation.[286]

An emerging field frequently classified as regenerative medicine is *gene therapy*, the use of DNA as a drug to treat disease by delivering therapeutic DNA into a patient's cells. The most common form of gene therapy involves using DNA that encodes a functional, therapeutic gene to replace a mutated gene. Other forms involve directly correcting a mutation or using DNA that encodes a therapeutic protein drug (rather than a natural human gene) to provide treatment. In gene therapy, DNA that encodes a therapeutic protein is packaged within a vector, which is used to get the DNA inside cells within the body. Once inside, the DNA becomes expressed by the cell machinery, resulting in the production of therapeutic protein, which in turn treats the patient's disease.

Gene therapy was first conceptualized in 1972 when Friedmann and Roblin published a paper in *Science* entitled "Gene therapy for human genetic disease?".[287] However, the authors urged caution before commencing gene therapy studies in humans. The first FDA-approved gene therapy experiment in the United States occurred in 1990, when Ashanti DeSilva was treated for adenosine deaminase–severe combined immunodeficiency (ADA-SCID), an autosomal (not in a sex chromosome) recessive metabolic disorder that causes immunodeficiency. ADA-SCID occurs in less than 1 in 100,000 live births worldwide but accounts for about 15% of all cases of SCID.

Although early clinical failures led many to dismiss gene therapy as overhyped, clinical successes since 2006 have bolstered new optimism in the promise of gene therapy. By January 2014, about 2000 clinical trials had been conducted or had been approved using a number of techniques for gene therapy.

DNA must be administered to the patient, get to the cells that need repair, enter the cell, and express a protein. Generally the DNA is incorporated into an *engineered virus* that serves as a vector, to get the DNA through the bloodstream, into cells, and incorporated into a chromosome.

However, with development of our understanding of the function of nucleases (enzymes capable of cleaving the phosphodiester bonds between the nucleotide subunits of nucleic acids) such as *zinc finger nucleases* (ZFNs) in humans, efforts have begun to incorporate genes encoding nucleases into chromosomes; the expressed nucleases then "edit" the chromosome and disrupt genes causing disease.

A zinc finger is a small protein structural motif that is characterized by the coordination of one or more zinc ions in order to stabilize the fold. Since their original discovery and the elucidation of their structure, these interaction modules have proven extremely useful in various therapeutic and research capacities. Engineering zinc fingers[288] to have an affinity for a specific DNA sequence is an area of active research, and ZFNs and zinc finger transcription factors are two of the most important applications of this to be realized to date. A recent review of Gene therapy can be found in the March 2014 issue of Scientific American. The author concludes that recent advances have moved the experimental approach closer to being a mainstream treatment for some disorders such as acute lymphoblastic leukemia (ALL), other blood cancers, and head and neck cancer.[289]

Finally, *immunotherapy* is defined as the treatment of disease by inducing, enhancing, or suppressing an immune response. Immunotherapies can either be designed to elicit or amplify an immune response (activation immunotherapies) or to reduce or suppress an immune response.

The active agents of immunotherapy are collectively called immunomodulators. They are a diverse array of recombinant, synthetic and natural preparations, often cytokines, small proteins that are important in cell signaling.

Cell-based immunotherapies are proven to be effective for some cancers. Immune effector cells such as lymphocytes, macrophages, dendritic cells, natural killer cells (NK cells), cytotoxic T lymphocytes (CTLs), etc., work together to defend the body against cancer by targeting abnormal antigens expressed on the surface of the tumor due to mutation.

We will encounter examples of gene therapy and immunotherapy when discussing some of the emerging biotech companies in those exciting and promising new areas of regenerative medicine.

With so many promising research results, it is no wonder that regenerative medicine offers strong investment opportunities.

Clearly, Japan is looking to build on its early lead in the iPSC area, started by Nobel laureate S. Yamanaka, by creating in April 2010 the Center for iPS Cell Research and Application at Kyoto University. Its goals for 2020 have been specified by CiRA director Yamanaka as follows:[290]

- Establish basic iPSC technology and secure the associated intellectual property (IP) rights
- Build a stock of iPSCs for use in regenerative medicine
- Carry out preclinical studies and work toward clinical studies

- Contribute to the development of therapeutic drugs using patient-derived iPSCs

Another Kyoto-based research organization is the Institute for Frontier Medical Sciences, currently the only institute in Japan bearing the name "saisei ikagaku" (regenerative medicine). It has approximately 150 staff members, who conduct research on studies related to regenerative medicine, such as basic biology, stem cell biology, and engineering, as well as applied studies on regenerative medicine.

By retooling the framework of the regulatory process for regenerative medicine products, Japan has shifted from a very conservative drug development environment to one of the most proactive countries, aiming to expedite growth in the stem cell industry to meet the needs of its aging population. By trying to streamline regulatory approvals without sacrificing safety, Japan's prime minister Shinzo Abe is clearly aiming at Japanese leadership position in this exciting new area of nanomedicine.

In addition to the United States, Europe, and Japan, another global power positioned to invest in regenerative medicine is China. Since 2005, Shenzhen-based Chinese biotech company Beike Biotechnology has been specializing on research, clinical translation, and technology support services of adult stem cells. Beike's CEO, Xiang Hu, presented the following findings by the Chinese Academy of Sciences at the Shenzhen International Biotech Summit 2014:

- By 2030, the Chinese population will be older than the US population and 83% of healthcare resources will be spent on chronic conditions
- Regenerative medicine may reduce this burden; it is an area of high priority for Chinese science.

Statistical data collected by the Chinese Academy of Sciences:

- Global statistics point to US and European dominance: Worldwide 418 companies, of which 268 are in North America, 100 in Europe, 39 in Asia, and 11 in Australia + New Zealand.
- 56% of companies are focused on cell-based therapies, gene therapies, tissue-engineered biomaterials and scaffolds,

implantable devices, and small-molecule- and biologic-based therapies.

- 19% are focused on stem cell–based drug discovery tools (e.g., preclinical in vitro assays, toxicity testing, etc.).
- 13% are focused on "banking" of stem cells, in particular cord blood.
- 12% provide services.
- There are over 500 therapeutical products in the pipeline, 37 of them marketed (as of September 2014).
- The total global financing of regenerative medicine is $4.74 B (stem cells $1.872 B, gene therapy $1.85 B, and cell-based immunotherapy $580.7 B).

The Alliance for Regenerative Medicine's[291] (ARM) mission is to advance regenerative medicine by representing, supporting, and engaging all stakeholders in the field, including companies, academic research institutions, patient advocacy groups, foundations, health insurers, financial institutions, and other organizations. Based in Washington, DC, ARM is a global advocacy organization that promotes legislative, regulatory, reimbursement, investment, technical, and other initiatives to accelerate the development of safe and effective regenerative medicine technologies. ARM also works to increase public understanding of the field and its potential to transform human healthcare.

ARM was formed in 2009 and has more than 170 members worldwide. Its website is very well organized and provides an overview of ongoing efforts in the various disease areas.

Table 5.4 lists stem cell gene therapy and immunotherapy companies, ordered by market capitalization (as of September 2014), adding up to ~$5.4 B without counting Celgene and other established biopharmaceutical companies with ongoing research programs.

Celgene Corporation, founded in 1986, is a leader in oncology and immunotherapeutics that is also active in regenerative medicine. The company was not included in Table 5.4 because its market capitalization of $73.7 B (see Table 4.1) is dominated by products (integrating both small-molecule and cell-based therapies) outside the area of regenerative medicine. However, the Organ &

Table 5.4 Regenerative medicine companies, ordered by Market Cap

	Company name and HQ's location	Year founded	SC-stem cell, GT-gene th, IT-immuno-therapy	Stock symbol	NASDAQ	Other Stock exchange	Market cap (M)
1	Celgene (Summit, New Jersey)	1986	SC, IT +++	CELG	Y		$73,700
2	Mesoblast (Melbourne, Australia)	2001	SC, Cardiov, Ophtalm.	MBLTY	N	OTC	$1,430
3	Sangamo BioSciences (Richmond, California)	1995	GT, Hemophilia	SGMO	Y		$781
4	Organovo (San Diego, California)	2007	3D Bioprinting	ONVO	Y		$490
5	Osiris Therapeutics (Columbia, Maryland)	1992	SC, Skin, Bone, Cartil.	OSIR	Y		$444
6	EVOTEC (Hamburg, Germany)	1993	SC, Diabetes, Alzheimer	EVTCY	N	TecDAX, OTC	$439
7	Medipost (Seoul, South Korea)	N/A	SC, Blood, Cartil., CNS	078160.KQ	N	KOSDAQ	$430
8	AGTC (Alachua, Florida)	2001	GT, Ophtalmology	AGTC	Y		$302
9	Advanced Cell Technology (Marlborough, Massachusetts)	1994	SC, Retinal Disease	ACTCD	N	OTC	$262
10	BioTime (Alameda, California)	1990	SC, CNS, SCI, Cancer	BTX	N	NYSE	$229
11	Pluristem Therapeutics (Haifa, Israel)	2001	SC, Cardiov, PAD	PSTI	Y		$203
12	NeoStem (New York, New York)	1980	SC, IT Cancer etc	NBS	Y		$190
13	Argos Therapeutics (Durham, North Carolina)	1997	IT Cancer, HIV	ARGS	Y		$190
14	Celladon Corporation (San Diego, California)	2005	GT Heart Failure	CLDN	Y		$183
15	UniQure (Amsterdam, Netherlands)	2012	GT Rare diseases	QURE	Y		$160
16	Fibrocell Science (Exton, Pennsylvania)	N/A	SC Skin	FCSC	Y		$121
17	TiGenix-Cellerix (Leuven, Belgium and Madrid, Spain)	2000	SC Autoimmune D	TIG.BR	N	Brussels	$120
18	Cellular Dynamics International (Madison, Wisconsin)	1999	SC Services	ICEL	Y		$120
19	Athersys (Cleveland, Ohio)	1995	SC Cardiov, CNS	ATHX	Y		$108
20	StemCells Inc (Newark, California)	1998	SC CNS, Liver	STEM	Y		$89
21	Capricor (Beverly Hills, California)	2005	SC Cardiov	CAPR	N	OTC	$50
22	Aastrom Biosciences (Ann Arbor, Michigan)	1989	SC Cartilage, Burns	ASTM	Y		$21
23	STEMCELL Technologies (Vancouver, Canada)	1993	SC Services		N	Private	N/A
24	Gamida Cell (Jerusalem, Israel)	1998	SC (Blood)	(Novartis)	N	Private	N/A
25	ViaCyte (San Diego, California)	1999	SC, Diabetes	(J&J)	N	Private	N/A
	TOTAL (without Celgene)						$6,362

Tissue Therapeutics (OTT) unit of Celgene Cellular Therapeutics (CCT) is dedicated to developing innovative engineered tissue and biomaterial approaches for the many challenges in regenerative medicine. Cell therapy research teams focus on the discovery and early development of novel cell-based therapeutics. Celgene has developed a portfolio of clinical candidates with broad therapeutic potential, including human placenta–derived stem cells (HPDSCs), their ex vivo "natural killer" expansion into cancer immunotherapy, and amnion-derived adherent cells (AMDACs). Note that the amnion is a membrane covering the embryo, becoming the amniotic sac, filled with amniotic fluid, providing a protective environment for the developing embryo. Celgene has achieved industry leadership in adult stem cell isolation, cell culture, characterization, and functional interpretation and collaborates with international expert groups to elucidate and apply the complex mechanisms that are associated with live cell-based therapeutics.

Mesoblast acquired Angioblast (founded in 2001) in 2010 and is conducting clinical programs in the areas of congestive heart failure (CHF), acute myocardial infarction (AMI), diabetes, macular degeneration, and bone marrow transplantation.[292] Mesoblast's cell-based core technologies include its highly purified, immunoselected mesenchymal precursor cells (MPCs), culture-expanded mesenchymal stem cells (MSCs), dental pulp stem cells (DPSCs), and expanded HSCs. Mesoblast's allogeneic MPCs have been shown in preclinical studies to be capable of repairing the structure of intervertebral discs. Mesoblast is currently in clinical development of a nonsurgical adult stem cell treatment for patients with early disc degeneration using a single intradisc injection of allogeneic MPCs. HSCs are generally used autologously, to avoid the potential for recipient immune reactions to unrelated donor cells. Mesoblast is partnered with Israel's Teva Pharmaceuticals for the development and commercialization of its MPC products in a number of fields, including CVD and neurologic conditions. It partners with Swiss Lonza for the clinical and long-term commercial manufacturing of its allogeneic MPC products. Finally, Mesoblast has a partnership in place with JCR Pharmaceuticals (market cap $\approx$ \$920 M, known for a human growth hormone [HGH] product, but with new focus on regenerative medicine) for the exclusive use in Japan of Mesoblast's culture-expanded MSC technology for the treatment of hematologic malignancies using HSC transplants.

Sangamo BioSciences started in 1995 and went public in 2000 and is a global leader in specific regulation of gene expression and gene modification. The basis of this platform is a naturally occurring class of transcription factors, zinc finger DNA-binding proteins (ZFPs), which can be engineered to drive desired therapeutic outcomes by activating or repressing gene expression to create ZFP transcription factors (ZFP TFs) capable of turning genes on or off. ZFPs can also be linked to nuclease domains to create ZFNs, which enable precise gene-editing in cells. In 2012, Sangamo established a therapeutic partnership with Shire AG to develop ZFP Therapeutics for hemophilia, Huntington's disease, and other monogenic diseases. Sangamo is also conducting phase 2 clinical trials and two phase 1 trials to evaluate the first therapeutic application of a ZFN-modified T cell product for the treatment of human immunodeficiency virus (HIV)/acquired immunodeficiency syndrome (AIDS). In addition, Sangamo's clinical collaborators at City of Hope (COH) have initiated a phase 1 clinical trial to evaluate a ZFN-based therapeutic for the treatment of glioblastoma multiforme, a type of brain cancer. Other preclinical development programs of ZFP Therapeutics are focused on hemophilia B, PD, neuropathic pain, and neuroregenerative programs in spinal cord injury (SCI), traumatic brain injury, and stroke. Via its scientific advisory board, Sangamo has ties to the universities of Berkeley, Harvard, and Cambridge, UK.

Organovo started in 2007 and went public in 2012. Organovo designs and creates functional human tissues using a proprietary 3D bioprinting technology. The goal is to build living human tissues that are proven to function like native tissues. With reproducible 3D tissues that accurately represent human biology, Organovo partners with biopharmaceutical companies and academic medical centers to design, build, and validate more predictive in vitro tissues for disease modeling and toxicology, thereby enabling drug toxicology tests on functional human tissues and bridging the gulf between preclinical testing and clinical trials. Organovo is further working to fulfill its vision of building human tissues for surgical therapy and transplantation.

Osiris Therapeutics, started in 1992 as a spinoff from Case Western University, primarily uses stem cells obtained from human bone marrow. Prochymal, its first approved stem cell product, is based on ex vivo cultured human mesenchymal stem cells (hMSCs) derived from the bone marrow of healthy volunteers. However, in October 2013, Osiris sold the Prochymal assets to Mesoblast for an initial $50 M and potential subsequent payments. Osiris' current product line includes Grafix for acute and chronic wounds, Cartiform, a viable cartilage mesh for cartilage repair, and OvationOS, a viable bone matrix used for bone repair and regeneration.

Evotec, started in 1993 in Hamburg Germany, by a group of scientists including the chemistry Nobel Prize (1967) laureate Manfred Eigen,[293] started out as a small-molecule drug discovery tools company but has recently added regenerative medicine to its portfolio of tools and technologies. Evotec is specializing on partnerships with both biopharmaceutical companies and leading academic institutions and programs. After teaming with Novartis, GSK, Bayer and Teva, Evotec is now working Roche on AD, with AstraZeneca/MedImmune on kidney disease and diabetes, and with Janssen (Johnson & Johnson) on Diabetes and AD. The diabetes project includes a collaboration with Harvard University and the Howard Hughes Medical Institute in diabetes research (CureBeta), kidney disease, cancer, and antibacterials. In 2013, Evotec and Yale University formed the Open Innovation Alliance aimed at drug development and commercialization partnerships with Pharma companies. Evotec is also working with Hamburg University on a treatment for multiple sclerosis (MS). MS is an inflammatory and neurodegenerative disease in which the insulating covers of nerve cells (myelin sheath) in the brain and spinal cord are damaged. This damage disrupts the ability of parts of the nervous system to communicate, resulting in a wide range of signs and symptoms, including physical, mental, and sometimes psychiatric problems.

Medipost provides services and develops products related to a readily available and noncontroversial source—human umbilical cord blood and human umbilical cord blood–derived mesenchymal stem cells (hUCB-MSCs),[294] It is focused on developing products related to regenerative or functional recovery of knee articular cartilage, the nervous system, the pulmonary system, and hematopoi-etic transplantation engraftment areas. Hematopoietic stem cell transplantation (HSCT) is the transplantation of multipotent HSCs, usually derived from bone marrow, peripheral blood, or umbilical cord blood. It may be autologous (patient's own stem cells) or allogeneic (the stem cells come from a donor). Note that Medipost market capitalization was obtained by converting Korean Wons to USD.

Applied Genetic Technologies Corporation (AGTC) was started in 2001 by Barry Birne (Johns Hopkins), by Terence Flotte (UMMS), by William Hauswirth and Nicholas Muzyczka (both University of Florida, Gainesville), and by Jude Samulski (University of ophthalmology. AGTC's lead product candidates, which are each in the preclinical stage, focus on X-linked retinoschisis, achromatopsia, and X-linked retinitis pigmentosa, which are rare diseases of the eye, caused by mutations in single genes, that significantly affect visual function and currently lack effective medical treatments.

Advanced Cell Technology (ACT) was formed already in 1994 but has only recently generated wide attention by—in 2010—receiving FDA approval to initiate the first-ever human clinical trial using ESCs to treat retinal diseases and—in August 2014—announcing a 1:100 reverse split of its previous "penny stock," listed on the OTC Markets Group. ACT is primarily developing stem cell–based technologies, both adult and human embryonic, and other methods and treatments in the area of regenerative medicine, in particular regenerative ophthalmology. ACT's research team has strong connections to Wake Forest University.

BioTime was started in 1990 as a Berkeley University spinoff and is focused on regenerative medicine, with technologies centered on pluripotent stem cells and with focus on neuroscience, oncology, orthopedics, and blood and vascular diseases. Already marketed products include Renevia^{TM}, a hydrogel device for cell delivery, and PanC-Dx^{TM}, a blood-based pancancer-screening diagnostic. In 2009, BioTime secured \$4.72 M in grant funding from the California Institute for Regenerative Medicine (CIRM) to expand its ACTCellerate program, which it had licensed the previous year from ACT. In 2013, BioTime acquired the stem-cell assets of Geron

Corporation, with the aim of restarting its ESC-based clinical trial for SCI.

Pluristem Therapeutics (Haifa, Israel), was founded in 2001 by Shai Meretzki of the Technion, who made use of stem cell patents he had developed with colleagues from the Weizmann Institute of Science. Pluristem is focused on clinical use of its PLacental eXpanded (PLX) cells to treat peripheral arterial disease (PAD). Its PLX cells are adherent stromal cells that are expanded using a proprietary process, termed PluriX. PluriX uses a system of stromal cell cultures and substrates to create an artificial 3D environment where placental-derived stromal cells (obtained after birth) could grow.

In 2003, the NASDAQ-listed shell company A1 Software acquired all shares and patents belonging to Pluristem and changed its name to Pluristem Life Systems. In 2007 the name was changed again, this time to Pluristem Therapeutics Inc. In 2007, Pluristem entered into a collaborative research agreement with the Charité medical school in Berlin, Germany, for a period of five years. In 2012, the agreement was renewed until 2017. Also in 2012, Pluristem was awarded two grants ($2.4 and $3.2 M) from the Office of the Chief Scientist of Israel.

NeoStem (New York, NY), was founded in 1980 and was formerly known as Cornich Group Inc. (until 2003) and Phase III Medical Inc. (until 2006). Since 2006, Neostem has been led by physician and business executive Robin L. Smith. As NeoStem's CEO, she has successfully completed six acquisitions, one divestiture and has raised over $190 M for research and development, expansion of business units, and strategic transactions. In 2011, Neostem acquired Progenitor Cell Therapy (PCT), a spinoff from New Jersey's Hackensack Medical Center, cofounded in 1997 by Andrew L. Pecora and Robert A. Preti to provide stem cell and tissue processing services for cancer patients. In 2000, PCT and Dendreon Corp. signed the first in a series of contracts for PCT to manufacture PROVENGE (the first FDA-approved immunotherapy for advanced prostate cancer) and other cell therapy cancer vaccines for phase 1, 2, and 3 clinical trials. In 2004, PCT established Amorcyte, also to be acquired by Neostem in 2011. Amorcyte is focused on cell therapy for the treatment of CVD. In 2014 NeoStem acquired California

Stem Cell, creating the company's Targeted Cancer Immunotherapy Program, a first step toward cancer therapeutics. NeoStem has built a strong intellectual property (IP) position focused on immunology, cardiology, orthopedic, wound healing, stem cell isolation and purification, and very small embryonic-like (VSEL) patent rights and know-how, in-licensed from the University of Louisville and covering isolation, purification, and therapeutic use of VSEL stem cells. In September 2014, NeoStem announced the licensing of Rockefeller University's IP in the area of cancer immunotherapy.

Argos Therapeutics started in 1997 as Merix Bioscience and changed its name to Argos Therapeutics, Inc. in October 2004. Argos traces its roots to Rockefeller University, where cofounder and Nobel laureate Dr. Ralph Steinman discovered the role of dendritic cells in the immune system and developed a method to generate dendritic cells. The other Argus cofounders worked at Duke University, where they developed a unique RNA-based dendritic cell technology. Dendritic cells are the master switch that can turn the immune system on or off. They capture, process and identify antigens, or foreign bodies, and alert the body's immune response, activating antibodies and T cells to destroy the invaders. Arcelis is a fully personalized immunotherapy technology that captures mutated and variant antigens that are specific to each patient's disease. It is designed to overcome tumor- and disease-induced immunosuppression in cancer and HIV by eliciting a durable memory T cell response. Argos' most advanced product candidates include AGS-003 for the treatment of metastatic renal cell carcinoma, or mRCC, and AGS-004 for the treatment of HIV. Argos is well positioned to be a leader in the immunotherapy field.

Celladon, started in 2005, is a clinical-stage biotechnology (BT) in the field of gene therapy and calcium dysregulation to develop novel therapies for diseases with unmet medical needs. Celladon targets SERCA enzymes that play an integral part in the regulation of intra-cellular calcium in all human cells. Calcium dysregulation is implicated in a number of important and complex medical conditions and diseases, such as heart failure, vascular disease, diabetes, and neurodegenerative diseases. MYDICAR, the company's most advanced product candidate, uses gene therapy to target

SERCA2a, which is an enzyme that becomes deficient in patients with heart failure.

UniQure, having acquired AMT in 2012, is focused on delivering single-gene therapy treatments with potentially curative results. The geneQure platform has four key elements: (1) therapeutic gene cassettes that carry a transgene that encodes, or provides the blueprint for the expression of, a therapeutic protein; (2) an adeno-associated virus (AAV)-based vector delivery system for delivering the gene cassette; (3) administration technologies to effectively deliver the relevant transgene into the tissues and organs; And (4) a scalable, proprietary manufacturing process to produce our AAV-based vectors.

Glybera and all of UniQure's current product candidates are based on AAV vector technologies. Researchers have used AAV-based vectors in preclinical research and over 80 clinical trials in which AAV-based vectors have demonstrated a good safety profile. AAVs have also demonstrated lasting therapeutic gene expression following a single treatment in preclinical and clinical studies. In addition to monogenic inherited diseases, there are opportunities to apply gene therapy in diseases caused by more complex pathology in which one particular protein plays a crucial role in the causation of the disease. In such indications, such as some liver diseases, disorders of the CNS and CVD, it may be possible to halt or eradicate the disease with a gene therapy that promotes the natural production or function of the relevant protein. Glybera has been approved by EMA to treat lipoprotein lipase deficiency, or LPLD, a rare metabolic disease. UniQure has partnered with Chiesi Farmaceutici.

In August 2014, UniQure announced the acquisition of InoCard GmbH, an early-stage BT company focused on the development of gene therapy approaches for cardiac disease. InoCard has developed a novel gene therapy to preclinical proof of concept, for the one-time treatment of CHF.

Fibrocell Science, based in Exton, PA, is at the forefront of the field of personalized biologics. By extracting cells from skin to create localized therapies that are compatible with the unique biology of each patient, Fibrocell is cultivating and harnessing autologous fibroblast cells, which fuels its personalized biologics approach.

Fibrocell's proprietary technology uses a person's own fibroblast cells—with or without genetic modification—to target rare and serious skin and connective tissue diseases and conditions caused by genetic defects that cause protein deficiencies. It has a licensing agreement with UCLA.

TiGenix, started in 2000, is a leading European cell therapy company with an advanced clinical stage pipeline of adult stem cell programs. The stem cell programs are based on a validated platform of allogeneic expanded adipose-derived stem cells (eASCs) targeting autoimmune and inflammatory diseases. Focus on Crohn's disease, rheumatoid arthritis, and severe sepsis. The commercialized product ChondroCelect®, for cartilage repair in the knee, was the first approved cell-based product in Europe and is marketed and distributed by Swedish Orphan Biovitrum AB and Finnish Red Cross Blood Service. TiGenix merged with Spanish Cellerix in 2011.

Cellular Dynamics International (CDI) was cofounded by stem cell pioneer James A. Thomson and develops and manufactures fully functioning human cells in industrial quantities to precise specifications. Dr. Thomson's derivation of human ESCs was featured as *Science* magazine's "Scientific Breakthrough of the Year" in 1999, and work from Dr. Thomson's laboratory has been cited in *TIME* magazine's "Top 10 Discoveries of the Year" on three separate occasions, including the isolation of human ESCs (1998, #1), the isolation of human iPSCs (2007, #1), and the collaborative mapping of the human epigenome (2009, #2). CDI's proprietary iCell Operating System (iCell O/S) includes true human cells in multiple cell types (iCell products), human induced pluripotent stem cells (iPSCs), and custom iPSCs and iCell products (MyCell products). CDI is a global leader in iPSC technology and the world's largest producer of terminally differentiated human cells derived from iPSCs. In addition to its portfolio of iCell® products (i.e., cardiomyocytes, neurons, endothelial cells, hepatocytes), CDI offers MyCell Products from donor samples of CDI customers, including iPSC reprogramming, genetic engineering, and cell differentiation.

Athersys, founded in 1995 by Gil van Bokkelen and John Harrington, is developing MultiStem, a patented, adult stem cell–derived product platform, for multiple disease indications in the areas of inflammatory and immune, neurological, and CVD. By 2014, Athersys is conducting two ongoing clinical studies—a phase 2 study in ulcerative colitis being conducted with Pfizer and a phase 2 study in ischemic stroke—and overall five clinical stage programs in inflammatory bowel disease, stroke, graft-versus-host disease, AMI, and solid organ transplant.

StemCells Inc., founded in 1998, is focusing on the CNS and the liver, seeking to address unmet medical needs through the development of stem cells as therapeutic agents to treat damage to or degeneration of major organ systems. Its scientific advisory board includes Irving Weissman, Fred Gage, and David Anderson, who were responsible for the discovery of the mammalian blood, brain and peripheral nervous system stem cells, respectively. They continue to make significant contributions to the company's research. Nobuko Uchida is vice president of stem cell biology and responsible for StemCell's discovery initiative focusing on identifying new stem or progenitor cells, as well as for characterizing human neural stem cell and candidate liver and pancreas stem cells. She was the first to identify, by a combination of cell surface markers, the human neural stem cell. Ann Tsukamoto, a codiscoverer of the human HSCs while at SyStemix, Inc. (acquired by Novartis in 1997), is vice president of scientific and strategic alliances. StemCells, Inc. claims ownership of the preeminent neural stem cell patent portfolio, covering broadly human neural stem cells irrespective of whether they were derived from embryonic, juvenile, or adult tissue or derived using presently known iPS technologies.

Capricor, established 2013 by merger with Nile Therapeutics (start-up year 2005), is a clinical-stage BT company developing cardiac-derived stem cell therapeutics. Capricor is the only stem cell company working to develop and commercialize cardiac stem cell therapies based on a proprietary mixture of 3D cells from the heart itself. Under the guidance of Eduardo Marban, previously at Johns Hopkins and now directing the Cedars-Sinai Heart Institute, Capricor's allogeneic cardiosphere-derived cell (CDC) product, CAP-1002, aims to attenuate and potentially improve damage to the heart caused by a heart attack. Capricor has received ~\$20 M of financial support by the CIRM and has signed a partnership agreement with Johnson & Johnson's Janssen Biotech[295] in January 2014. Janssen agreed to pay Capricor $12.5 M up front, and up to $325 M if Janssen exercises option rights, plus royalties on commercial sales of CAP-1002.

Aastrom Biosciences was founded in 1989 and is dedicated to the development of patient-specific expanded cellular therapies applied to severe diseases and conditions. Aastrom markets two autologous cell therapy products in the United States—Carticel, an autologous chondrocyte implant for the treatment of articular cartilage defects in the knee, and Epicel, a permanent skin replacement for full thickness burns greater than or equal to 30% of total body surface area. Aastrom is also developing MACI, a third-generation autologous chondrocyte implant for the treatment of cartilage defects in the knee, and ixmyelocel-T, a patient-specific multicellular therapy for the treatment of advanced heart failure due to ischemic dilated cardiomyopathy (DCM).

STEMCELL Technologies is a privately held, global BT company headquartered in Vancouver, British Columbia, Canada. The company develops and markets specialized cell culture media, cell separation products, instruments, and other reagents for life sciences research. Founded by Allen Eaves in 1993, STEMCELL Technologies currently has more than 500 employees and distributes over 1500 products.

In August 2014, Cytori Therapeutics (San Diego, CA) halted trials of its experimental stem cell therapy for heart failure after three patients developed blood flow problems. As a consequence its stock suffered a significant decline and the company has to go through a restructuring phase.

Cytori began operations under the name MacroPore Biosurgery (Frankfurt Stock Exchange) in 1996, following the invention of a line of bioresorbable plates and screws for craniofacial surgery—based on polylactic acid polymer technology. By 2000, a commercialization partnership had been formed with Medtronic and two more product lines based on the same technology were developed, one for spinal implants and another for adhesion barriers. Between 2000 and 2002, MacroPore began looking at alternative opportunities in regenerative medicine, including a start-up stem cell company, StemSource, Inc., which commercialized the processing and banking of adult stem cells originating from adipose (body fat) tissue.

In 2002, MacroPore sold commercialization rights to the craniofacial product line to Medtronic. Later in 2002, MacroPore acquired StemSource and began the development of the Celution System, based on autologous adipose-derived stem and regenerative cells (ADRCs), to treat CVD and repair soft tissue defects along with research of clinical applications for the device's stem and regenerative cell output. In 2005, MacroPore changed its name to Cytori and entered into a joint venture with Olympus Corporation of Japan.

Other promising start-up companies in the field of regenerative medicine are as follows.

Gamida Cell (Jerusalem, Israel), founded in 1998, started out with stem cell technology licensed from and jointly developed with Hadassah University Medical Center, Jerusalem, Israel. Cofounder Tony Peled is still Gamida Cell's chief scientific officer. She has made the discoveries that have led to Gamida Cell's key patents and is leading the development of Gamida Cell's pipeline of products. She has 15 years of experience in hematopoiesis and stem cell research.

One of the great limitations of therapies based on stem cells derived from umbilical cord blood is the small number of stem/progenitor cells available in any given tissue. This small number severely limits the ability to manufacture a therapeutic dose. To overcome this limitation, Gamida Cell has developed proprietary platform technologies to increase the number of stem cells in culture with limited differentiation—the goal being to create new stem cells identical to those in the original population. Gamida Cell's epigenetic technologies for expansion of populations of functional hematopoietic (blood) progenitor cells (HPCs) utilize small molecules (high-affinity copper chelator tetraethylpentamine [TEPA] and nicotinamide [NAM], a form of vitamin B_3) to modulate differentiation, homing, and the engraftment abilities of cultured cells. Gamida Cell has partnered with Teva and Amgen and is majority owned by Elbit Imaging Ltd., a holding company with diverse business interests, including real estate, medical imaging, hotels, shopping malls, and retail. In August 2014, Gamida Cell's CEO Yael Margolin announced an agreement with Novartis, starting with an initial investment by Novartis to acquire 15% of the company for

$35 M, with an option to acquire the remaining 85% for another $165 M by year-end 2015.

Viacyte (San Diego, California), was founded in 1999 as NovoCell and merged with Cythera and Bresagen in 2004. Cythera was an ESC research company and Bresagen was the first Australian Biotech company to clone a pig in 2001. In 2006, Viacyte published first results of engineered insulin-secreting islet cells. In 2008, Viacyte started two important collaborations, (i) with Pfizer on drug discovery and (ii) with Nobel laureate S. Yamanaka to work on human iPSCs. The name change to Viacyte occurred in 2010. ViaCyte is a private venture-backed company focused on profoundly improving the way patients with diabetes are treated.

ViaCyte's diabetes therapy, VC-01, combines a pancreatic endoderm cell product (PEC-01) with a proprietary encapsulation medical device (Encaptra drug delivery system). The VC-01 combination product is aimed at providing a superior solution to insulin therapy, avoiding the problems of frequent glucose monitoring and insulin injections, and the risk of unsafe hypoglycemic events caused by too high doses of insulin. In preclinical experiments, PEC-01 cells produce insulin and other functions characteristic of endocrine cells, including the production of glucagon, amylin, somatostatin, and other factors. Encaptra is an immune-protecting and retrievable encapsulation medical device. The VC-01 combination product is expected to be implanted under the skin of the patient through a simple outpatient surgical procedure. The cells are then expected to further differentiate to produce mature pancreatic cells that will synthesize and secrete insulin and other factors, thereby regulating blood glucose levels. Figure 5.3 (Section 5.1) above illustrated how a healthy pancreas is regulating blood glucose levels via insulin and glucagon. Preclinical studies with mice have been successful. In August 2014, ViaCyte received FDA approval to begin evaluation in human clinical trials. On August 21, ViaCyte entered into an agreement with Janssen R & D (a part of Johnson & Johnson) that provides Janssen with a future right to evaluate a transaction related to the VC-01™ combination product that ViaCyte is developing for T1D.

Below are a few more interesting regenerative medicine start-ups.

Voyager Therapeutics (Cambridge, MA) was founded in 2014 by the following world leaders in the fields of AAV gene therapy, expressed RNA interference, and neuroscience: Krystof Bankiewicz, professor of neurological surgery and neurology, University of California San Francisco (UCSF); Guangping Gao, professor and director of the Gene Therapy Center at the University of Massachusetts Medical School (UMMS); Mark Kay, professor and head of Human Gene Therapy, Stanford University; and Philipp Zamore, professor and codirector of the RNA Therapeutics Institute, UMMS. Led by CEO Stephen Paul, a distinguished leader for 35 years in the field of neuroscience, including the presidency of Lilly Research Labs, Voyager will develop life-changing gene therapies for fatal and debilitating diseases of the CNS, including PD, a monogenic form of amyotrophic lateral sclerosis (ALS), and Friedreich's ataxia, an autosomal recessive inherited disease that causes progressive damage to the nervous system and results from the degeneration of nerve tissue in the spinal cord. Voyager has formed a broad strategic collaboration with UMMS and has entered into license and other agreements with UMMS, UCSF, and Stanford to access relevant technology and data. Voyager is a private company with $45 M in Series A venture capital financing from Third Rock Ventures.

Juno Therapeutics (Seattle, WA), was founded in 2013 by Fred Hutchinson Cancer Research Center and Memorial Sloan-Kettering Cancer Center, along with pediatric partner Seattle Children's Research Institute, to advance a broad pipeline of immunotherapies. Backed by two large investments ($134 M in April 2014 and $176 M in August 2014) adding up to a total of $210 M, Juno will develop immunotherapy platforms that harness the potency of memory T cells, redirecting them to targets expressed on or in cancer cells. Using synthetic receptors and/or augmented natural antigen receptors, Juno's T cell reprogramming technologies enable the creation of a powerful antitumor immune response built from the patient's own immune system. This transformative approach has the potential of reducing or eliminating the need for debilitating surgery, radiation, and chemotherapy.

Two of Juno's scientific founders, Drs. Michel Sadelain and Renier J. Brentjens of Memorial Sloan Kettering Cancer Center, have won the New York Intellectual Property Law Association (NYIPLA)

"Inventor of the Year" award for their design of chimeric antigen receptors (CARs). The CAR technology targets cell surface antigens that are expressed on cancer cells. In addition, the high-affinity TCR technology can also detect alterations in intracellular proteins present in tumor cells. These treatments have the potential to reduce longer-term toxicities associated with chemotherapeutics. Juno's goal is to drive multiple product candidates in select hematologic and solid tumor cancers to FDA licensure. Each product candidate has the potential to treat a variety of high-risk cancers.

Altor Bioscience Corporation (Miami, FL) was formed in 2002 by Hing C. Wong, as a spin-off from Sunol Molecular Corporation (Sunol), and is engaged in the discovery and development of high-value, targeted immunotherapeutic agents for the treatment of cancer, viral infections, and inflammatory and autoimmune diseases. Altor is backed by leading venture capital funds, including Sanderling Ventures and the Florida Growth Fund. Altor currently has three products in clinical development:

(1) ALT-801, a T cell receptor-targeted immunotherapeutic developed from the STAR™ platform technology for cancer, concluded a phase 1/2a clinical trial in patients with metastatic malignancies in 2009 and a phase 2 trial for ALT-801 in combination with cisplatin for patients with metastatic melanoma in 2012

(2) Altor's second product ALT-836, a mab-based tissue factor antagonist, completed a second 90-patient, multi-center, randomized, placebo-controlled, phase 2 trial for treatment of patients with acute respiratory distress syndrome and acute lung injury, life-threatening, systemic inflammatory diseases.

(3) Altor's ALT-803, a novel IL-15 superagonist protein complex, has entered first-in-human phase 1 clinical trials for treatment of metastatic melanoma, refractory multiple myeloma, BCG-naive nonmuscle invasive bladder cancer and relapse of hematologic malignancy after allogeneic stem cell transplantation.

The decisions by Celgene, Teva, Amgen, Novartis, Pfizer, Johnson & Johnson, and other major global biopharmaceuticals to invest in regenerative medicine provides validation to the promise of this exciting new field.

Chapter 6

Impact of Nanomedicine

In this chapter we will discuss the impact that nanomedicine is expected to have on specific disease areas and selected areas of importance for medicine.

While DNA and RNA sequencing are expected to have the biggest impact as we are entering the era of genomic medicine, other nanomedical advances such as targeted drug delivery, nanodiagnostics, and regenerative medicine will not be far behind, in particular when combined with medical devices and micro-/nanoprocessor-enabled implants.

6.1 The Gut Microbiome

Dis-moi ce que tu manges, je te dirai ce que tu es.

(Tell me what you eat and I will tell you what you are.)

(Der Mensch ist, was er ißt.)

> —Anthelme Brillat-Savarin in *Physiologie du Gout* (1826)

In 2001, Nobel laureate Joshua Lederberg[296] coined the term "human microbiome," thereby emphasizing the importance of

Nanomedicine: Science, Business, and Impact
Michael Hehenberger
Copyright © 2015 Pan Stanford Publishing Pte. Ltd.
ISBN 978-981-4613-76-7 (Hardcover), 978-981-4613-77-4 (eBook)
www.panstanford.com

nutrition for human health and acknowledging the fact that the average human body contains 10 trillion cells but houses 10 times more cells in the gut, the skin, the mouth, etc.[297]

Lederberg received the 1958 Nobel Prize for physiology or medicine (at the young age of 33 years) for his discoveries concerning genetic recombination and the organization of the genetic material of bacteria.

Estimates of the weight of the human microbiome are ranging between 0.2 and 1.6 kg compared to the brain's ~1.4 kg. About 100 mg are located at the tongue and at our dental plaque.

About 80 billion bacteria are swallowed per day in our saliva.

Finally, there are about 3.3 million genes in the microbiome, more than 150 times the number of genes in the human genome.

Increasingly, it is believed that the human microbiome is playing an important role in autoimmune diseases like juvenile (type I) diabetes, rheumatoid arthritis, muscular dystrophy, multiple sclerosis (MS), fibromyalgia, and perhaps some cancers. The predisposition to such diseases may be inherited, but the triggering factor may be microbiome related. A poor mix of microbes in the gut may also aggravate common obesity. Since some of the microbes in our body can modify the production of neurotransmitters known to be found in the brain, the composition of the gut microbiome may also have an effect on schizophrenia, depression, bipolar disorder and other neurochemical imbalances.

It is well known that certain components of our diet can lead to responses that can cause allergies or even serious medical conditions. We also know that certain conditions occur more frequently in developed countries with high levels of hygiene, comprehensive immunization programs and frequent use of antibiotics to fight infectious disease. Could significant changes in the gut microbiome be at least partly responsible for the increase in autoimmune diseases?

Celiac disease provides a good example: While this condition is thought to affect about one in 1750 worldwide, it is ~17 times much more prevalent in the United States. Celiac (or "Coeliac") disease is caused by a reaction to gliadin, a gluten protein found in wheat, and similar proteins found in other common grains, such as barley and rye. Upon exposure to gliadin, the enzyme transglutaminase

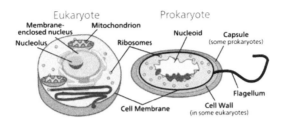

Figure 6.1 Eukaryotes and prokaryotes (archaea and bacteria). Source: Wikipedia.

modifies the protein, and the immune system cross-reacts with the small-bowel tissue, causing an inflammatory reaction, which in turn destroys the small intestine lining (called villous atrophy) and interferes with the absorption of nutrients. If not diagnosed, celiac disease causes vitamin deficiencies, weight loss and failure to grow in children, fatigue, iron deficiency, and ultimately inability to even walk. So far, the only known effective treatment is a lifelong gluten-free diet. Fortunately, unprocessed meats, poultry, fish, beans, nuts, vegetables, and fruits are naturally gluten free and food supermarkets have loaded store shelves with gluten-free foods, labeled as such.

To elucidate the human microbiome, we have to identify the members of a microbial community that includes bacteria, eukaryotes, and viruses, as shown in Fig. 6.1.

And to do so, we have to rely heavily on nanomedicine, in particular sequencing technologies that were developed as part of the Human Genome Project (HGP), as described above in Chapter 3.

Archaea are single-celled microorganisms without cell nucleus or any other membrane-bound organelles in their cells. As opposed to bacteria, archaea possess genes and several metabolic pathways, in particular related to enzymes involved in transcription and translation. Archaea reproduce asexually by binary fission, fragmentation, or budding. They are particularly numerous in the oceans, the archaea in plankton being one of the most abundant groups of organisms on the planet. Methanogens are microorganisms that produce methane as a metabolic by-product. They are common in wetlands, where they are responsible for marsh gas, and in the digestive tracts of plant-eating animals and humans, where they are

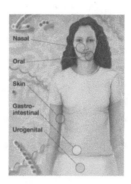

Figure 6.2   HMP analysis of microbial genomes from five sites in the human body. Source: Wikipedia.

responsible for the methane content of flatulence (flatus expelled through the anus) in humans.

Identification of the human microbiome members is done primarily using DNA-based studies, complemented with studies of RNA, proteins, and metabolites. DNA-based microbiome studies typically are *metagenomic* studies (see definition in Section 3.10).

The Human Microbiome Project (HMP) is a US National Institutes of Health (NIH) initiative[298] with the goal of identifying and characterizing the microorganisms that are found in association with both healthy and diseased humans (the human microbiome). The first HMP phase (2007–2012) characterized the composition and diversity of microbial communities (see Fig. 6.2) and evaluated their genetic metabolic potential.

The second HMP phase (2013–2015) is focused on the creation of the first integrated dataset of biological properties from both the microbiome and host from cohort studies of microbiome-associated diseases. The HMP can be considered as an extension of the HGP and HMP goals have been defined in similar ways:

- To develop a reference set of microbial genome sequences and to perform preliminary characterization of the human microbiome
- To explore the relationship between disease and changes in the human microbiome

- To develop new technologies and tools for computational analysis
- To establish a resource repository
- To study the ethical, legal, and social implications of human microbiome research

Another HMP goal is to break down the pre-existing barriers between medical and environmental microbiology (see the Earth Microbiome Project [EMP] below). It is hoped that the HMP will not only identify new ways to determine health and predisposition to diseases but also define the parameters needed to design, implement, and monitor strategies for intentionally manipulating the human microbiota, to optimize its performance in the context of an individual's physiology. In other words, we can expect a new focus on nutritional science.

Organized characterization of the human microbiome is also being done internationally under the auspices of the International Human Microbiome Consortium.[299]

Finally, the Earth Microbiome Project (EMP)[300] is an initiative to collect natural samples and to analyze the microbial community around the globe. The EMP's steering committee includes representatives from Argonne National Laboratory and University of Chicago, University of California at Davis, University of Southern California, Lawrence Berkley National Laboratory, University of Colorado at Boulder, Center for Ecology and Hydrology (UK), and the Netherlands Institute of Ecology.

The EMP's key goal is to survey microbial composition in many environments across the planet, across time as well as space, using a standard set of protocols. The EMP is developing the Global Environmental Sample Database (GESD). This database will hold information about the types of environmental samples that are currently available globally for the EMP.

It remains to be seen if enough EMP funding will be available to realize this ambitious project.

In conclusion, evidence is mounting that our microbial guests are playing an important role for our wellness and can significantly influence the outcome of therapeutic interventions, across disease areas such as cancer, obesity and type II diabetes, rheumatoid arthritis, possibly also type 1 diabetes (T1D), and even malnutrition.[301]

6.2 Central Nervous System: Brain and Spinal Cord

The human body is the best picture of the human soul.

—Ludwig Wittgenstein (1889–1951)

The central nervous system (CNS) is the part of the nervous system consisting of the brain and the spinal cord. As shown in Fig. 6.3, the peripheral nervous system (PNS) is composed of nerves leading to and from the CNS.

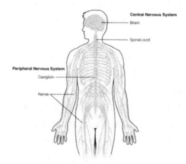

Figure 6.3 Central (pink) and peripheral nervous systems. Source: Wikipedia.

The CNS is so named because it integrates information it receives from, and coordinates and influences the activity of, all parts of the human body. Included in the CNS are also the retina and the optic nerve, as well as the olfactory nerves (1st) and olfactory epithelium inside the nasal cavity, connecting (synapsing) directly with brain tissue without intermediate ganglia. Hence the olfactory epithelium is the only central nervous tissue in direct contact with the environment, a fact that is of relevance for therapeutic treatments.

Before focusing on CNS diseases and the impact of nanomedicine in diagnosing them and potentially developing cures, let's briefly review a few facts about the brain. It is safe to say that understanding the brain will be the next big frontier of life sciences and medicine.

Several Grand Science projects, very similar in their level of ambition to the 15-year HGP 1988–2003, have been started:

- *Human Brain Project*:[302] This European Union initiative is a large 10-year scientific research project, established in 2013, directed by the École polytechnique fédérale de Lausanne (EPFL), Switzerland, and aimed at simulating the complete human brain on supercomputers. Among other goals, the project aims to build a full computer model of a functioning brain to simulate drug treatments. The project is coordinated by Professor Henry Markram, EPFL.

- *BRAIN*:[303] The U.S.-based Brain Research through Advancing Innovative Neurotechnologies (BRAIN) initiative was announced by the Obama administration on April 2, 2013, with the goal of mapping the activity of every neuron in the human brain. BRAIN has been projected to cost more than $300 M per year for 10 years.

- *Human Connectome Project* (HCP):[304] The HCP is a 5-year project sponsored by 16 departments of the NIH, split between two consortia of research institutions. The project was launched in July 2009 as a Grand Challenge for Neuroscience Research. The goal of the HCP is to build a network map that will shed light on the anatomical and functional connectivity within the healthy human brain, as well as to produce a body of data that will facilitate research into brain disorders such as dyslexia, autism, Alzheimer's disease (AD), and schizophrenia. HCP will greatly advance the capabilities for imaging and analyzing brain connections and will lead to major advances in our understanding of what makes us uniquely human.

Here is a glimpse of our current knowledge, soon to be augmented by new insights as the above-mentioned significant global investments will start producing results.

6.2.1 *The Human Brain*

The brain is what makes humans special. There are three important facts that have made humans what we are, and they have shaped our brain:

- The opposable thumb in conjunction with our ability to walk upright on two legs has made it possible to develop complicated tools and weapons, and to construct objects with high precision.
- Our 3D visual system, although inferior in acuity to some highly specialized birds of prey, has made it possible for humans to effectively hunt, to estimate distances, and to quickly recognize shapes.
- Our ability to communicate by speech and develop language has made it possible to develop socially, to teach and to learn, and to develop "culture," the capacity to classify and represent experiences with symbols, and to act imaginatively and creatively.

The brain is the organ serving as the center of the nervous system. The brain is the most complex organ in the human body. Its largest part, the cerebral cortex, is estimated to contain 15–33 billion neurons, each connected by synapses to several thousand other neurons. These neurons communicate with one another by means of long protoplasmic fibers called axons, which carry trains of signal pulses called action potentials to distant parts of the brain or body. The function of the brain is to exert centralized control over the other organs of the body. The brain acts on the rest of the body both by generating patterns of muscle activity and by driving the secretion of chemicals called hormones. This centralized control allows rapid and coordinated responses to changes in the environment. Some basic types of responsiveness such as reflexes can be mediated by the spinal cord or peripheral ganglia, but sophisticated purposeful control of behavior based on complex sensory input requires the information integrating capabilities of a centralized brain.

6.2.1.1 Brain function

Historically, the mind was thought to be separate from the brain. However, we have now proof that there is a close relationship between brain activity and mind activity or conscience, leading neuroscientists to be materialists, believing that mental phenomena are ultimately reducible to chemistry and biology. The operations of individual brain cells are now understood in considerable detail, but we still need to learn more about the way they connect and cooperate. The most promising approaches treat the brain as a biological computer that acquires information from the surrounding world, stores it, and processes it in a variety of ways. Specific brain functions have been found to be localized in various part of the brain. What distinguishes the human brain is the highly developed cerebral cortex, a thick layer of neural tissue that covers most of the brain. This layer is folded in a way that increases the amount of surface that can fit into the available volume. The pattern of folds is similar across individuals, although there are many small variations. The cortex is divided into four lobes, called the frontal lobe, parietal lobe, temporal lobe, and occipital lobe, as shown in Fig. 6.4a. Within each lobe are numerous cortical areas, each associated with a particular function, including vision, motor control, and language.

6.2.1.2 Left and right brain

The left and right sides of the cortex are broadly similar in shape, and most cortical areas are replicated on both sides. Some areas, though, show strong lateralization, particularly areas that are involved in language. In most people, the left hemisphere is dominant for language, with the right hemisphere playing only a minor role. There are other functions, such as spatiotemporal reasoning, for which the right hemisphere is usually dominant. Each hemisphere of the brain interacts primarily with one half of the body, but the connections are crossed: the left side of the brain interacts with the right side of the body, and vice versa. Motor connections from the brain to the spinal cord, and sensory connections from the spinal cord to the brain, both cross the midline at the level of the brainstem.

The brain's cortical areas are shown in Fig. 6.4b.

6.2.1.3 The cerebral cortex

The functions of the cortex are divided into three functional categories of regions. One consists of the primary sensory areas, which receive signals from the sensory nerves and tracts by way

Lateral View of the Brain

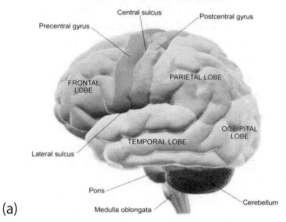

(a)

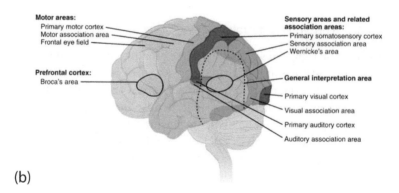

(b)

Figure 6.4 (a) Four lobes of the cerebral cortex. (b) Functional areas of the cerebral cortex. Source: Wikipedia.

of relay nuclei in the thalamus (a midline symmetrical structure of two halves, situated between the cerebral cortex and the midbrain). Primary sensory areas include the visual area of the occipital lobe, the auditory area in parts of the temporal lobe and insular cortex, and the somatosensory cortex in the parietal lobe. A second category is the primary motor cortex, which sends axons down to motor neurons in the brainstem and spinal cord. This area occupies the rear portion of the frontal lobe, directly in front of the somatosensory area. The third category consists of the remaining parts of the cortex, which are called the association areas. These areas receive input from the sensory areas and lower parts of the brain and are involved in the complex processes of perception, thought, and decision making.

6.2.1.4 Language

The Austrian philosopher Ludwig Wittgenstein has been quoted as saying that "language is a part of our organism and no less complicated than it." The production of language has been linked to Broca's area since Pierre Paul Broca reported impairments to speak in two patients after injury to the posterior inferior frontal gyrus of the brain. Functional magnetic resonance imaging (fMRI) studies have confirmed via activation patterns in Broca's area that it is associated with various language tasks. Patients with Broca's aphasia are individuals who are typically able to comprehend words and sentences with a simple syntactic structure but are more or less unable to generate fluent speech.

Wernicke's (speech) area, on the other hand, is involved in the understanding of written and spoken language. It is located in the posterior section of the superior temporal gyrus (STG) in the dominant cerebral hemisphere, which is the left hemisphere in about 95% of right-handed individuals and 60% of left-handed individuals. Recent research[305] largely confirms Broca's and Wernicke's observations but adds additional insights extending our ability to speak and to understand speech into left hemisphere brain regions remote from the narrowly defined traditional Broca and Wernicke areas.

6.2.1.5 Thalamus

The thalamus is located in the forebrain superior to the midbrain, near the center of the brain, with nerve fibers projecting out to the cerebral cortex in all directions. Some of its functions are the relaying of sensory and motor signals to the cerebral cortex, and the regulation of consciousness, sleep, and alertness. Fatal familial

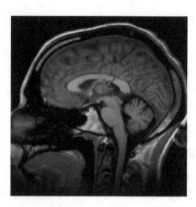

Figure 6.5 MRI cross section of the brain with the thalamus marked in red. Source: Wikipedia.

insomnia is a hereditary prion disease in which degeneration of the thalamus occurs, causing the patient to gradually lose his ability to sleep and progressing to a state of total insomnia, eventually leading to death. Prions[306] are infectious agents composed of protein in a misfolded form. The thalamus plays a major role in regulating the level of awareness and activity. Damage to the thalamus can lead to permanent coma. The thalamus is functionally connected to the hippocampus as part of the extended hippocampal system. A cerebrovascular accident (stroke) can lead to the thalamic syndrome, involving a one-sided burning or aching sensation often accompanied by mood swings.

Figure 6.5 shows a cross section obtained by MRI with the thalamus clearly marked by a red arrow.

6.2.1.6 Hippocampus

The hippocampus (named after its resemblance to the seahorse) is a major component of the human brain and of the vertebrate (species with backbones) brain. We have two hippocampi, one in each side of the brain. The hippocampus has a great deal to do with the formation of memories and plays important roles in the consolidation of information from short-term memory to long-term memory. Our improved and deeper understanding of its significant

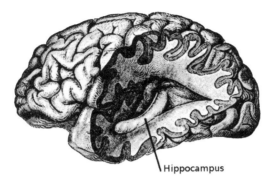

Hippocampus

Figure 6.6 The hippocampus is located under the cerebral cortex. Source: Wikipedia.

role in managing spatial navigation has been recognized by the Nobel Prize for physiology or medicine in 2014.[307] John O'Keefe, May-Britt Moser, and Edvard I. Moser received this award for their discoveries of cells that constitute a positioning system in the brain. As shown in Fig. 6.6, the hippocampus is located under the cerebral cortex, underneath the cortical surface.

Hippocampal place cells, which fire action potentials when the head passes through a specific part of its environment, interact extensively with head direction cells, whose activity acts as an inertial compass, and conjecturally with grid cells in the neighboring entorhinal cortex. It is an interesting fact that the navigation mechanisms used by rodents and by humans are very similar. It is therefore possible to study rats as model systems for neurophysiology. The form of neural plasticity known as long-term potentiation (LTP) was first discovered to occur in the hippocampus and is believed to be one of the main neural mechanisms by which memory is stored in the brain.

In Alzheimer's disease (AD), the hippocampus is one of the first regions of the brain to suffer damage; memory loss and disorientation are included among the early symptoms. Damage to the hippocampus can also result from oxygen starvation (hypoxia), encephalitis, or medial temporal lobe epilepsy. People with extensive, bilateral hippocampal damage may experience anterograde amnesia—the inability to form or retain new memories.

6.2.1.7 Brain damage and disease

Despite being protected by the thick bones of the skull, suspended in the cerebrospinal fluid and isolated from the bloodstream by the blood–brain barrier (BBB) (see Section 4.3), the human brain is susceptible to damage and disease. The most common forms of physical damage are closed head injuries such as a blow to the head, a stroke, or poisoning by a variety of chemicals that can act as neurotoxins. Infection of the brain, though serious, is rare due to the biological barriers that protect it. Degenerative disorders and other anomalies, such as Parkinson's disease (PD), multiple sclerosis (MS), AD, schizophrenia, depression, etc., are studied extensively but still not well enough understood to develop effective therapeutics.

6.2.1.8 Brain metabolism

The brain consumes up to 20% of the energy used by the human body, more than any other organ. Brain metabolism normally relies upon blood glucose as an energy source, but during times of low glucose (such as fasting, exercise, or limited carbohydrate intake), the brain will use ketone bodies for fuel with a smaller need for glucose. The brain can also utilize lactate during exercise. Long-chain fatty acids cannot cross the BBB, but the liver can break these down to produce ketones. However, the medium-chain fatty acids octanoic and heptanoic acids can cross the barrier and be used by the brain. The brain stores glucose in the form of glycogen, albeit in significantly smaller amounts than that found in the liver or skeletal muscle. Although the human brain represents only 2% of the body weight, it receives 15% of the cardiac output, 20% of total body oxygen consumption, and 25% of total body glucose utilization.

6.2.1.9 Dementia and Alzheimer's disease: a global health priority

In 2008, the WHO declared dementia as a priority condition. The total number of people with dementia in 2010 was estimated at ~36 million and is projected to reach ~66 million in 2030 and ~115 million in 2050. Caring for dementia patients costs the world more than \$604 B per year, of which only 15% is for medical care and the rest for informal and formal social care.

AD is responsible for 60%–80% of all dementia cases. Vascular dementia (also known as poststroke or multi-infarct dementia) is responsible for ~10% of cases. In the past, evidence of vascular dementia was used to exclude a diagnosis of Alzheimer's (and vice versa). That practice is no longer considered consistent with the pathological evidence, which shows that the brain changes of both types of dementia commonly coexist. When two or more types of dementia are present at the same time, the individual is considered to have mixed dementia.

Characteristic symptoms of the condition are:

- Difficulty remembering recent conversations, names, or events is often an early clinical symptom.
- Apathy and depression are also often early symptoms.
- Later symptoms include impaired communication, disorientation, confusion, poor judgment, behavior changes, and, ultimately, difficulty speaking, swallowing, and walking.

At this point, nanomedicine can offer diagnostic options for early recognition of AD via brain imaging and other emerging tests, including genetic tests.

The best-known genetic risk factor is the inheritance of the *e4* allele of the apolipoprotein E (APOE). Between 40% and 80% of people with AD possess at least one *APOE-e4* allele. The *APOE-e4* allele increases the risk of the disease by three times in heterozygotes and by 15 times in homozygotes.

Mutations in the *TREM2* gene have been associated with a three to five times higher risk of developing AD. A suggested mechanism of action is that when *TREM2* is mutated, white blood cells in the brain are no longer able to control the amount of beta-amyloid present.

Despite many ongoing research projects, it has so far not been possible to stop or reverse the onset of AD.

6.2.1.10 Parkinson's disease

PD is a degenerative disorder of the CNS. The motor symptoms of PD result from the death of dopamine-generating cells in the substantia nigra, a region of the midbrain; the cause of this cell death is unknown. Early in the course of the disease, the most obvious symptoms are movement related; these include shaking, rigidity, slowness of movement, and difficulty with walking and gait. In addition, speech is inhibited. Later, thinking and behavioral problems may arise, with dementia commonly occurring in the advanced stages of the disease. Other symptoms include sensory, sleep and emotional problems.

Diagnosis of PD is mainly based on symptoms, with tests such as neuroimaging used for confirmation.

PD traditionally has been considered a nongenetic disorder; however, around 15% of individuals with PD have a first-degree relative who has the disease.

Several therapeutic options are pursued, so far without breakthrough results. Deep brain stimulation (DBS) has shown some positive results. DBS is a neurosurgical procedure involving the implantation of a medical device called a brain pacemaker, which sends electrical impulses, through implanted electrodes, to specific parts of the brain (brain nucleus) for the treatment of movement and affective disorders. DBS has provided therapeutic benefits for PD, essential tremor, dystonia, chronic pain, major depression, and obsessive compulsive disorder (OCD).

Cognitive dysfunction is a difficult area for therapeutic intervention. Hopefully, the ongoing brain research project will help us understand deeper mechanisms and lead to more effective treatments.

Let us now leave the brain and turn to the spinal cord and conditions associated with it.

6.2.2 *The Spinal Cord*

A *spinal cord injury* (SCI) refers to any injury to the spinal cord of the vertebral column (see Fig. 6.7) that is caused by trauma instead of disease. Damage to any part of the spinal cord or nerves at the end of the spinal canal often causes permanent changes in strength, sensation, and other body functions below the site of the injury.

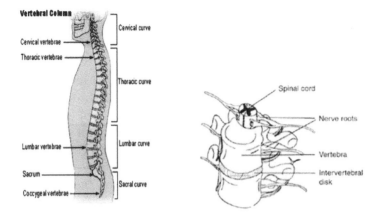

Figure 6.7 Vertebral column and spinal cord and detailed view of the sacrum (bottom). Source: Wikipedia.

The American Spinal Injury Association (ASIA) first published an international classification of SCIs in 1982, called the International Standards for Neurological and Functional SCI Classification.

- A indicates a complete SCI where no motor or sensory function is preserved in the sacral segments S4–S5.
- B indicates an incomplete SCI where sensory but not motor function is preserved below the neurological level and includes the sacral segments S4–S5. This is typically a transient phase and if the person recovers any motor function below the neurological level, that person essentially becomes a motor incomplete, that is, ASIA C or D.
- C indicates an incomplete SCI where motor function is preserved below the neurological level and more than half of key muscles below the neurological level have a muscle grade of less than 3, which indicates active movement with full range of motion against gravity.
- D indicates an incomplete SCI where motor function is preserved below the neurological level and at least half of the key muscles below the neurological level have a muscle grade of 3 or more.
- E indicates a normal level where motor and sensory scores are normal. Note that it is possible to have an SCI and

neurological deficits with completely normal motor and sensory scores.

There are three potential areas of impact for nanomedicine, namely:

- **Diagnostics**: There is a strong need for imaging techniques to monitor the changes in the spinal cord tissue in vivo. SCI is caused by a mechanical insult to the cord (i.e., primary injury), followed by a myriad of biological events such as ischemia, inflammation, etc., that further damage the cord (i.e., secondary injury). By means of advanced imaging techniques it could be possible to elucidate the mechanisms of tissue injury in the spinal cord. Varying degrees of demyelination, white and gray matter injury, and spatial distribution of tissue damage will be dependent upon the mechanism of injury and should be identifiable using fMRI. For instance, a research team led by Tom Oxland at Univ of British Columbia, Vancouver, Canada, has developed a novel device to produce three different mechanisms of SCI in a rat model. These cord injury mechanisms (i.e., contusion, dislocation, distraction) mimic three common spinal column injury patterns (i.e., burst fracture; fracture-dislocation; flexion, distraction, and hyperextension) and should be diagnosed as such.
- **Minimally invasive surgery (MIS) of the spine**: Most types of MIS rely on a thin telescope-like instrument, called an endoscope, or on a portable X-ray machine, called a fluoroscope, to guide the surgeon during the operation. The endoscope is inserted through small incisions in the body. The endoscope is attached to a tiny video camera that projects an internal view of the patient's body onto television screens in the operating room. Small surgical instruments are passed through one or more incisions, later to be closed with sutures and covered with surgical tape. The fluoroscope is positioned around the patient to give the surgeon the best X-ray views from which to see the anatomy of the spine.

Benefits of MIS include:

- a few tiny scars instead of one large scar
- shorter hospital stays—usually only a couple of days
- reduced postoperative pain
- shorter recovery time and quicker return to daily activities, including work
- less blood loss during surgery
- reduced risk of infection
- opportunity to administer local treatment to improve outcomes

- **Regenerative medicine (see also Section 5.6)**: The Wings for Life Foundation[308] has classified SCI research areas in the following way:

- Prevention of secondary damage and protection of intact cells: An SCI is followed by a massive breakdown of neuronal and supporting cells (known as glial cells) around the site of injury. The goal is to prevent the secondary damage and preserve neuronal cell function by, for example, bone marrow stromal cell (BMSC) transplantation in the lesioned spinal cord; by local administration of drugs to treat inflammation and tissue damage; by looking for genetic predictors of inflammatory response; and by strengthening the local immune response via increase of immune cells.
- Plasticity research: SCI is accompanied by the release of substances—known as natural growth inhibitors—that block the renewed growth of nerves. Plasticity research aims at finding, analyzing, and eliminating these sub-stances. In addition, it is important to understand the mechanisms underlying the reorganization of the spinal circuitry and interconnections.
- Regeneration: When an adult nerve fiber in the CNS is completely severed, its ability to regrow is very limited. Can nerves be stimulated to regenerate and regrow? Perhaps by using hydrogels for targeted addition of the glial cell line–derived neurotrophic factor (GDNF) or by knocking down phosphatase (Inpp5f) in neurons to improve the

regeneration of axons or by finding out if dual leucine zipper kinase (DLK) activates the pro-regenerative program in the cell or by local administration of cancer drugs such as Gleevec and Taxol.

- Neural reconstruction and remyelination (insulation of nerve fibers): Replace destroyed tissue by the transplantation of cells (focus on stem cells) and/or (prosthetic) biomaterials. There is a lot of hope surrounding therapies with stem cells as these are able to form tissue scaffolds, release growth factors, form new circuits, and promote the regrowth of the protective myelin sheath. Similar to an electric wire that loses its insulation, the demyelinated nerve fibers lose their ability to properly transmit signals. Another promising approach may be to target the protein Nogo, a myelin associated inhibitor. Substances known to counteract the effect of Nogo are already being tested.

6.3 Cancer and Immunology

That which does not kill us makes us stronger.

—Friedrich Nietzsche (1844–1900)

In 2008, according to WHO,[309] 7.6 million people died of cancer— 13% of all deaths worldwide. In 2012, the number had increased to 8.2 million deaths. It is expected that annual cancer cases will rise from 14 million in 2012 to 22 million within the next two decades.

About 30% of cancer deaths are due to the five leading behavioral and dietary risks: obesity (high body mass index), low fruit and vegetable intake, lack of physical activity, tobacco use, alcohol use.

Tobacco use is the most important risk factor for cancer, causing over 20% of global cancer deaths and about 70% of global lung cancer deaths.

Cancer causing viral infections such as hepatitis B virus (HBV)/hepatitis C virus (HCV) and human papillomavirus (HPV) are responsible for up to 20% of cancer deaths in low- and middle-income countries.

More than 60% of world's total new annual cases occur in Africa, Asia, and Central and South America. These regions account for 70% of the world's cancer deaths.[310]

"Cancer" is a generic term for a large group of diseases that can affect any part of the body. Other terms used are malignant tumors and neoplasms. One defining feature of cancer is the rapid creation of abnormal cells that grow beyond their usual boundaries (i.e., metastasize) and that can then invade adjoining parts of the body and spread to other organs. Metastases are the major cause of death from cancer.

Cancer is a group of diseases involving abnormal cell growth with the potential to invade or spread to other parts of the body. There are at least 200 forms of cancer and many more subtypes. Each of these is caused by errors in DNA that cause cells to grow uncontrolled.

The most common type of cancer is skin cancer, accounting for at least 40% of cases. In females, the most common types are breast cancer, colorectal cancer, lung cancer, and cervical cancer. In males, the most common types are lung cancer, prostate cancer, colorectal cancer, and stomach cancer. In children acute lymphoblastic leukemia and brain tumors are most common.

Cancer is fundamentally a disease of tissue growth regulation failure. For a normal cell to transform into a cancer cell, the genes that regulate cell growth and differentiation must be altered.

The affected genes are divided into two broad categories. Oncogenes are genes that promote cell growth and reproduction. Tumor suppressor genes are genes that inhibit cell division and survival. Malignant transformation can occur through the formation of novel oncogenes, the inappropriate overexpression of normal oncogenes, or by the underexpression or disabling of tumor suppressor genes. Typically, changes in many genes are required to transform a normal cell into a cancer cell.

Nanomedicine is applied both to cancer diagnostics and treatment. An important contribution to diagnostics is the direct sequencing of primary tumor tissue, adjacent normal tissue, the tumor micro environment such as fibroblast/stromal cells, or metastatic tumor sites. By studying the cancer genome, we can discover what mutations are causing a healthy cell to become a cancer cell. The genome of a cancer cell can also be used to tell one type of cancer from another. In some cases, studying the genome in a cancer can help identify a subtype of cancer within that type, such as HER2+ breast cancer. Understanding the cancer genome will also help an oncologist select the best treatment for each patient.

Two main cancer sequencing projects, the *Cancer Genome Project*, initiated by Wellcome Trust Sanger Institute, and *The Cancer Genome Atlas* (TGCA), funded by the National Cancer Institute (NCI) and the National Human Genome Research Institute (NHGRI), have been launched during the past decade.

In 2004, the Cancer Genome Project at the Sanger Institute (Hinxton near Cambridge, UK) launched "Catalogue Of Somatic Mutations In Cancer" (COSMIC), an online database of somatically acquired mutations found in human cancer. Somatic mutations are changes to the genetics of a multicellular organism that are not passed on to its offspring through the germline. Many cancers are somatic mutations. COSMIC curates data from papers in the scientific literature and large-scale experimental screens from the Sanger Institute. The first four somatically mutated cancer genes included in COSMIC were *HRAS, KRAS2, NRAS,* and *BRAF.* By 2010, COSMIC added mutation data from tumor suppressor p53, a protein that in humans is encoded by the *TP53* gene. The p53 protein is crucial in multicellular organisms, where it regulates the cell cycle. By August 2014, the COSMIC database[311] contained over 28,735 genes and 2,002,811 mutations associated with over 1 million samples. It covered the findings of 19,703 scientific publications.

TCGA[312] was established in 2006 as a joint effort of the NCI and the NHGRI, two of the 27 institutes and centers of the NIH. TCGA's goal is to accelerate our understanding of the molecular basis of cancer through the application of genome analysis technologies, including large-scale genome sequencing. TCGA criteria for the selection of cancers for study include poor prognosis and overall public health impact and availability of human tumor and matched-normal tissue samples that meet TCGA standards for patient consent, quality, and quantity. As of May 2014, the cancers (C.) selected by TCGA are breast C., CNS C. (glioblastoma, glioma), endocrine C., gastrointestinal (GI) C., gynecologic C., head and neck C., hematologic C. (leukemia), skin C. (melanoma), soft tissue C. (sarcoma), thoracic C. (including lung C., mesothelioma), and urologic C. (including prostate C.).

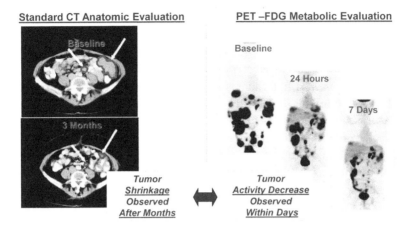

Standard CT Anatomic Evaluation

PET –FDG Metabolic Evaluation

Tumor
Shrinkage
Observed
After Months

Tumor
Activity Decrease
Observed
Within Days

Figure 6.8 Comparison of tumor activity decrease and associated tumor shrinkage.[314]

The Cancer Genome Project and TCGA are also collaborating along with the global cancer research community in the International Cancer Genome Consortium (ICGC).[313] The ICGC's stated goal is to "obtain a comprehensive description of genomic, transcriptomic and epigenomic changes in 50 different tumor types and/or subtypes that are of clinical and societal importance across the globe."

The ICGC is committed to generating comprehensive catalogs of genomic abnormalities (somatic mutations, abnormal expression of genes, epigenetic modifications) in tumors from 50 different cancer types and/or subtypes that are of clinical and societal importance, and make the data available to the entire global research community as rapidly as possible, and with minimal restrictions.

The basic approach to cancer diagnostics can vary widely between various cancer types and many diagnostic tools have already been discussed in Sections 5.4 and 5.5. Figure 6.8 shows how separate imaging modalities (here computed tomography [CT] and positron emission tomography [PET]) can provide complementary information, with PET capturing a significant decrease in glucose uptake (tumor metabolic rate) already after a few days, far ahead of CT, which needs three weeks to show the subsequent structural tumor shrinkage.

Until recently, cancer therapy has been dominated by surgery followed by administration of chemotherapy (preferably administered by means of nanomedical encapsulation—see Section 5.2.3) and/or radiation. However, monoclonal antibody (mab) technology (see Section 4.5) has changed cancer therapy by delivering drugs that could inhibit oncogenic proteins or survival factors selectively expressed by tumors.

Recently, in addition to genomics and other "omics" introduced in Section 3.10, nanomedicine may have even more impact on cancer therapeutics via immunotherapy, hailed as "breakthrough of the year" by *Science* magazine in December 2013.[315]

The story behind the breakthrough started in the late 1980s when French scientists[316] identified a new protein receptor on the surface of T cells, called cytotoxic T-lymphocyte antigen 4, or CTLA-4. T cells lead the cellular immune attack on antigens. The T cell attack can be turned on by stimulating the CD28 receptor on the T cell and turned off by stimulating the CTLA-4 receptor, which acts as an off switch. Its crystal structure is shown in Fig. 6.9.

CTLA4 is a member of the immunoglobulin superfamily, which is expressed on the surface of Helper T cells and transmits an inhibitory signal to T cells. It is encoded by the CTLA4 gene. Mutations in this gene have been associated with insulin-dependent diabetes mellitus (IDDM or type I/juvenile diabetes) and several other, less

Figure 6.9 Cytotoxic T-lymphocyte antigen 4 (CTLA-4). Source: Wikipedia.

known, autoimmune diseases such as Graves' disease, Hashimoto's thyroiditis, celiac disease, systemic lupus erythematosus, thyroid-associated orbitopathy, and primary biliary cirrhosis.

To *reduce immune activity*, fusion proteins (created through the joining of two or more genes that originally coded for separate proteins) of CTLA4 and antibodies (CTLA4-Ig) have been developed by Bristol–Myers Squibb (BMS) into a commercially available drug, *Orencia (abatacept)*, to treat rheumatoid arthritis. Abatacept has also been proposed for the treatment of T1D, for patients in the "honeymoon phase" of the disease, when Abatacept may protect surviving beta cells from autoimmune attack. A second-generation form of CTLA4-Ig known as *belatacept* was recently approved by the FDA for immunosuppression after renal transplantation.

Cancer immunologist James Allison, then working at University of California at Berkeley, had another idea: He wondered whether blocking the off switch—the CTLA-4 molecule—would set the immune system free to destroy cancer.[317] Together with his student M.F. Krummel, he decided to use antagonistic antibodies against CTLA-4. The result was the monoclonal antibody *ipilimumab* which erased tumors in mice and which he succeeded in licensing to biotech company Medarex in Princeton, New Jersey, in 1999. It took then 11 years to get clinical results. In February 2011, the FDA approved Ipilimumab (also known as MDX-010 and MDX-101), now marketed as Yervoy,[318] for the treatment of melanoma, a type of skin cancer. Already in 2010, BMS acquired Medarex for more than $2B. In addition to melanoma, BMS is targeting ipilimumab for the treatment of non–small cell lung carcinoma (NSCLC), small-cell lung cancer (SCLC), bladder cancer, and metastatic hormone-refractory prostate cancer.

In the United States, BMS is charging $120,000 for a course of treatment, based on the fact that Yervoy helps patients with advanced metastatic melanoma to significantly improve outcomes. Average survival was extended from 6 to 10 months, and nearly a quarter of participants in a randomized trial survived at least two years. In 2013, BMS reported that of 1800 melanoma patients treated with ipilimumab, 22% were alive three years later. By combining ipilimumab and another antibody, PD-1 (called "programmed death" because the molecule was discovered as expressed in dying

T cells), deep and rapid tumor regression was reported in almost a third of melanoma patients.

Engineered T cells are now pursued for cancer therapy by at least five major drug companies.

For oncologists accustomed to lose hope when cancer patients enter the advanced metastatic phase, there is finally a chance that immunotherapy may significantly improve the odds. Time will tell if it will be possible to build on this initial success.

6.4 Cardiovascular Disease

If we could give every individual the right amount of nourishment and exercise, not too little and not too much, we would have found the safest way to health.

—Hippocrates (460–371 BC)

According to the World Health Organization (WHO),[319] cardiovascular diseases (CVDs) are a group of disorders of the heart and blood vessels and they include:

- coronary heart disease—disease of the blood vessels supplying the heart muscle;
- cerebrovascular disease—disease of the blood vessels supplying the brain;
- peripheral arterial disease—disease of blood vessels supplying the arms and legs;
- rheumatic heart disease—damage to the heart muscle and heart valves from rheumatic fever, caused by streptococcal bacteria;
- congenital heart disease—malformations of heart structure existing at birth; and
- deep vein thrombosis and pulmonary embolism—blood clots in the leg veins, which can dislodge and move to the heart and lungs.

Heart attacks and strokes are usually acute events and are mainly caused by a blockage that prevents blood from flowing to the heart or brain. The most common reason for this is a build-up of fatty deposits on the inner walls of the blood vessels that supply the heart or brain. Strokes can also be caused by bleeding from a blood vessel in the brain or from blood clots.

Risk factors for CVD include unhealthy diet, physical inactivity, tobacco use, and harmful use of alcohol. Behavioral risk factors are responsible for about 80% of coronary heart disease and cerebrovascular disease. Most CVDs can either be prevented by addressing the risk factors listed above or reduced by preventing/treating hypertension, diabetes, and raised blood lipids.

CVDs are the number one cause of death globally: more people die annually from CVDs than from any other cause.

- An estimated 17.3 million people died from CVDs in 2008, representing 30% of all global deaths. Of these deaths, an estimated 7.3 million were due to coronary heart disease and 6.2 million were due to stroke. By 2030, the number of people expected to die from CVDs will increase to reach 23.3 million. CVDs are projected to remain the single leading cause of death.

- Low- and middle-income countries are disproportionally affected: over 80% of CVD deaths take place in low- and middle-income countries and occur almost equally in men and women.

- 9.4 million deaths each year, or 16.5% of all deaths, can be attributed to high blood pressure. This includes 51% of deaths due to strokes and 45% of deaths due to coronary heart disease.

Cessation of tobacco use, reduction of salt in the diet, consuming fruits and vegetables, regular physical activity, and avoiding harmful use of alcohol have been shown to reduce the risk of CVD.

Now let's take a look at the possible impact of nanomedicine on diagnosis and therapy of CVDs, as covered in a recent review by Cicha, Garlichs, and Alexiou.[320] Applications range from plaque imaging and thrombus (blood clot) detection to targeted drug delivery, stent endotheliazation, and blood vessel regeneration.

In Section 5.4, we already touched on imaging techniques that can be applied to the detection and characterization of cardiovascular disorders. The modalities commonly used are MRI, PET, and optical fluorescence imaging (suitable for plaque endothelium). Suitable nanoparticles to be used as MRI contrast agents have already been covered extensively in Section 5.4.

Targeted drug delivery and regenerative medicine are also expected to have significant impact on cardiovascular therapy, as covered in Sections 5.2.3 and 5.6, respectively. Seven of the twenty-five companies listed in Table 5.4 are targeting cardiovascular therapies.

In addition, the cardiovascular field has attracted many medical device companies. They are all trying to include nanotechnology in their next generation products.

6.5 Diabetes

Learning never exhausts the mind.

—Leonardo da Vinci (1452–1519)

Extrapolating from a recent study[321] published in the *Lancet*, close to 400 million people worldwide have diabetes. In 2004, an estimated 3.4 million people died from consequences of fasting high blood sugar. A similar number of deaths has been estimated for 2010. More than 80% of diabetes deaths occur in low- and middle-income countries.[322]

Diabetes is a chronic disease that occurs either when the pancreas does not produce enough insulin or when the body cannot effectively use insulin. Insulin is a hormone that regulates blood sugar. As shown in Fig. 5.3, it is produced in the beta cells of the pancreas.

Hyperglycemia, or raised blood sugar, is a common effect of uncontrolled diabetes and over time leads to serious damage to many of the body's systems, especially the nerves and blood vessels.

The opposite condition, hypoglycemia, can occur as a complication of diabetes treatment with insulin or oral medications. It can cause serious issues such as seizures, unconsciousness, and even permanent brain damage or death.

T1D (also called insulin-dependent, juvenile, or childhood-onset diabetes mellitus) is characterized by deficient insulin production and requires daily administration of insulin.

Type II diabetes (also called non-insulin-dependent or adult-onset diabetes mellitus) results from the body's ineffective use of insulin. Type II diabetes comprises 90% of people with diabetes around the world, and is largely the result of excess body weight and physical inactivity.

The consequences of diabetes are severe. Over time, diabetes—even if treated and controlled—can damage the heart, blood vessels, eyes, kidneys, and nerves.

- Diabetes increases the risk of heart disease and stroke. 50% of people with diabetes die of CVD
- Combined with reduced blood flow, neuropathy (nerve damage) in the feet increases the chance of foot ulcers, infection, and eventual need for limb amputation.
- Diabetic retinopathy is an important cause of blindness, and occurs as a result of long-term accumulated damage to the small blood vessels in the retina. One percent of global blindness can be attributed to diabetes. Diabetes is the biggest reason for blindness in working-age people.
- Diabetes is among the leading causes of kidney failure, requiring either dialysis or transplantation.

What can nanomedicine do for patients suffering from diabetes?

As to type II diabetes, behavioral changes (diet and exercise) are the most important actions to be taken, both to prevent and possibly reverse the condition. For patients severely affected by the condition, the therapeutic options are similar to the ones developed for T1D patients, discussed below.

In Section 5.1 we already covered the Artificial Pancreas project. When fully developed and approved by the FDA, the combination of continuous glucose monitoring (CGM), advanced software to feed back results and regulate insulin (and perhaps even glucagon) injections into the bloodstream, could maintain healthy blood glucose levels and protect patients from the above-mentioned consequences of diabetes.

In Section 5.6, we discussed the use of stem cells to regenerate beta cell function as provided by a healthy pancreas. Several projects are moving forward with this idea, including San Diego–based Viacyte's VC-01 diabetes therapy and Johnson & Johnson's Betalogics[323] stem cell project. Recently, the Harvard Stem Cell Institute has discovered a way to produce large quantities of insulin-secreting beta cells from stem cells. The team led by Doug Melton,[324] with support from JDRF and the Helmsley Charitable Trust, has demonstrated that these in vitro cells have morphological and functional similarity to human beta cells and are secreting insulin in a glucose-responsive manner. They have already been found to rescue rodent models of T1D from hyperglycemia. These cells could benefit patients that are dependent on insulin, and could also help screen new drug candidates on a massive scale. However, for clinical use it will be necessary to encapsulate the cells to protect them against autoimmune attack. Harvard, Johnson & Johnson, and Evotec are engaged in a three-way partnership. The encapsulation or immune therapies are likely to be more difficult to get right vs the cells! The commercial model is also difficult to build out.

Another important area of activity is the prevention and possible cure of T1D, based on a deeper understanding of the autoimmune response responsible for the disease. There is a genetic disposition for T1D, which can be diagnosed by detection of antibodies, but there are also environmental reasons that are not yet fully understood. Increasingly, the gut microbiome[325] (see also Section 6.1) is suspected of playing an important role in triggering the unfortunate response of the patient's immune system that causes destruction of the beta cells in the pancreas.

Ethically, it is difficult to decide whether we should allow diagnosis of all children when we do not have prevention available, and not all parents of children with a predisposition to develop diabetes will do so. As there is a higher incidence and prevalence in certain geographic areas, such as Scandinavia, we may expect more Swedish, Danish, and Finnish research into environmental (food, diet, infectious disease, etc.) reasons for diabetes triggers, in the child or even the mother.

6.6 Infectious Disease

Soap and water and common sense are the best disinfectants.

—William Osler (1849–1919)

Worldwide, infectious diseases are the leading cause of death of children and adolescents. About 16% of all deaths each year are caused by infectious diseases.[326] Most of these deaths are in low- and middle-income countries and are attributable to preventable or treatable diseases such as diarrhea, lower respiratory infections, human immunodeficiency virus (HIV)/acquired immunodeficiency syndrome (AIDS), tuberculosis (TB), and malaria.

While significant advances have been made in interventions to prevent and treat most of these diseases, they are not always available to the populations most in need.

Infection is the invasion of a host organism's body tissues by disease-causing agents, their multiplication, and the reaction of host tissues to these organisms and the toxins they produce.

Infectious diseases are also known as transmissible diseases or communicable diseases. They are associated with the presence and growth of pathogenic biological agents in an individual host organism.

Most infections are caused by infectious agents such as viruses, and microorganisms such as bacteria, but also by ticks, mites, fleas, lice, fungi such as ringworm, etc. Figure 6.10 is showing two examples of infectious agents, the mosquito and the deer tick.

Figure 6.10 Agents of infection: mosquito (malaria) and deer tick (Lyme disease). Source: Wikipedia.

Humans fight infections using the immune system and react with an innate response, often involving inflammation, followed by an *adaptive response*.

The adaptive, or acquired, immune system is a subsystem of the overall immune system that is composed of highly specialized, systemic cells and processes that eliminate or prevent pathogen growth. One of the two main immunity strategies (the other being the innate immune system), adaptive/acquired immunity creates immunological memory after an initial response to a specific pathogen, leading to an enhanced response to subsequent encounters with that same pathogen. This process of *acquired immunity is the basis of vaccination*. Like the innate system, the adaptive system includes both humoral immunity components and cell-mediated immunity components. Humoral immunity, also called antibody-mediated immune system, is the aspect of immunity that is mediated by macromolecules (as opposed to cell-mediated immunity) found in extracellular fluids such as secreted antibodies, complement proteins and certain antimicrobial peptides. We will return to this fact when—further below—we will discuss the possible impact of antimicrobial polymer nanoparticles on infectious disease.

In acquired immunity, pathogen-specific receptors are acquired during the lifetime of the organism (whereas in innate immunity pathogen-specific receptors are already encoded in the germline). The acquired response is said to be adaptive because it prepares the body's immune system for future challenges. However, it can actually also be maladaptive when it results in autoimmunity.

Vaccines save millions of lives each year and are among the most cost-effective health interventions ever developed. Immunization has led to the eradication of smallpox, a 74% reduction in childhood deaths from measles over the past decade, and the near-eradication of polio.[327] Unfortunately, the global polio eradication program, intended for completion by 2000, has fallen victim to political and civil unrest in Nigeria, Afghanistan, Pakistan, and India.

Despite these great strides, there remains an urgent need to reach all children with lifesaving vaccines. One in five children worldwide is not fully protected with even the most basic vaccines. As a result, an estimated 1.5 million children die each year from vaccine-preventable diseases such as diarrhea and pneumonia. Tens of thousands of other children suffer from severe or permanently disabling illnesses.

Vaccines are often expensive for the world's poorest countries, and supply shortages and a lack of trained health workers are challenges as well. Unreliable transportation systems and storage facilities also make it difficult to preserve high-quality vaccines that require refrigeration.

Increased routine vaccination for measles, bacterial meningitis, tetanus, diphtheria, polio, pertussis, yellow fever, and rotavirus greatly improved with better coordination, discrete budget sources and additional outside funding from groups like the Global Alliance for Vaccines and Immunization (GAVI).[328] For newly introduced vaccines and old, efforts must focus on further increasing routine coverage of immunization through the broad array of strategies that have proved themselves successful, including targeted community campaigns, child health days, and immunization weeks.

The branch of medicine that focuses on infections and pathogens is infectious disease medicine. In addition to vaccines as preventive measures, doctors may use specific pharmaceutical drugs such as antibiotics and antivirals to treat infections.

Since bacterial and viral infections can cause the same kinds of symptoms, it can be difficult to distinguish between them. However, it is important to find out because viral infections cannot be cured by antibiotics.

6.6.1 *Pathogenic Bacteria and Antibiotics*

Classic symptoms of a bacterial infection are localized redness, heat, swelling, and pain. A characteristic of a bacterial infection is *local pain*, pain that is in a specific part of the body. A cut that produces pus and milky-colored liquid is most likely infected.

Most bacteria are harmless or even beneficial. However, some are pathogenic.

For instance, *Mycobacterium tuberculosis* is still killing about 2 million people a year, mostly in sub-Saharan Africa. Other pathogenic bacteria contribute to other important diseases, such as pneumonia, which can be caused by bacteria such as *Streptococcus* and *Pseudomonas*. There are also food-borne illnesses that can be caused by bacteria such as *Shigella, Campylobacter,* and *Salmonella.* Urinary tract infection is almost exclusively caused by bacteria. The main causal agent is *Escherichia coli,* which can also cause bacterial gastroenteritis. Skin infections may be caused by *Staphylococcus aureus* or by *Streptococcus.* Pathogenic bacteria also cause infections such as tetanus, typhoid fever, diphtheria, syphilis, and leprosy. Lyme disease is caused by *Borrelia burgdorferia.*

Bacterial infections may be treated with antibiotics, classified as bacteriocidal if they kill bacteria or bacteriostatic if they just prevent bacterial growth. There are many types of antibiotics but they all have to discriminate between pathogen and host. For example, the antibiotics chloramphenicol and tetracyclin inhibit the bacterial ribosome but not the structurally different eukaryotic ribosome. They exhibit selective toxicity. Antibiotics are used both in treating human disease, but also in intensive farming to promote animal growth. Both uses may be contributing to the rapid development of antibiotic resistance in bacterial populations.

In 1928, Alexander Fleming, London, noticed that a number of disease-causing bacteria were killed by a fungus of the genus *Penicillium.* Fleming postulated that the effect is mediated by an antibacterial compound he named penicillin, and that its antibacterial properties could be exploited for therapy. The first sulfonamide and first commercially available antibacterial, Prontosil, was developed by a research team led by Gerhard Domagk in 1932 at the Bayer Laboratories in Germany. In 1939, Rene Dubos (New York) reported the discovery of the first naturally derived antibiotic, tyrothricin. It was one of the first commercially manufactured antibiotics universally and was very effective in treating wounds and ulcers during World War II. Walter Florey and Ernst Chain succeeded in purifying the first penicillin, penicillin G, in 1942, and Dorothy Crowfoot Hodgkin determined the structure, shown in Fig. 6.11, in 1945. Purified penicillin displayed potent antibacterial activity against a wide range of bacteria and had low toxicity in humans.

Fleming,[329] Domagk,[330] Florey,[331] Chain,[332] and Hodgkin[333] all won Nobel Prizes for their respective contributions. Hodgkin, an outstanding pioneer of protein crystallography, also determined the structures of vitamin B_{12} and insulin.

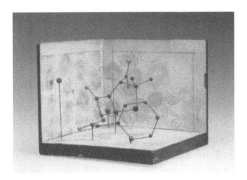

Figure 6.11 Dorothy Crowfoot Hodgkin's model of penicillin. Source: Wikipedia.

6.6.1.1 Diagnosis

Microbiological culture is a principal tool used to diagnose infectious disease. In a microbial culture, a solid medium that supplies carbohydrates and proteins necessary for growth of a bacterium, along with generous amounts of water, most pathogenic bacteria are grown to form a colony. The size, color, shape, and form of a colony is characteristic of the bacterial species, its specific genetic makeup (its strain), and the environment that supports its growth. Another principal tool in the diagnosis of infectious disease is microscopy. Virtually all of the culture techniques discussed above rely, at some point, on microscopic examination for definitive identification of the infectious agent. Samples obtained from patients may be viewed directly under the light microscope and can often rapidly lead to identification.

6.6.1.2 Antibiotic resistance

Antibiotic resistance is a form of drug resistance where subpopulations of a microorganism, usually a bacterial species, are able to survive after exposure to one or more antibiotics; pathogens resistant to multiple antibiotics are considered *multidrug resistant* (MDR), or, more colloquially, *superbugs*.

Antibiotic resistance is a serious and growing phenomenon and has emerged as one of the pre-eminent public health concerns of the 21st century.

Staphylococcus (S.) aureus is one of the major resistant pathogens. Its occurrence is pervasive and it is extremely adaptable to antibiotic pressure. It was one of the earlier bacteria in which penicillin resistance was found—in 1947. Methicillin was then the antibiotic of choice, but has since been replaced by oxacillin due to significant kidney toxicity. Methicillin-resistant *S. aureus* (MRSA) was first detected in Britain in 1961, and is now quite common in hospitals. MRSA was responsible for 37% of fatal cases of sepsis in the UK in 1999, up from 4% in 1991. In the United States, half of all *S. aureus* infections are resistant to penicillin, methicillin, tetracycline, and erythromycin.

Other superbugs:

- *Streptococcus pneumoniae* resistance to penicillin and other beta-lactams is increasing worldwide. The major mechanism of resistance involves the introduction of mutations in genes encoding penicillin-binding proteins. *S. pneumoniae* is responsible for *pneumonia* (inflammatory condition of the lung affecting primarily the microscopic air sacs known as alveoli), *bacteremia* (bacteria in blood), *otitis media* (middle ear inflammation), *meningitis* (inflammation of the protective membranes covering the brain and spinal cord), *sinusitis* (inflammation of the paranasal sinuses, a group of four paired air-filled spaces that surround the nasal cavity), *peritonitis* (inflammation of the peritoneum, the thin tissue that lines the inner wall of the abdomen), and *arthritis* (a form of joint disorder that involves inflammation of one or more joints).
- *Enterococcus faecalis* and *E. faecium* are associated with *nosocomial* infections (hospital-acquired infection [HAI] such as one acquired by a patient during a hospital visit or one developing among hospital staff). Three strains have been observed: penicillin-resistant *Enterococcus*, vancomycin-resistant *Enterococcus* (VRE), and linezolid-resistant *Enterococcus*.

- Pseudomonas aeruginosa has low antibiotic susceptibility and can infect compromised immune systems such as those of patients taking immunosuppressants after organ transplantation.
- *Clostridium difficile* is a nosocomial pathogen that causes diarrheal disease in hospitals worldwide.
- *Escherichia (E.) coli* and *Salmonella* infections can result from the consumption of contaminated food and water. Both can cause nosomical (hospital-linked) infections.
- *Mycobacterium (M.) tuberculosis*: Tuberculosis (TB) is again on the rise. Globally, MDR TB causes 150,000 deaths annually. TB did not have a cure until the discovery of streptomycin by Selman Waksman in 1943. However, the bacteria soon developed resistance. Since then, drugs such as isoniazid and rifampin have been used. *M. tuberculosis* develops resistance to drugs by spontaneous mutations in its genomes. Extensively drug-resistant tuberculosis (XDR TB) is TB that is also resistant to the second line of drugs.[334]

What can be done to deal with this serious problem? As to pharmaceutical R&D investments, antibiotics have a poor return on investment because they are taken for a short period of time and cure their target disease. There is therefore a lack of incentive for companies to develop antibiotics. Only five major biopharmaceutical companies—GlaxoSmithKline, Novartis, AstraZeneca, Merck, and Pfizer—still had active antibacterial discovery programs in 2008, according to an article published in the journal *Clinical Infectious Diseases* in January 2009.

As new ideas are needed could nanomedicine be of help? Perhaps!

Jim Hedrick et al. at IBM's Almaden research lab in San Jose, CA, have developed biodegradable in vivo applicable antimicrobial polymer nanoparticles that have been successfully tested by the Institute of Bioengineering and Nanotechnology (IBN) in Singapore and by the State Key Lab for Diagnosis and Treatment of Infectious Diseases in Hangzou, China. They attack and kill MRSA and VRE on the basis of a mechanism that differs from antibiotics.[335]

Clinical studies will have to be done to validate the claims made by Hedrick's team.

6.6.2 *Viral Pathogens and Therapies*

A virus is a biological agent that reproduces inside the cells of living hosts. When infected by a virus, a host cell is forced to produce many thousands of identical copies of the original virus, at an extraordinary rate. Unlike most living organisms (including bacteria), viruses do not have cells that divide; new viruses are assembled in the infected host cell. But viruses contain genes, which gives them the ability to mutate and evolve.

Viral infections can cause disease in humans, animals and even plants. However, they are usually eliminated by the immune system, conferring lifetime immunity to the host for that virus. Antibiotics have no effect on viruses, but antiviral drugs have been developed to treat life-threatening infections. Vaccines that produce lifelong immunity can prevent some viral infections.

Viruses spread in many ways. Plant viruses are often spread from plant to plant by insects and other organisms, known as vectors (a person, animal or microorganism that carries and transmits an infectious pathogen into another living organism). Some viruses of animals, including humans, are spread by exposure to infected bodily fluids. Viruses such as influenza are spread through the air by droplets of moisture when people cough or sneeze. Viruses such as norovirus are transmitted by the fecal–oral route, which involves the contamination of hands, food, and water. Rotavirus is often spread by direct contact with infected children. HIV is transmitted by bodily fluids transferred during sex. Others, such as the Dengue virus, are spread by blood-sucking insects.

Among the human infecting families there are a number of rules that may assist physicians and medical microbiologists/virologists.

6.6.2.1 Diagnosis

For viral diagnosis, a regular light microscope may no longer be sufficient. Therefore, microscopy is often used in conjunction with biochemical staining techniques, applied in combination with antibody-based techniques. For example, the use of antibodies made artificially fluorescent (fluorescently labeled antibodies) can be directed to bind to and identify a specific antigens present on a pathogen. A fluorescence microscope is then used to detect fluorescently labeled antibodies bound to internalized antigens within clinical samples or cultured cells. This technique is especially useful in the diagnosis of viral diseases, where the light microscope is incapable of identifying a virus directly.

6.6.2.2 Classification

The *Baltimore classification*, developed by Nobel laureate David Baltimore,[336] is a virus classification system that groups viruses into families, depending on their type of genome (DNA, RNA, single-stranded, double-stranded, etc.) and their method of replication.

As a general rule, DNA viruses replicate within the nucleus while RNA viruses replicate within the cytoplasm.

- DNA virus-related diseases: herpes, chickenpox, shingles, mononucleosis, smallpox, hepatitis B, etc. More than 90% of adults have been infected with at least one of these, and a latent form of the virus remains in most people.
- RNA virus-related diseases: West Nile virus, dengue virus, tick-borne encephalitis virus, yellow fever virus, and several other viruses that may cause encephalitis. The Ebola and Marburg virus families are also included. Infection with both can have fatal consequences.

An important concept for classification is the "viral envelope." Many viruses (e.g., influenza) have viral envelopes covering their protective protein *capsid*, that is, the protein shell of a virus, enclosing the genetic material of the virus. The envelopes typically are derived from portions of the host cell membranes (phospholipids and proteins) but include some viral glycoproteins (see Section 2.4, Fig. 2.44). Viral envelopes are essential to entry into host cells; they may help viruses avoid the host immune system. Glycoproteins on the surface of the envelope serve to identify and bind to receptor sites on the host's membrane. The viral envelope then fuses with the host's membrane, allowing the capsid and viral genome to

Scheme of a CMV virus

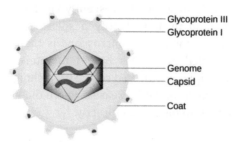

Figure 6.12   Viral envelope for cytomegalovirus. Source: Wikipedia.

enter and infect the host. Figure 6.12 shows the viral envelope for cytomegalovirus (CMV), a member of the viral family known as Herpesviridae, or herpesviruses. Herpesviruses like CMV share a characteristic ability to remain latent within the body over long periods. CMV infections are frequently associated with the salivary glands.

The cell from which the virus itself buds will often die or be weakened, and shed more viral particles for an extended period. The lipid bilayer envelope of these viruses is relatively sensitive to desiccation, heat, and detergents. Therefore these viruses are easier to sterilize than nonenveloped viruses, have limited survival outside host environments, and typically must transfer directly from host to host. Enveloped viruses possess great adaptability and can change in a short time in order to evade the immune system. Enveloped viruses can cause persistent infections.

Due to recent events, the Ebola virus is receiving a lot of attention. Ebola virus symptoms show up 2 to 21 days after infection and usually include: High fever, headache, joint and muscle aches, sore throat, weakness, stomach pain, and lack of appetite.

Ebola virus disease (EVD), formerly known as Ebola hemorrhagic fever, is a severe, often fatal illness in humans. Here are some facts as communicated by WHO:[337]

- The virus is transmitted to people from wild animals (fruit bats) and spreads in the human population through human-to-human transmission through body fluids.

- The average EVD case fatality rate is around 50%. Case fatality rates have varied from 25% to 90% in past outbreaks.
- The first EVD outbreaks occurred in remote villages in Central Africa, near tropical rainforests, but the most recent outbreak in West Africa (2014) has involved major urban as well as rural areas.
- Community engagement is key to successfully controlling outbreaks. Good outbreak control relies on applying a package of interventions, namely, case management, surveillance and contact tracing, good laboratory service, safe burials, and social mobilization.
- Early supportive care with rehydration, symptomatic treatment improves survival. There is as yet no licensed treatment proven to neutralize the virus but a range of blood, immunological and drug therapies are under development. Inactivated Ebola virus vaccines were shown to not promote enough of an immune response to the real pathogen. Ebola virus subunit vaccines are showing promise in lab animals for protecting against Ebola infection.

EVD first appeared in 1976 in two simultaneous outbreaks, one in Nzara, Sudan, and the other in Yambuku, the Democratic Republic of Congo. The latter occurred in a village near the Ebola River, from which the disease takes its name.

The 2014 outbreak in West Africa (first cases notified in March 2014) is the largest and most complex Ebola outbreak since the Ebola virus was first discovered in 1976. There have been more cases and deaths in this outbreak than all others combined. It has also spread between countries, starting in Guinea, then spreading across land borders to Sierra Leone and Liberia, by air (one traveler only) to Nigeria, and by land (one traveler) to Senegal.

The most severely affected countries, Guinea, Sierra Leone, and Liberia have very weak health systems, lacking human and infrastructural resources, having only recently emerged from long periods of conflict and instability. On August 8, the WHO director general declared this outbreak a public health emergency of international concern.

The influenza virus family is also RNA based. The type A influenza viruses are the most virulent human pathogens among the three influenza types and cause the most severe disease. The serotypes that have been confirmed in humans, ordered by the number of known human pandemic deaths (see further below), are as follows:

- H1N1 caused Spanish flu in 1918 and swine flu in 2009.
- H2N2 caused Asian flu.
- H3N2 caused Hong Kong flu.
- H5N1 is a pandemic threat.
- H7N7 has unusual zoonotic potential.
- H1N2 is endemic in humans and pigs.
- H9N2, H7N2, H7N3, and H10N7.

Typically, influenza is transmitted from infected mammals through the air by coughs or sneezes, creating aerosols containing the virus, and from infected birds through their droppings. Influenza can also be transmitted by saliva, nasal secretions, feces, and blood. Flu viruses can remain infectious for about one week at human body temperature but indefinitely at very low temperatures (such as lakes in northeast Siberia). They can be inactivated by disinfectants and detergents.

The viruses bind to a cell through interactions between its hemagglutinin glycoprotein and sialic acid sugars on the surfaces of epithelial cells in the lung and throat. Influenza hemagglutinin (HA) is a glycoprotein found on the surface of the influenza viruses. It is responsible for binding the virus to cells with sialic acid on the membranes, such as cells in the upper respiratory tract or erythrocytes. The name "hemagglutinin" comes from the protein's ability to cause red blood cells (erythrocytes) to clump together (agglutinate) in vitro.

Influenza viruses have 2 surface glycoproteins, hemagglutinin and *neuraminidase*. Influenza neuraminidase (also named *sialidase*) exists as a mushroom-shape projection on the surface of the influenza virus. Sialidase cleaves sialic acid to form new particles. This cleavage releases new viruses which can later invade new host cells. Neuraminidase actions include assistance in the mobility of

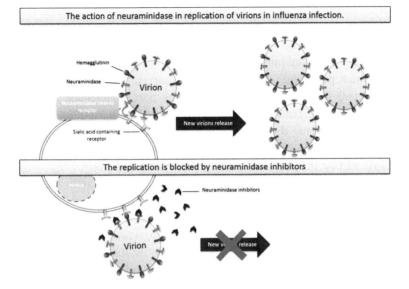

Figure 6.13 Influenza infection and neuraminidase inhibitors. Source: Wikipedia.

virus particles through the respiratory tract mucus. Neuraminidase is therefore a target for antiviral drugs. Neuraminidase inhibitors are useful for combating influenza infection: *zanamivir*, administered by inhalation; *oseltamivir*, administered orally; and under research is *peramivir*, administered parenterally, that is through intravenous or intramuscular injection. Figure 6.13 illustrates the mechanism of action.

Oseltamivir was the first orally administered neuraminidase inhibitor, commercially developed by Gilead Sciences, which licensed the exclusive rights to Roche in 1996. Marketed under the trade name *Tamiflu*, it is on the World Health Organization's List of Essential Medicines,[338] a list of the most important medication needed in a basic health system.

Retroviruses are a family of enveloped viruses that replicate in a host cell through the process of reverse transcription. A retrovirus is a single-stranded RNA (ssRNA) virus that stores its nucleic acid in the form of a messenger RNA (mRNA) genome (including the 5′ cap and 3′ poly A tail) and targets a host cell as an obligate parasite.

Once inside the host cell cytoplasm, the virus uses its own reverse transcriptase enzyme to produce DNA from its RNA genome, the reverse of the usual pattern, thus retro (backward). This new DNA is then incorporated into the host cell genome by an integrase enzyme, at which point the retroviral DNA is referred to as a provirus. The host cell then treats the viral DNA as part of its own genome, translating and transcribing the viral genes along with the cell's own genes, producing the proteins required to assemble new copies of the virus. It is difficult to detect the virus until it has infected the host. At that point, the infection will persist indefinitely.

Retroviruses are proving to be valuable research tools in molecular biology and have been used successfully in gene delivery systems. When retroviruses have integrated their own genome into the germ line, their genome is passed on to a following generation. These endogenous retroviruses (ERVs), now make up 5%–8% of the human genome. Most insertions have no known function and have previously been referred to as "junk DNA." However, many endogenous retroviruses control gene transcription, cell fusion during placental development in the course of the germination of an embryo, and resistance to exogenous retroviral infection. Endogenous retroviruses have also received special attention in the research of immunology-related pathologies, such as autoimmune diseases like MS.

Because reverse transcription lacks the usual proofreading of DNA replication, a retrovirus mutates very often. This enables the virus to grow resistant to antiviral pharmaceuticals quickly, and impedes the development of effective vaccines and inhibitors for the retrovirus.

Antiretroviral drugs are medications for the treatment of infection by retroviruses, primarily HIV. Different classes of antiretroviral drugs act on different stages of the HIV life cycle. Combination of several (typically three or four) antiretroviral drugs is known as highly active antiretroviral therapy (HAART).

6.6.3 *Pandemic Risk*

A pandemic (or global epidemic) is a disease that affects people over an extensive geographical area. Here is a historic overview:

- The plague of Justinian, from 541 to 750, killed between 50% and 60% of Europe's population.[32]
- The Black Death of 1347 to 1352 killed 25 million in Europe over five years. The plague reduced the world population from an estimated 450 million to between 350 and 375 million in the 14th century.
- The introduction of smallpox, measles, and typhus to the areas of Central and South America by European explorers during the 15th and 16th centuries caused pandemics among the native inhabitants. Between 1518 and 1568 disease pandemics are said to have caused the population of Mexico to fall from 20 million to 3 million.
- The first European influenza epidemic occurred between 1556 and 1560, with an estimated mortality rate of 20%.
- Smallpox killed an estimated 60 million Europeans during the 18th century (approximately 400,000 per year). Up to 30% of those infected, including 80% of the children under five years of age, died from the disease, and one-third of the survivors went blind.
- In the 19th century, TB killed an estimated one-quarter of the adult population of Europe; by 1918 one in six deaths in France was still caused by TB.
- The influenza pandemic of 1918 (or the Spanish flu) killed 25–50 million people (about 2% of the world population of 1.7 billion). Today influenza kills about 250,000 to 500,000 worldwide each year.

Could nanomedicine help us to reduce the risk of pandemic outbreaks? Yes, by developing vaccines based on antibodies and by delivering vaccines. In addition, there may be ways to use computer simulations, both to anticipate virus mutations[339] with potentially highly contagious mutants that easily infect large parts of the population, and to simulate the path a pandemic[340] could take along public transportation infrastructures.

In the mid-19th century, important steps were taken to significantly improve public health and reduce pandemic risk, starting with water quality: Since contamination of water by cholera and typhoid infected others, it was important to take action like adding chlorine (introduced around 1900). Other important milestones in the fight against infectious diseases are listed below:

- Louis Pasteur (1822–1895) proved beyond doubt that certain diseases are caused by infectious agents, and developed a vaccine for rabies.
- Robert Koch (1843–1910) provided the study of infectious diseases with a scientific basis known as Koch's postulates. His students Behring, Ehrlich, Wernicke, and Kitasato developed vaccines against diphtheria, tetanus, meningitis, and pneumonia.
- Edward Jenner, Jonas Salk (1914–1995), and Albert Sabin developed effective vaccines for smallpox and polio, which would later result in the eradication and near-eradication of these diseases, respectively.
- The discovery of antibiotics (see above) greatly reduced the threat of bacterial infections.

However, despite major advances there remains a constant threat as demonstrated by the recent Ebola outbreak. Despite annual influenza vaccinations of a large part of the population in the Western world, even the flu is still causing severe problems, in particular among the very young and the weak and old.

6.7 Tissue and Organ Transplantation

I love those who yearn for the impossible.

—Johann Wolfgang von Goethe (1749–1832)

Tissue and organ transplantation is the moving of tissue or an organ from one body to another or from a donor site to another location on the patient's own body. The goal is to replace the recipient's damaged or absent tissue and/or organ. When tissues are transplanted within the same person's body they are called *autografts*. Transplants received from a donor are called *allografts*. Allografts can either be from a living or cadaveric source.

Tissues that can be transplanted include bones, tendons (both referred to as musculoskeletal grafts), cornea, skin, heart valves, nerves, and veins. Organs that can be transplanted are the heart, kidneys, liver, lungs, pancreas, intestine, and thymus. The kidneys are the most commonly transplanted organs, followed by the liver and then the heart. Cornea and musculoskeletal grafts are the most commonly transplanted tissues. They outnumber organ transplants by more than tenfold.

Unlike organs, most tissues (with the exception of corneas) can be preserved and stored for up to five years, meaning they can be "banked."

Transplantation raises a number of bioethical issues, including the definition of death, when and how consent should be given for an organ to be transplanted, and payment for organs for transplantation. In compensated donation, donors get money or other compensation in exchange for their organs. This practice is common in some parts of the world, whether legal or not, and is one of the many factors driving medical tourism.

In the United States, the National Organ Transplant Act of 1984 made organ sales illegal. In the UK, the Human Organ Transplants Act 1989 made organ sales illegal but has since been superseded by the Human Tissue Act 2004. Monetary compensation for organ donors is in the process of being legalized in Australia and Singapore. Iran has had a legal market for kidneys since 1988, and the market price is below $2000 for the recipient.[341]

Transplantation medicine is one of the most challenging and complex areas of modern medicine. To avoid transplant rejection, the body's immune response to the transplanted tissue or organ, the patient is given immunosuppressant drugs. When possible, transplant rejection can be reduced by determining the most appropriate donor–recipient match.

We have already touched upon the promise of regenerative medicine for patients waiting for organ transplantation (Section 5.6) and the promise of treating diabetes with beta cell transplants. Growing numbers of biotech start-ups (see Table 5.4) are working on tissue generation for cardiovascular regeneration, skin, muscular and skeletal conditions, spinal cord injuries, 3D printing of organs, etc.

The regrowing of tissues and organs from stem cells would be a blessing for patients with failing organs that are currently waiting for donors.

The field of tissue and organ transplantation is therefore a particularly promising area of impact for nanomedicine. In addition to transplantation, nanomedicine also offers engineering solutions like the Artificial Pancreas project (Section 5.1), the bionic eye (Section 5.6), DBS, and other uses of implants.

Finally, mabs, as discussed in Section 4.5, are offering new ways to develop immunosuppressants with less side effects than current standard medical practice based on 30-year old drugs like Tacrolimus (trade names Prograf, Advagraf, Protopic) that, while reducing the activity of a patient's immune system, also generate severe side effects.[342]

What we are looking for is to preserve the organ function, while not hurting the immune system that is constantly trying to reject the foreign tissue.

Chapter 7

The Healthcare Ecosystem and Biomedical Research Funding

Rulers need three things: food, weapons, and trust. If one of those three things has to be given up it must be weapons. If two of those three things have to be given up, food must go as well. Trust cannot be given up; there can be no governing without trust!

—Confucius

What Confucius said about government applies to an even higher degree to wellness and health. There is a never-ending need to develop new "weapons", i.e., new scientific and technological breakthroughs that can be translated into patient benefits.

The role of food is played by the healthcare ecosystem where the stakeholders are working together to provide clinical care to patients.

However, to really make it work, to prevent diseases, and to help patients recover from illness, we cannot achieve our goals without trust—trust between the various stakeholders of the healthcare ecosystem and ultimately trust between the patient and the caregiver.

Nanomedicine: Science, Business, and Impact
Michael Hehenberger
Copyright © 2015 Pan Stanford Publishing Pte. Ltd.
ISBN 978-981-4613-76-7 (Hardcover), 978-981-4613-77-4 (eBook)
www.panstanford.com

Without trust we have to spend more and more resources on bureaucratic control, on regulation, and on administrative overhead related to financial matters, fraud protection, matters of privacy, etc.

A society based on trust will achieve its goals with less effort and a minimum of administrative overhead and regulatory protection.

If there is no trust in the ability of scientists to improve our understanding of disease mechanisms, there will be no funding of their efforts.

If there is no trust in the ability of doctors and nurses to do what's in the best interest of patients and to have the proper skills and experience to achieve positive health outcomes, there may not be a willingness by payers to insure patients.

The translation of current and future advances in biomedical research and nanotechnology into patient benefits requires a huge effort; it requires funding, risk taking, and proper regulatory oversight. It will always be difficult to achieve positive results, but if translational medicine projects go well they have the potential to address important unmet needs and to transform healthcare.

To make it happen requires strong coordination and trust among the stakeholders in the healthcare ecosystem.

7.1 It's All about the Patient/Consumer

To transform healthcare, its stakeholders must be aligned across this ecosystem. To adopt new ideas such as genomic medicine and nanomedicine, the stakeholders must agree that it is in their common interest. There is no point in making a difficult change unless it saves lives, prevents and cures disease and, above all, more than balances the cost of implementation.

Let's quickly review the respective roles of the ecosystem stakeholders, also shown in Fig. 7.1.

7.2 Providers of Medical Care

Healthcare is provided by physicians, nurses, and other medical professionals in hospitals, community medical centers, or family

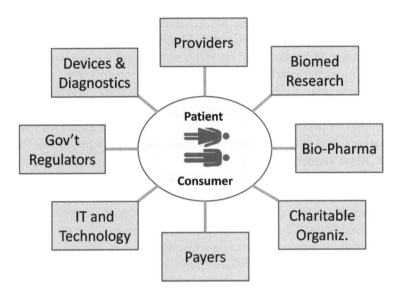

Figure 7.1   Healthcare ecosystem around patients/consumers.

practices. However, diagnostic devices are increasingly operated by specialized technicians and geneticists are expected to be consulted in the new era of genomic medicine, starting with oncology. As we learn more about the importance of diet and exercise to prevent metabolic and cardiovascular disease (CVD) as well as cancer, we are further expecting fitness experts and nutritionists to join the medical teams in charge of wellness and patient care.

7.2.1 *Biomedical Research: Academic Medical Research Centers and Government Research*

Academic medical research centers (AMRCs) are conducting medical research, while also providing medical care to patients seeking nonstandard treatments that may involve leading-edge procedures and medications. Research may include both independent academic research activities and sponsored clinical research for the testing of new drugs and/or devices. Sponsors are usually commercial drug and device companies that want to conduct clinical trials aimed at the generation of data needed for regulatory approval.

AMRCs are competing for research grants and for clinical research studies, often on a global level. There is competition for relevance and volume of scientific output and for leadership in selected disease areas. AMRCs are also competing for the best students who study for medical degrees. AMRCs are therefore ranked by their quality of teaching and research and by their reputation in providing clinical care, often by disease area. There are national and global rankings of AMRCs published in places like the US News and World Report.[343,344]

The US National Institutes of Health (NIH) are still the world's leading provider of government research funding. However, as reported in *Science* magazine,[345] recent budget decisions have reduced the NIH budget as adjusted for inflation and forced leading academic research groups to look elsewhere for additional funding.

Between 1998 and 2003, the NIH's biomedical research budget soared from \$14 B per year to \$27 B. From 2004 to 2008 it stayed largely flat. In 2009, as part of the "Obama Stimulus" package after the recession of 2008, the NIH received an additional \$10.4 B, followed by first stagnation and then a sizable 5% cut in 2013 due to the politically motivated "Sequester." In 2014, the NIH budget was back to \$30 B, which, adjusted for inflation, is still about \$5 B smaller than during its peak in 2003. Grant approval rates dropped from about 30% in the late 1990s to below 17% in 2013.

US government agencies like the Department of Energy (DOE) and the Department of Defense (DOD), in particular via the Defense Advanced Research Projects Agency (DARPA), have also contributed to biomedical research funding. The important role of the DOE in the start-up phase of the Human Genome Project (HGP) is well known. It may be less well-known that the DOD has spent $\sim$\$3 B since 1992 on breast cancer research[346] via the National Breast Cancer Coalition (NBCC). Only the National Cancer Institute (NCI), the largest institute of the NIH, which is spending about \$600 M a year on breast cancer, has funded more breast cancer research studies in the United States than the NBCC. DARPA, less known for investments in medical research than for its significant role in, for example, the creation of the Internet, has just set aside \$70 M to advance innovative neurotechnologies,[347] with particular focus on deep brain stimulators applied to neurological disorders such as posttraumatic stress, depression and chronic pain. Finally, the National Science Foundation (NSF) is an independent US federal agency that supports fundamental research and education across all fields of science and engineering. In fiscal year 2014, its budget was $7.2 B across all fields of science, distributed to nearly 2000 colleges, universities, and other institutions.

In Europe and AP there are numerous government programs funding medical research, often with special focus on cancer.

The European Union (EU) is active via its framework programs[348]—estimated to provide funding for about 10% of all European research projects—and EU member countries typically also fund their own national biomedical research funding efforts. To name a few:

- UK Medical Research Council (MRC),[349] established in its current form as early as 1919. In 2012/2013, the MRC spent £766.9 M on research.
- Several German Max–Planck Institutes (MPIs),[350] going back to 1911, support a significant portion of biomedical and life sciences research, divided into eight focus areas. The individual Max–Planck Institutes are independent and autonomous, they have their own, internally managed budgets, and they are free to attract third-party project funds. An interesting example is the MPI for neuroscience in Jupiter, FL, which was established 2008 and supported by $94 M of US public funding. It can be considered as an attempt by the Max–Planck Society to go global and to tap into private philanthropic support from individuals, corporations, and foundations across the United States. Other German research organizations worth mentioning are the Fraunhofer Society (mostly focused on applied research) and the Helmholtz Society. The German Cancer Research Center (DKFZ) in Heidelberg has an annual budget of ~$250 M and is part of the Helmholtz Association of German research centers, which have a total budget of ~$5 B, most of which spent on research outside life sciences and medicine.

- In France, the Institut Gustave Roussy (IGR),[351] founded 1926, is one of the biggest health centers dedicated to oncology in Europe together with the European Institute of Oncology in Milan, Italy. It is located south of Paris, France.
- The Karolinska Institute (KI)[352] in Stockholm, Sweden, founded in 1810 and financed by the Swedish federal government (total annual budget ∼$900 M), has established a strong reputation for medical research excellence, consistently ranked in the very top of Europe and in the top 10 globally. Since 1901 the Nobel Assembly at KI has selected the Nobel laureates in physiology or medicine.

In the Asia Pacific region, several governments are investing heavily in biomedical and life sciences research. To name a few:

- Japan's National Cancer Center (NCC)[353] was founded in 1962 and has been on the cutting edge of research to unravel the cause of cancer and to develop medical treatment. Founded in 1917, RIKEN[354] is Japan's largest government-funded (∼$1 B) research center covering many scientific disciplines, among them chemistry, biology, and medicine.
- Established in 1991, the Agency for Science, Technology and Research (A*STAR)[355] was established with the primary mission to raise the level of science and technology in Singapore. Its Biomedical Research Council (BMRC) oversees seven research institutes and several other research units that focus on both basic and translational and clinical research. A*STAR has entered into several global research collaborations and attracted outstanding guest scientists and life sciences industries to Singapore.
- In 2002, Taiwan's Academia Sinica launched the National Research Program for Genomic Medicine (NRPGM)[356] as a national priority program, focused on disease-oriented topics dedicated to lung and liver cancer, infectious diseases, and highly heritable diseases.
- The Commonwealth Scientific and Industrial Research Organisation (CSIRO)[357] is the federal government agency for scientific research in Australia. It was founded in 1926

and includes animal, food, and health sciences as a focus area. The state of Victoria has recently created a world class life sciences center[358] at the University of Melbourne.

- India established the Council of Scientific and Industrial Research (CSIR)[359] to "maximize the economic, environmental and societal benefits for the people of India." Focus areas include medical and life sciences research with emphasis on serious Indian food and health challenges.
- China has invested heavily in Genomics and the Beijing Genome Institute (BGI)[360] is now the world's leading provider of DNA sequencing services.

7.2.2 Drug Developers and Manufacturers and Clinical Research Organizations

Biopharmaceutical are important participants in the healthcare ecosystem as they develop and manufacture lifesaving medicines. They rely on clinical research organizations (CROs) to administer the global multicenter clinical trials that are required to bring new drugs to market. CROs play the role of an interface between pharma companies and AMRCs where patients are recruited.

7.2.3 Charitable Organizations: Advocacy Groups and Foundations

In many specific disease areas, patients and their families as well as dedicated doctors and medical professionals have formed support groups and organized themselves into advocacy groups and foundations that collect donations and work hard to make a difference for people affected from the disease. Examples of such organizations will be provided below. As they are gaining visibility and collecting resources, they make an impact on health policy, quality of care, prevention, and research.

7.2.4 Public and Private Payers of Health Insurance

Most developed countries have implemented single-payer systems, sometimes complemented with optional private insurance add-ons for people willing and able to pay extra for additional services like shorter wait times for surgical procedures, single rooms in hospitals, extended drug formularies, etc. The US health system is an exception as it relies mostly on private payers. Even in government programs such as Medicare (for retirees), Medicaid (for low-income households), and the Veteran's Administration (for soldiers and war veterans), private payers are involved. Payers have a strong influence on medical care because they decide about reimbursements of diagnostic tests, doctor's visits, hospital stays, surgical procedures, and approved medications. There are also combined payer–provider organizations (such as Kaiser Permanente[361] in the United States) that may have an advantage in offering preventive care, simplified administrative processes and lower overall costs.

7.2.5 *IT and Technology Solution Providers*

In today's healthcare ecosystem, information technology (IT) companies play a significant role in providing computer hardware, software and services. From information systems supporting medical centers and hospitals, electronic medical records, and IT support of drug developers and medical instruments, the ecosystem is critically dependent on IT. Stakeholder interactions are facilitated by global standards such as Health Level Seven International (HL7)[362] and CDISC.[363]

Founded in 1987, HL7 is a standards-developing organization dedicated to the exchange, integration, sharing, and retrieval of electronic health information that supports clinical practice. Similarly, the CDISC is a global nonprofit organization that has established standards to support the acquisition, exchange, submission, and archive of clinical research data and metadata.

7.2.6 *Government Regulators*

The US Food and Drug Administration (FDA)[364] and equivalent government regulatory authorities such as the European Medicines Agency (EMA)[365] and the Japanese Ministry for Health, Labor and Welfare (MHLW) play a key role in regulating food, drugs, diagnostics, and devices. It is one of the most important functions of governments around the world to protect patients and consumers against health risks. Government functions such as the US Centers for Disease Control (CDC) should also be mentioned as helping to prevent and manage outbreaks of infectious diseases and as organizers of vaccination campaigns for prevention and protection.

7.2.7 *Medical Devices and Diagnostics Companies*

Similar to the biopharmaceutical industry, medical diagnostics and device companies are also conducting clinical research and using the results to gain regulatory approval. However, the clinical studies are generally less time consuming and do not quite involve the high numbers of patients as needed to win the approval of new drugs.

After regulatory approval, the new diagnostic tests and new medical devices have to be introduced into medical practice, first by AMRCs and then by all other providers.

Examples of diagnostic tests that are part of standard medical practice are listed below:

- Blood tests
- Urine tests
- Medical imaging including X-rays, magnetic resonance imaging (MRI) scans, positron emission tomography (PET) scans, colonoscopies, mammograms, etc.
- Genetic tests (single-nucleotide polymorphisms [SNPs], DNA)

Examples of medical devices:

- Medical imaging equipment
- Hip and knee replacements, bone and tooth implants
- Cardiovascular stents, pacemakers
- Device technologies to enable minimally invasive surgery (MIS)
- Deep brain stimulators
- Continuous glucose measurement implants and insulin pumps

7.3 The Future of Funding for Basic Biomedical and Translational Research

As government budgets are increasingly scrutinized after the big recession of 2008, a new trend—the privatization of American science—is emerging. As covered in a story published by *The New York Times* on March 15, 2014, "Billionaires with Big Ideas"[366] are trying to fill the gaps left by reductions in the funding of scientific research budgets. A subsequent paper in *Science* examined NIH funding volatility and pointed to an emerging "venture philantropy" trend.[367]

The creation of charitable foundations enjoying preferred tax status in the United States is not a recent phenomenon. For example, the Howard Hughes Medical Institute (HHMI),[368] founded by the famous aviator and industrialist Howard Hughes, has been around since 1953. In 2013, the HHMI spent $727 M on medical research and another $80 M on scientific education.

More recently, in 2000, Microsoft cofounder Bill Gates started the Bill & Melinda Gates Foundation (BMGF)[369] with a total endowment exceeding $40 B (as of year-end 2013). The BMGF, based in Seattle, WA, is controlled by its three trustees: Bill Gates, Melinda Gates, and Warren Buffett. In 2006, Buffet pledged to give the foundation approximately 10 million Berkshire Hathaway Class B shares spread over multiple years through annual contributions, with the first year's donation of 500,000 shares being worth ~$1.5 B. The BMGF has made grant payments of more than $30 B since inception. Grant payments during 2012–2013 were $3.5 B on average.[370] The BMGF focuses mainly on global public health and is driving initiatives such as fighting malaria, eradicating polio, significantly reducing the incidence of human immunodeficiency virus (HIV) and improving the life of people living with HIV, and fighting pneumonia, TB, neglected infectious diseases, etc.

Former New York City mayor and founder of Bloomberg LP, Michael Bloomberg, has donated a total of $1.1 B to his alma mater, Johns Hopkins University. A large fraction went to its School of Public Health.

The "Giving Pledge"[371] is a new movement that gives hope to future scientific funding: The campaign, announced in 2010 by Bill and Melinda Gates along with investor Buffet, aims at encouraging the wealthiest people in the world to make a commitment to give most of their wealth to philanthropic causes. As of February 2014, 122 wealthy individuals and couples have signed the pledge. It can be assumed that more than $100 B will be donated to science and that the movement will spread globally.

What is already happening is that a number of important areas of medical research are currently receiving boosts provided not only by the past and current fortunes of wealthy individuals but also by the passion and enthusiasm of individuals that have developed effective fund-raising organizations and depend on a large number of small donors. Here are a few illustrative examples sorted by disease area:

1. Basic Biomedical, Bioengineering, and Genomic Research

- In 1936, American born Sir Henry Wellcome created the Wellcome Trust that has since championed science and biomedical research and influenced health policy across the globe. The trust's £16.4 B investment portfolio provides the income for its funding, currently around £600 M per year. With headquarters in London,[372] the Wellcome Trust supports research of the highest quality with the aim of improving human and animal health. In 1992, the Wellcome Trust and the UK Medical Research Council founded the Sanger Centre/Institute, named after Frederick Sanger, recipient of two Nobel Prizes for chemistry, for protein sequencing (1958) and DNA sequencing (1980).[373] The institute is located on the Wellcome Trust Genome Campus in Hinxton near Cambridge. Twenty years later it has grown to over 900 people and is one of the world's largest genomic research centers.
- In 1929, C. C. Little ended his term as president of the University of Michigan and founded the Roscoe B. Jackson Memorial Laboratory in Bar Harbor, ME. Support came from Detroit industrialists, such as Edsel Ford and Jackson, president of the Hudson Motorcar Company, with land donated by family friend George B. Dorr, the philanthropist who organized the gifts of land that established Maine's Acadia National Park. Today, the Jackson Lab is an independent,

nonprofit biomedical research institution with more than 1500 employees in Bar Harbor, ME, Sacramento, CA, and a new genomic medicine institute in Farmington, CT. Jackson Lab is a National Cancer Institute (NCI) and the world's source for more than 7,000 strains of genetically defined mice, the home of the mouse genome database. Its annual budget is ~$230 M, it receives public funding of ~$60 M and generates annual services revenues of ~$160 M, depending on its dominating position and expertise in the breeding of mice used for medical research. Jackson Lab can claim 26 Nobel Prizes that are linked to its mouse-related research activities.[374] Jackson Lab could also be mentioned as an example of philanthropic beginning turned into a self-sustaining operation, having achieved dominance in a special niche area, namely, mouse genetics and breeding.

- The SALK Institute[375] was established in 1962 by Dr. Jonas Salk, the developer of the polio vaccine. His goal was to explore questions about the basic principles of life and to make it possible for biologists and others to work together in a collaborative environment. For San Diego mayor Charles Dail, a polio survivor, bringing the Salk Institute to San Diego was a personal quest. Dail showed Salk 27 acres on a mesa in La Jolla, just west of the proposed site for the new University of California campus then planned for San Diego. In June 1960, in a special referendum, the citizens of San Diego voted overwhelmingly to give the land for Salk's dream. With initial financial support from the National Foundation/March of Dimes, Salk and architect Louis Kahn were able to proceed. The first Nonresident Fellows selected were a group of eminent scientists including Nobel laureates Francis Crick,[376] Salvador Luria,[377] and Jacques Monod.[378] Today the major areas of study at Salk are: molecular biology and genetics, neurosciences, and plant biology. Funding has been provided by private donors and by the NIH. Today's annual SALK Institute budget is about $130 M.

- Located in Cambridge, MA, Whitehead Institute[379] was founded in 1982 by businessman and philanthropist Edwin C. "Jack" Whitehead, who provided $35 M to construct and equip a new building, as well as $5 M per year in guaranteed income and a substantial endowment in his will (for a total gift of $135 M). The successful Whitehead Institute concept—tight association with a major research university—was codeveloped by Whitehead and MIT professor and Nobel laureate David Baltimore. In the 1990s, Whitehead Institute scientists shaped the emerging field of genomics by making the single largest contribution to the HGP. Current research programs are focused on cancer, immunology, developmental biology, stem cell research, regenerative medicine, genetics, and genomics. The annual operating budget is ~$75 M.
- In 2003, Eli Broad (who made a fortune in housing and insurance) provided seed funding of $200 M to the Eli and Edythe L Broad Institute of Harvard and MIT,[380] to "transform medicine by using systematic approaches in the biological sciences to dramatically accelerate the understanding and treatment of disease." In 2008, the Broads gave another gift of $400 M and in 2013 another $100 M to permanently endow the institute, providing long-term sustainability for its unique model of collaborative, interinstitutional research. The annual operating budget has reached $270 M in 2013. Directed by Eric Lander,[381] the Broad Institute aims at:
  - assembling a complete picture of the molecular components of life, identifying the functional elements in the human genome and revealing how they vary in humans and other species
  - defining the biological circuits that underlie cellular responses in order to understand how this complex circuitry functions in human health and disease
  - uncovering the molecular basis of major inherited diseases through comprehensive studies of genetic variation: Examples of disease areas are type 2 diabetes, leukemia, and Crohn's

  – unearthing all the mutations that underlie different cancer types by creating systematic catalogues of these changes across different types of tumors
  – discovering the molecular basis of major infectious diseases, defining components in both pathogens and hosts, to develop effective vaccines, rapid diagnostics, and new therapeutics
  – transforming biopharmaceutical discovery and development by innovating the drug discovery process, including chemical synthesis of unprecedented diversity, testing of candidate drugs on living cells and tissues, developing new methods to identify drug targets, optimize drug efficacy and safety, and increase accuracy and efficiency of clinical trials

- In 2009, Swiss billionaire Hansjorg Wyss (founder and CEO of Synthes, a medical device company sold to Johnson & Johnson for $20 B in 2012) donated $125 M to help establish the Wyss Institute for Biologically Inspired Engineering[382] at Harvard University, a cross-disciplinary institute with alliances to other leading academic and clinical institutions in the Boston area. The Wyss Institute applies biological principles to develop new engineering solutions for medicine, industry, the environment, and many other fields that have previously not been touched by the biology revolution. Wyss, a graduate of Harvard Business School, doubled his contribution in 2013 to a total of $250 M.

- The Ellison Medical Foundation was founded in 1997 and is located in Bethesda, MD. The foundation supported research in the following discipline areas: biomedical research on aging, age-related diseases, and disabilities. Its major philanthropic support comes from Oracle founder and CEO Larry Ellison who has donated hundreds of millions of dollars. Until 2013, $40 M per year were given to 25 senior scholars and 25 new scholars. The senior scholars received $1 M each and the new scholars received $400,000 each for four years of research. Ellison whose net worth is estimated to exceed $40 B keeps details private but has declared that he intends to "give billions more over time" to medical research.

2. Neuroscience (Brain Research)

- Launched in 2003 and located in Seattle, WA, with a seed contribution of $200 M and an additional $300 M in 2012 from Microsoft cofounder and philanthropist Paul G. Allen, the mission of the Allen Institute for Brain Science[383] is to accelerate the understanding of how the human brain works in health and disease.
- Fred Kavli (1929–2013) was a Norwegian born entrepreneur who established the Kavli Foundation[384] for the advancement of science, starting in December 2000. He donated $200 M distributed across 17 Kavli Institutes, each with a starting endowment of $7.5 M and a goal to reach $10 M. Kavli also established $1 M "Kavli Prizes" for astrophysics, neuroscience, and nanoscience, awarded since 2008 by the King of Norway. The Kavli Institutes[385] in Neuroscience are located at the following universities: Columbia, NY (Brain Science—led by Nobel laureate Eric Kandel,[386] famous for his research into molecular mechanisms of memory storage in Aplysia and mice and, more recently, with animal models of memory disorders and mental illness); UC San Diego (Brain and Mind); Yale, CT (Neuroscience);and Trondheim, Norway (Systems Neuroscience). Harvard is hosting a Kavli Bio-Nanoscience Institute.
- The McGovern Institute for Brain Research at MIT[387] was established in February 2000 by Patrick J. McGovern (1937–2014), founder and chairman of the International Data Group (IDG),[388] and his wife Lore Harp McGovern. Their donation will total $350 M over 20 years.
- The Stanley Medical Research Institute (SMRI)[389] is supporting research on the causes of, and treatments for, schizophrenia and bipolar disorder. Since it began in 1989, the SMRI has supported more than $300 M in research

in over 30 countries around the world. It is the largest nongovernmental source of funds for research on these diseases in the United States. It supports brain research in Rockville, MD, and the Center for Psychiatric Research[390] as part of the Broad Institute in Cambridge, MA (see above).

- The Wings for Life Foundation[391] is a European initiative started in 2003 by Red Bull founder Dietrich Mateschitz[392] and two-time motocross world champion Heinz Kinigadner. In 2003, Kinigadner's son Hannes had a tragic accident, which left him tetraplegic, prompting Kinigadner and Mateschitz to set up a research foundation with the goal of finding ways to cure spinal cord injuries (SCIs). It is not known how much billionaire Mateschitz has contributed personally to the SCI research foundation Wings for Life, which he cofounded. However, in 2012 he donated ~$100 M to the Paracelsus Medical University (PMU)[393] in Salzburg, Austria, to build a new facility for research into paraplegia.

- Since 2000, the Michael J. Fox Foundation[394] is dedicated to finding a cure for Parkinson's disease (PD) through an aggressively funded research agenda and to ensuring the development of improved therapies for those living with PD today. By 2012, the Michael J. Fox Foundation has funded more than $450 M to speed a cure for PD. Google founder Sergey Brin and his wife Anne Wojcicki challenged the Michael J. Fox Foundation to raise $50 M by December 31, 2012, and ended up matching $53 M of donations.[395]

- Since 1980, the Alzheimer's Association (initially named Alzheimer's Disease and Related Disorders Association or ADRDA) with headquarters in Chicago, IL, is focused on care, support and research for AD. Its mission[396] is "to eliminate AD through the advancement of research; to provide and enhance care and support for all affected; and to reduce the risk of dementia through the promotion of brain health." The Alzheimer's Association invests in Alzheimer research through a peer-reviewed research grants program. Since 1982, it has committed more than

$220 M to best-of-field research proposals. It's southwest affiliate, Banner Alzheimer's Institute (BAI), in Arizona[397] was established in 2006 and has made a name for itself as a research center focused on brain imaging and genomics and as a site for clinical trials focused on prevention of AD in high-risk populations.[398] In 2012, its Banner Alzheimer's Foundation announced a $40 M campaign to advance promising AD treatment and prevention. In 2013, the BAI announced a major prevention trial funded by a $33 M NIH grant to evaluate a treatment in cognitively healthy older adults at the highest-known genetic risk for developing AD at older ages. In July 2014, the BAI announced a partnership with Novartis in a pioneering medical trial to determine whether two investigational antiamyloid drugs—an active immunotherapy and an oral medication—can prevent or delay the emergence of symptoms of AD.

3. Cancer Research

- The David H. Koch Institute for Integrative Cancer Research[399] at MIT was launched in October 2007 with a $100 M grant from David Koch, the executive vice president of the oil conglomerate Koch Industries. It continues activities started as early as 1974 by MIT Professor and Nobel laureate Salvador Luria,[400] who created MIT's Center for Cancer Research (CCR), one of the NCI centers supported by the NIH. Recently, Koch has contributed additional funds for cancer research, estimated to add up to a total of $150 M.

- Jon Huntsman,[401] who made a fortune in the chemical industry (polystyrene, plastic containers, etc.), has donated at least $1.2 B to philanthropic causes, among them the Jon & Karen Huntsman Cancer Foundation and the Huntsman Cancer Institute at the University of Utah. This cancer center was conceived by the Huntsmans in 1993 with a donation of $10 M. They donated $100 M in 1995 and another $125 M shortly thereafter. The annual contributions by the Huntsman Foundation amount to over \$20 M and the Huntsman Cancer Center is employing over 1400 people. In 2013, the Huntsmans added another \$50 M for a new research building dedicated to children's cancer.

- In 1957, Danny Thomas—a successful movie and television entertainer of Lebanese origin—founded the American Lebanese Syrian Associated Charities (ALSAC) to raise the money needed to build and to finance the ongoing operating expenses of St. Jude Children's Research Hospital[402] in Memphis, TN. Sustained by a highly sophisticated fundraising organization, St. Jude has established a reputation of excellence in cancer research and the clinical care of children with cancer. St. Jude has the world's best survival rates for the most aggressive childhood cancers.[403] St. Jude's operating budget for cancer research and patient care is ~\$700 M per year, entirely funded by donations and by investment income from a ~\$3 B asset base.

- The Milken Family Foundation was established in 1982 to develop educational K–12 and medical research programs. In 1993, Mike Milken started the Prostate Cancer Foundation (PCF).[404] By the end of 2013, the PCF had raised a total of ~\$600 M to support prostate cancer research. In 2007, Milken started the Myeloma Research Alliance (MRA), focused on finding and funding the most promising translational melanoma research. By 2013, MRA has awarded more than \$51 M to 121 research programs.[405] Another Milken initiative is Faster Cures,[406] a nonprofit organization that works to improve the medical research system and make it faster.

- In September 2013,[407] Nike cofounder Phil Knight and his wife Penny pledged \$500 million to kick-start a \$1 B cancer research initiative at the Knight Cancer Institute at Oregon Health & Science University (OHSU). Knight said the pledge is contingent on OHSU's success in raising at least \$500 M more for cancer within two years. OHSU has been in ongoing conversations with the Knights about the possibility of additional investment in the Knight Cancer Institute, which the couple supported with a $100 M gift in 2008. OHSU President Joe Robertson accepted this historic challenge. Knight Cancer Institute Director Brian Druker is known for his leadership in the development of the first molecularly targeted anticancer drug, Gleevec, which strongly contributed to today's revolution in personalized cancer medicine. His achievements earned him the prestigious Lasker Award.[408] The OHSU Knight Cancer Institute is a designated NCI and offers the latest treatments and technologies as well as hundreds of research studies and clinical trials.

- The American Cancer Society[409] was founded in 1913 as the American Society for the Control of Cancer (ASCC) by 15 prominent physicians and business leaders in New York City. It is today among the United States' largest foundations with a budget of ~$1 B per year, spent on patient support (38%), mission support (28%), research (16%), prevention (15.5%), and detection/treatment (10%).
- Several nonprofit foundations are raising funds for breast cancer research, among them the Susan G. Komen Foundation[410] (invested $58 M in 2012 toward research), the National Breast Cancer Foundation, the American Breast Cancer Foundation, the Breast Cancer Research Foundation, etc.

4. Immunology (Diabetes)

- The Phillip T. and Susan M. Ragon Institute was established in 2009 at Massachusetts General Hospital, MIT, and Harvard with the dual mission to contribute to the accelerated discovery of an HIV/acquired immunodeficiency syndrome (AIDS) vaccine and to establish a world leader in the collaborative study of immunology. The Ragons donated $100 M. The Ragon Institute is directed by Prof. Bruce Walker of Harvard University.

- Since 1970, the Juvenile Diabetes Research Foundation (JDRF) is dedicated to research into type 1 diabetes (T1D). By 2013, the JDRF had awarded more than $1.7 B to diabetes research. The JDRF's efforts have helped to significantly advance the care of people with this disease and have expanded the critical scientific understanding of T1D. The JDRF provided more than $110 M for T1D research in 2012 and is also trying to facilitate innovation by working with pharmaceutical and medical device companies. Major donors include Woody Johnson,[411] the great-grandson of Robert Wood Jonson, cofounder of Johnson & Johnson.
- The American Diabetes Association (ADA)[412] was founded in 1940 and is the United States' largest voluntary health organization leading the fight to stop (both type 1 and type 2) diabetes. The ADA funds research to manage, cure and prevent diabetes, delivers services to hundreds of communities, and provides information for both patients and healthcare professionals. According to the ADA's annual report[413] for 2012, its total research investments add up to ~$640 M across over 4000 projects.

5. Genetic Diseases

- The Cystic Fibrosis Foundation (CFF)[414] was established in 1955. In 1989, the CFF-supported scientist Francis Collins[415] discovered the defective gene that causes cystic fibrosis. In 2012, the CFF provided ~$140 M of funding for medical and community services. The CFF is the world's leader in the search for a cure for cystic fibrosis and is funding more cystic fibrosis research than any other organization. Nearly every cystic fibrosis drug available today was made possible because of CFF support.
- The Huntington's Disease Society of America (HDSA) and the Huntington's Disease Assistance Foundation (HDAF) are voluntary health organization dedicated to improving the lives of people with Huntington's disease and their families. Their mission is to promote and support research and medical efforts to eradicate Huntington's disease.

Table 7.1   Examples of private funding of biomedical research

1st year	Field	Total (US M$)	Annual ($M)	Name	Sponsor(s)
1929	B/G	N/A	230	Jackson Lab	C.C. Little / G.B. Dorr, R.B. Jackson
1936	B/G	26,400	960	Wellcome Trust	H. Wellcome (UK)
1953	BMR	N/A	800	HHMI	Howard Hughes
1957	Ca	3,000	700	St. Jude Children's Res. Hosp.	Danny Thomas, ALSAC
1962	BMR	N/A	130	Salk Institute	Jonas Salk + Community
1970	T1D	1,700	110	JDRF	Community
1982	B/G	135	75	Whitehead Inst.	E.C. "Jack" Whitehead
1982	Ca	650	N/A	Prostate Ca and Myeloma Res	Mike Milken + Community
1989	Psy	300	N/A	Stanley Medical Res.Inst.	Stanley / Broad
1993	Ca	650	50	Huntsman Cancer Inst.	Jon Huntsman / Univ. Utah
1997	BR	600	40	Ellison Medical Found.	Larry Ellison
2000	PH	40,000	3500	BMGF	Bill & Melinda Gates
2000	Br	350	N/A	McGovern Inst. Brain Res	Patrick J. McGovern (MIT)
2000	BR	200	N/A	Kavli Institutes	Fred Kavli
2000	PD	450	65	Michael J. Fox Found.	Michael J. Fox + Community
2003	B/G	700	270	Broad Inst.	Eli & Edythe L. Broad
2003	Br	500	60	Allen Inst. Brain Science	Paul G. Allen
2007	Ca	150	N/A	David H. Koch Inst.	David H. Koch / MIT
2009	BE	250	N/A	Wyss Institute	Hansjörg Wyss (Harvard)
2009	HIV	100	N/A	Ragon Inst.	Phillip & Susan Ragon / Harvard+MIT
2013	Ca	500	N/A	Knight Cancer Inst. Oregon	Phil & Penny Knight / Oregon HSU
2014	PH	350	N/A	Harvard T.H. Chan School of Public Health	Morningside Foundation (Ronnie & Gerald Chan, Hongkong)

B/G Basic Research and Genomics; BMR Basic Medical Research; Ca Cancer; BR Basic Research; Psy Psychic Disorders, Brain Research; PD Parkinson's Disease; PH Public Health Focus, Vaccines etc.; Br Brain; BE Biologically Inspired Engineering

Foundations dedicated to a single cause/disease area have made and are continuing to make a strong impact—not only on research but also as communities providing education and support.

Table 7.1 summarizes the ongoing private funding efforts. It is by no means complete but clearly demonstrates an ongoing trend.

Here is another interesting fact: When President Obama announced his "BRAIN Initiative" in April 2013, he was able to commit public funding of $100 M in 2014 and he is proposing $200 M (NIH: $100 M, DARPA: $80 M; NSF: $20 M) in his budget for 2015. He was also able to announce[416] that the Allen Institute for Brain Science, HHMI, Kavli Institute, and the SALK Institute had committed total annual investments of about $100 M into his initiative, hence more or less matching the US government investment of 2014.[417] Looking into the future, the total BRAIN budget could well approach $3 B and therefore closely match the previous public investment into the Human Genome Project (HGP), but with perhaps one third contributed by private donors, such as Paul Allen.

A noteworthy side phenomenon is the strong push to turn scientific results into benefit for patients, the emphasis on "translational research." It is understandable that donors with an interest in a particular disease—often caused by affliction within the family—want to see tangible results and returns of their investments.

According to research carried out by MIT Professor Fiona Murray,[418] "science philantropy" is accounting for 30% of funding (~$4 B per year) of the leading 50 research-based universities in the United States. When combined with endowment income, the funding is reaching $7 B per year.

For the top 50 public US universities the decade 2000–2009 saw a dramatic growth in university R&D expenditures from an average of $47 M in 2000 to $79 M per university in 2009. At the same time, the other 900 or so US universities outside the top 50 saw only moderate growth. Among the top 10 public science and engineering universities (Johns Hopkins University, University of Michigan, University of Wisconsin, University of California, San Francisco, University of California, Los Angeles, UC San Diego, Duke University, University of Washington, Penn State University, University of Minnesota), average spending has increased to ~$1 B, while the top 11–50 spend ~$480 M. Obviously, the leading private universities such as Harvard, Yale, and Stanford are even further ahead. Another interesting comparison by Murray is made in Table 3 of her paper, where she compares federal and philanthropic funding per scientific discipline for the period 1999–2009.

Combining life sciences and medicine, federal funding to academia for the 10-year period added up to $18.7 B, whereas science philanthropy contributed about 5%. Whereas federal funding to life sciences and medicine represented 61% of the total funding, 72% of all private donations were aimed at this research area.

It can be expected that the private portion of funding will increase and that the existing gap in funding between the very top institutions and the rest will widen even more. As in other areas of life such as professional sports, there may be a handful of places and a few scientific superstars per medical research specialty that will attract most of the attention and funding.

Another consequence of science philantropy is the protection of leaders such as Harvard from drastic policy changes: When the George W. Bush administration stopped funding of stem cell research projects in 2001, Harvard scientists turned to wealthy individuals and formed the Harvard Stem Cell Institute (HSCI), which quickly attracted over $40 M of private funding.

As seen from the list of biomedical research initiatives listed above, the Boston area (Harvard, MIT, etc.) seems to be an exceptional target of biomedical science philantropy.

No wonder that the biopharmaceutical industry is establishing research labs close by: The list of Boston-based industry labs is extensive and includes Pfizer, Sanofi/Genzyme, Novartis, AstraZeneca, Merck, Biogen, Vertex, Takeda/Millennium, etc.

Other hotbeds of activity are the San Francisco Bay area, Southern California around Los Angeles and San Diego, Washington State around Seattle, Washington, DC, St. Louis, Missouri, New York and the tri-state area including New Jersey and Connecticut.

Similar activities are emerging around the world's leading academic medical research centers.

Despite the new biomedical science philantropy trends, the US government would be well advised to continue (and step up) NIH funding. According to a recent study by Everett Ehrlich,[419] the NIH spent $26.6 B in 2010 on extramural activities, supporting directly and indirectly 487,900 jobs nationwide in the United States. And 15 states experienced job growth of 10,000 or more. When the NIH funding was cut in 2011 to only $23.7 B of extramural spending (partly caused by the end of supplementary investment provided by the American Recovery and Reinvestment Act—also known as "Obama-Stimulus"), the number of NIH-supported jobs decreased to 432,094. Leading beneficiaries among the states were California, Massachusetts, New York, and Texas with 63,196, 34,598, 33,193, and 25,878 jobs, respectively. The NIH spending in 2011 produced $62.132 B in new economic activity (compared to $69.190 B in 2010). Other statistical details show that a dollar of NIH support for research leads to an increase of private medical research of roughly 32 cents. A study by Tokyo University's Research Center for Advanced Science and Technology (RCAST) demonstrated further that NIH-funded entities produce new pharmaceutical discoveries that are more advanced, more likely to be considered "novel" by the FDA, and more likely to be "orphan drugs" targeted at smaller populations than generally considered by major biopharmaceutical companies. Several case studies confirm the beneficial complementarity between public and private sectors.

In 2010, 5 of the top 20 bestselling drugs were monoclonal antibody (mab) therapeutics, namely, Remicade, Humira, Avastin, Rituxan, and Herceptin, already discussed in Section 4.5. The mouse tumor cell line—critical for the invention of the hybridoma technique that is at the core of mab development—was funded by the NIH. mabs are, at their root, a product of NIH funding.

Another area of nanomedicine, DNA and RNA sequencing, has benefitted greatly from NIH funding. As explained in Sections 3.5–3.7, the amazing pace of advancement in sequencing technologies was driven not only by the NIH (and the DOE and other countries') investments 1988–2003 during the HGP, but also subsequently by continued NIH support. It should be noted that the benefits of NIH investments are also stimulating economic development on a global level, in line with the acknowledged leadership position of the United States.

Through its 27 institutes, funding ~6000 in-house scientists, and ~325,000 extramural researchers at over 3000 universities and medical schools, the NIH acts as a powerful economic engine.

Chapter 8

Public Health and Global Health Economics

Make a habit of two things: to help; or at least to do no harm.

—Hippocrates (460–371 BC)

In 2012, the World Health Organization (WHO) adopted the goal of reducing global noncommunicable disease (NCD) mortality by 25% in 2025— "25 by 25"—among adults 30 to 70 years of age.[420] NCDs include cardiovascular disease (CVD), cancer, chronic lung diseases, and diabetes. They kill three in five people worldwide. The greatest reductions in NCD mortality will come from population-wide interventions that address the risk factors of:

- tobacco use;
- unhealthy diet;
- lack of physical activity; and
- harmful use of alcohol.

Member states are encouraged to consider the development of national NCD targets and indicators, building on the global framework. By 2013, 50% of WHO member countries had an operational NCD policy with a budget for implementation.

Nanomedicine: Science, Business, and Impact
Michael Hehenberger
Copyright © 2015 Pan Stanford Publishing Pte. Ltd.
ISBN 978-981-4613-76-7 (Hardcover), 978-981-4613-77-4 (eBook)
www.panstanford.com

Some high-income countries have made progress in tobacco control and moderation of alcohol, but those health gains are threatened by dramatic increases in overweight and obesity.

The four risk factors reflect human behaviors and such behaviors are difficult to change. Population-level impact can best be achieved by changing environments to make healthy choices easier.

For instance, public policies such as taxes, worksite and public-space smoking bans, advertising bans, and information campaigns have been effective in reducing tobacco use.

Although such antismoking campaigns have worked well in Scandinavia and in the United States, it is never politically easy to force change in countries such as South Korea, where male smoking rates still are among the world's highest.[421]

Weight control seems to be even more difficult than tobacco control. However, as shown by population-based actions in Finland, it is possible to improve unhealthy dietary habits by means of national salt-labeling regulations (reduction of sodium intake), increasing fresh fruit and vegetable consumption, and substituting healthier fats.[422]

One of the few positive studies of obesity prevention, conducted in schoolchildren in the Netherlands, showed that blind replacement of sugar-containing beverages with sugar-free beverages modestly reduced weight gain and fat accumulation in normal-weight children.

Public health has important consequences for public finances. The more a country has to spend on the management of NCDs, the fewer resources will be available for other health-related initiatives such as biomedical research. Preventive population-based actions seem to be good investments and should therefore receive high priority.

What about global health funding? Poor and middle-income countries are receiving significant investments in health from the following major donors: US government, UK and European governments, including the European Commission, WHO, UNICEF, the GAVI Vaccine Alliance, the Global Fund to fight AIDS, Tuberculosis and Malaria, the Bill & Melinda Gates Foundation, other foundations, US and international nongovernmental organizations (NGOs), and the World Bank and regional development banks.

According to data compiled by the Institute for Health Metrics and Evaluation,[423] annual spending grew from an initial level of ~$5 B in 1990 to ~$10 B in 2000 and reached $30 B in 2011. The top 10 countries benefitting from global health donations were India, Nigeria, Ethiopia, Tanzania, Kenya, South Africa, Uganda, Mozambique, Congo, and Mexico. For instance, India received $2.53 B. In terms of disease area distribution, HIV/AIDS, malaria, tuberculosis (TB), and maternal/newborn and child health received the highest contributions. Not surprisingly, NCDs only received quite a small fraction.

Despite those positive actions and investments by rich countries, the world's readiness for emergencies such as outbreaks of infectious diseases (such as avian flu, Ebola, etc.) is still not at a sufficient level. When such problems occur, coordinated actions are often slow and not forceful enough. Even here, preventive measures should be put in place by organizations such as WHO and the United Nations. The science is available, what is needed are resources that can be readily deployed, even in poor countries with insufficient public health infrastructures.

Below we will take a look at the state of public health and associated current healthcare spending with particular focus on the biggest spender, the United States.

8.1 Global View of Healthcare Costs, Infant Mortality, and Life Expectancy

You can always count on Americans to do the right thing—after they've tried everything else.

—Winston Churchill (1874–1965; a similar quote sometimes attributed to Israeli politician and diplomat Abba Ebban)

According to WHO[424] and many other sources such as *The Economist*,[425] the United States is spending by far the highest percentage of its gross domestic product (GDP) on healthcare. With a GDP of ~$14.6 trillion, 17.9% (~$2.6 trillion) is spent on the health of its citizens. Countries with a comparable standard of living are spending much less but achieve similar or even better outcomes. The following table (Table 8.1) illustrates this fact.

Among the eight developed countries listed in Table 8.1, the United States is spending about 50% more per capita on healthcare than Canada or Germany but is trailing all other countries in infant mortality and life expectancy. The leaders in infant mortality as well as "Mortality of children up to 5 years," Sweden and Japan, are achieving their statistical leads by spending only 9.6% and 9.7% of their GDP, respectively. Physician density is also higher in five other countries. Switzerland is ahead of the United States by over 60% in physician density by only spending 10.8% of its GDP on healthcare. The only area where the United States is clearly leading is dental care, where no other country can point to such a high density of dental professionals.

To illustrate the significant differences in healthcare quality and coverage between the developed world (represented by the United States, Northern Europe, Japan, and Australia) and countries like China and India, those two countries have been added as well. As seen, healthcare spending per capita is still very low and infant mortalities are at very high levels. However, we expect significant progress as both China and India will grow their respective GDPs per capita.

What are the reasons for the discrepancy between very high spending and average measurable results on healthcare in the United States? Although there are no simple answers, a few observations can be made from Table 8.1:

(1) Compared to Canada, the UK, Sweden, Germany, and Japan, where healthcare expenses are carried by the government at a rate of 77%–82.5%, the US government is only funding 53.1% of the costs. Apparently, the governments in other countries are more efficient in running healthcare systems than private health insurance companies ("payers") in the United States. The payers are responsible for 46.9% of healthcare funding in the United States, compared to only 16.1% in the UK. On the other hand, the Swiss model, where as much as 41% of healthcare is funded privately, seems to indicate that high efficiencies can also be achieved outside government. Unlike the US system, in

Table 8.1 Global health statistics

Country	USA	Canada	UK	Sweden	Switzerland	Germany	Australia	Japan	China	India
Population [M]	310.384	34.017	62.036	9.380	7.812	82.302	22.268	126.536	1,348.932	1,224.614
GDP/Capita [$]	47,310	38,370	35,840	39,730	49,960	38,100	36,910	34,610	7,640	3,400
HC Spend/Cap [$]	8,362	4,404	3,480	3,820	5,394	4,332	3,441	3,355	379	132
HC/GDP [%]	17.7%	11.5%	9.7%	9.6%	10.8%	11.4%	9.3%	9.7%	5.0%	3.9%
% funded by Govt	53.1	81.1	83.9	81.1	59	77.1	68	82.5	53.6	29.2
% funded privately	46.9	18.9	16.1	18.9	41	22.9	32	17.5	46.4	70.8
Infant Mortality (per 1000)	7	5	5	2	4	3	4	2	16	48
Mortality up to 5 yrs (per 1000)	8	6	5	3	5	4	5	3	18	63
Life Expectancy	79	81	80	81	82	80	82	83	74	65
Physician Density per 10k	25	19.75	27.43	37.7	40.7	36.01	29.91	21.42	14.15	6.49
(Nurses + Midwives) per 10k	94.30	102.4	98.8	115.3	153.6	111.5	90.4	42.0	13.8	9.4
Dental Professionals per 10k	14.9	12.3	5.2	8.0	5.2	7.8	6.5	7.5	0.4	0.8
BMI > 25 (Male)	80.5	66.9	67.8	57.0%	56.5%	67.2%	75.7%	29.8%	45.0%	20.1%
BMI > 25 (Female)	76.7	59.5	63.8	47.2%	58.9%	57.1%	66.5%	16.2%	32.0%	18.1%

Source: WHO 2010

which employers or governments select health insurance, in Switzerland, it is the consumers themselves who purchase their health insurance. As proposed by Harvard's Regina Herzlinger, a study of the Swiss model could possibly lead to insights that may be applied to reduce US healthcare costs.[426]

(2) The US population has the highest values of the body mass index (BMI), a somewhat arbitrary but widely used parameter defined as (weight in kg)/(length in cm)^2. According to international guidelines, 80.5% of men and 76.7% of women in the United States were overweight in 2010. Since the BMI is linked to heart disease, diabetes, and other health complications, the United States has a health disadvantage explaining at least a fraction of the elevated cost per capita.

(3) US infant mortality is more than twice as high as that in Sweden, Japan, and Germany. Partly due to insufficient health coverage of the poor in the United States, the health systems in other countries are doing a better job in performing more preventive care with expecting mothers, as well as more comprehensive health controls and preventive measures with babies and small children.

(4) The high degree of fragmentation among US healthcare payers and providers is another contributing factor, leading to excessive paperwork, administrative overhead, and inefficiencies. According to a study by Citigroup documented in a recent Ernst & Young whitepaper,[427] healthcare businesses are wasting a huge amount on invoicing, payment processing and debt collection. Administrative overhead caused by repeated paperwork is huge, much higher than in single payer systems.

8.2 US Health Statistics

In a country well governed, poverty is something to be ashamed of.

—Confucius

Whereas WHO is the authoritative source for global health statistics, the Centers for Disease Control (CDC)[428] provides a lot of important

Current cigarette smoking

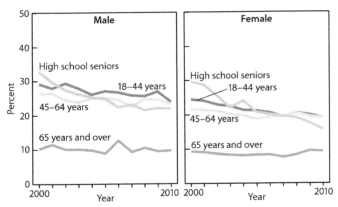

SOURCE: CDC/NCHS, *Health, United States, 2011*, Figure 8. Data from the National Health Interview Survey and the National Institutes of Health/National Institute on Drug Abuse, Monitoring the Future Study.

Figure 8.1 Cigarette smoking in the United States (2000–2010).

information helpful in understanding the situation in the United States.

For example, there has been a positive trend regarding smoking habits in the United States during the first decade of the 21st century, as illustrated in Fig 8.1 (CDC Fig. 8).

Unfortunately, the positive news about cigarette smoking is counterbalanced by already mentioned negative trends in the areas of body weight/obesity, shown below in Fig. 8.2 (CDC Fig. 11).

CDC data also suggest that healthcare costs have been growing at significant rates during the first decade of the 21st century, as shown in Fig. 8.3 (CDC Fig. 20).

Obviously, privately funded expenditures have been growing very fast, actually faster than the government-funded Medicaid and Medicare expenses. State-controlled Medicaid expenses have not kept up at all with Medicare and, in particular, private health insurance, indicating a growing discrepancy between healthcare for the poor and for the well-off.

Figure 8.4 (CDC Fig. 40) is further strengthening this observation. The facts are quite alarming: US citizens living below the poverty line only have ~40% chance of health insurance coverage, and

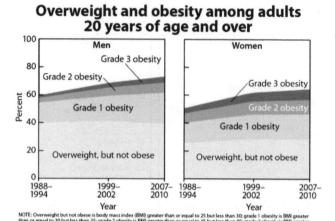

Figure 8.2 Overweight and obesity in the United States (1988–2010).

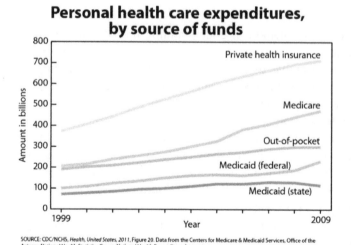

Figure 8.3 Growth of healthcare expenditures by source of funds.

Hispanics are the ethnic group facing the lowest odds of receiving good healthcare. Even among people living at an economic level above the poverty line and twice the poverty line (expressed as 200% in the figure), the coverage by health insurance is only

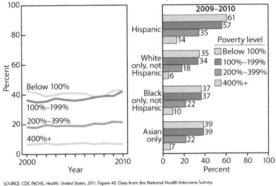

Figure 8.4 Percentage of no health insurance coverage as a function of the poverty level in the United States.

slightly higher. To reach acceptable levels of health insurance coverage, US citizens must earn incomes of more than four times the poverty line, currently defined by the US government as follows (Table 8.2).[429]

Table 8.2 US poverty line by number of persons in the household

2012 poverty guidelines for the 48 contiguous states and the district of Columbia	
Persons in family/household	Poverty guideline
1	$11,170
2	15,130
3	19,090
4	23,050
5	27,010
6	30,970
7	34,930
8	38,890

For families/households with more than 8 persons, add $3960 for each additional person.

There is no doubt that all measures of US health statistics are heavily distorted by the lack of health insurance coverage for citizens living below or close to the poverty line.

The Patient Protection and Affordable Care Act (PPACA), commonly called the Affordable Care Act (ACA) or "Obamacare,"[430] is a US federal statute signed into law by President Barack Obama on March 23, 2010. It represents the most significant regulatory overhaul of the US healthcare system since the passage of Medicare and Medicaid in 1965.

The ACA was enacted with the goals of increasing the quality and affordability of health insurance, lowering the uninsured rate by expanding public and private insurance coverage, and reducing the costs of healthcare for individuals and the government. It introduced a number of mechanisms—including mandates, subsidies, and insurance exchanges—meant to increase coverage and affordability. The law also requires insurance companies to cover all applicants within new minimum standards and offer the same rates regardless of pre-existing conditions or sex. Additional reforms are aimed to reduce costs and improve healthcare outcomes by shifting the system toward quality over quantity through increased competition, regulation, and incentives to streamline the delivery of healthcare. The Congressional Budget Office projected that the ACA will lower both future deficits and Medicare spending.

It will be interesting to assess the impact of the ACA[431] on the numbers shown in Fig. 8.4. An early assessment published in *The Economist*[432] shows the following results after the second quarter of 2014, when the first wave of enrolment ended:

- The proportion of US adults who were uninsured dropped by 22%–26%; in other words, some 8–10.3 million Americans have gained cover.
- In states that accepted the expansion of publicly funded Medicaid, only 10% remain uninsured.
- On the other hand, in states (all governed by Republicans who oppose the ACA) that did not expand Medicaid, nearly 20% of adults are still uninsured.

Growth in health spending per person slowed from a very high 7.4% per year from 1980 to 2009 to only 3% per year from 2009 to 2012.

The Economist is partly blaming health coverage provided by large US companies for their employees for the lack of transparency and apparent waste that is leading to the United States' high fraction of over 17% of the GDP devoted to healthcare. Employer-sponsored health coverage is tax exempt, costing the US government at least $200 B a year. Doctors have been paid for every test or invasive procedure, rather than for keeping patients well, and patients don't know what costs are incurred because expenses are settled above their heads by a third party.

However, an element of "Obamacare," applied to Medicare, is already making a difference: As groups of doctors and hospitals become accountable care organizations (ACOs) that are rewarded for keeping Medicare patients' costs below a set limit, the first success stories are emerging. As published in *The New York Times* on September 24, 2014,[433] the small border city of McAllen, TX, identified in June 2009 as the most expensive place for healthcare in the United States, has joined the Rio Grande Valley ACO and was able to save $20 M from its Medicare baseline between April 2012 to the end of 2013. By making preventive care a central focus, by advising patients about diet and lifestyle change, and by developing personal care plans for patients with uncontrolled diabetes, health outcomes improved and money was saved.

In addition to the high number of uninsured, there is a significant inequality of healthcare resources across the United States, illustrated in Fig. 8.5 (CDC Fig. 19) that shows physician density by state, reaching a maximum in the northeast, with Massachusetts at the top.

Comparison with Table 8.1 shows that the richest country in Europe, Switzerland, is approximately matching Massachusetts in physician density per 10,000 citizens. Despite the high quality of healthcare in places like Houston and Dallas, Texas has quite a low physician density associated with the fact that it has a higher share of uninsured citizens than any other state in the United States.

The growth of healthcare expenditures by type of expenditure 1999–2009 is shown in Fig. 8.6.

Clearly, "Hospital care", "Physician and clinical services" and "Prescription drugs" expenses have been growing the fastest during the past decade, outpacing all other categories.

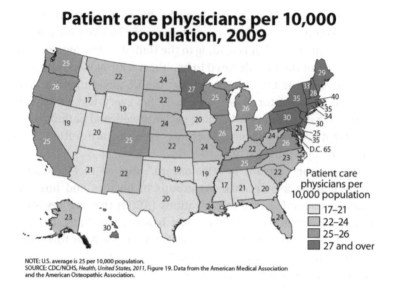

Patient care physicians per 10,000 population, 2009

NOTE: U.S. average is 25 per 10,000 population.
SOURCE: CDC/NCHS, *Health, United States, 2011*, Figure 19. Data from the American Medical Association and the American Osteopathic Association.

Figure 8.5 Variation of patient care coverage across the United States.

"Obamacare," if managed properly and fully implemented across the United States, should make a difference by significantly increasing health insurance coverage, and to control costs by emphasizing clinical patient outcomes in payments to healthcare providers, and by shifting more resources to preventive care.

Employer-sponsored healthcare may also become the subject of future political discussions. The Swiss model, where individuals decide about their health insurance, may be a valid alternative.

Compared to healthcare systems (such as, for example, in the UK) where governments prioritize resources and control/restrict access to some of the most expensive procedures and treatments, the US system has so far been rather uncontrolled and freely accessible to insured patients willing to pay for leading-edge service and resources.

In the past, investments in innovative new medical treatments and devices have therefore been made to a higher degree in the United States than elsewhere.

However, if healthcare expenses are consuming too high a share of available total public resources, there will be a risk that

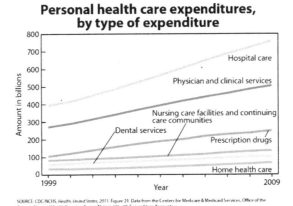

Personal health care expenditures, by type of expenditure

Figure 8.6 Growth of US healthcare expenses in 1999–2009 by type of expenditure.

government sponsored research will have to be cut, leading to less medical research and less opportunity for translational medicine.

Time will tell if the mixed private–public US model will remain to be the most attractive for academic and industrial research or if healthcare models, where governments control most resources, have a chance to compete for leadership in translational research and nanomedicine.

Chapter 9

Conclusions

Martin Caidin's novel *Cyborg* became a bestseller when it was first published in 1972. In 1974, it was turned into the TV series *The Six Million Dollar Man*, followed by *The Bionic Woman* in 1976.

When astronaut Steve Austin is severely injured in the crash of an experimental aircraft, he is "rebuilt" in an operation that costs $6 M. His right arm, both legs, and left eye are replaced with bionic implants, which enhance his strength, speed, and vision far above human norms: he can run at speeds of 60 mph, and his eye has a 20:1 zoom lens and infrared capabilities, while his bionic limbs all have the power of a bulldozer. The show's opening catch-phrase "We can rebuild him . . . we have the technology" could serve as a description of the future of nanomedicine.

Forty years later, we have not yet achieved the bionic improvements enjoyed by Steve Austin, but we have made the first baby steps toward that goal. The bionic eye of 2014 is not yet superior to the human eye, but it already provides basic capabilities helping blind persons. Above all, nanomedicine shows promise in using science and technology to—sometime in the not so distant future—help patients with heart disease, cancer, diabetes, cognitive dysfunction, spinal cord injuries, etc.

Nanomedicine: Science, Business, and Impact
Michael Hehenberger
Copyright © 2015 Pan Stanford Publishing Pte. Ltd.
ISBN 978-981-4613-76-7 (Hardcover), 978-981-4613-77-4 (eBook)
www.panstanford.com

Another important conclusion is the realization that break-throughs in nanomedicine require a truly interdisciplinary approach, involving physics, chemistry, biology, botany, zoology, anthropology, materials science, information technology (IT), etc. We need to keep an open mind and we need to keep studying what works and what does not. Botany and zoology are included because there is so much we can learn by studying plants and animals. Mendel arrived at his genetic laws by studying peas. We know that birds of prey have superior vision, that our dogs have a far superior sense of smell, that lizards can regenerate their broken tails, and that overweight pigs don't develop diabetes despite a serious lack of exercise. We need to keep studying the world around us and perhaps catch new ideas, while staying true to bioethical rules.

By covering both science and business in one book, I hope to have made a convincing case for public and private investment in both basic and applied biomedical research that can be translated into nanomedical breakthroughs. The numbers showing the impact of the Human Genome Project (HGP) and the impressive and stimulating role played by the US National Institutes of Health (NIH) and similar organizations in other countries should convince everyone that public investment in research is money well spent. Those investments must continue and public–private partnerships should be encouraged to drive innovation.

Big research budgets—such as the US NIH budget—must be managed in an effective and transparent way. If not, there will be a risk that opportunistic politicians will step in and make up new rules that invariably will shift the balance from long-term vision and funding decisions based on scientific merit to narrow-mindedness, political bias, and short-term thinking. It is the nature of science that breakthroughs cannot be planned—but the impact of such work can be managed. By referring frequently to the work done by outstanding individuals that earned awards such as the Nobel Prize, the important role of basic research is emphasized. However, to create true innovation it is equally important to translate basic research results into benefits for patients and society.

There is a current trend to partly offset or complement budget cuts with donations made by wealthy individuals. Although there may be a risk that personal tragedy shifts the focus away from the greatest needs for humanity to the greatest needs for the wealthy donor, contributions to worthy causes are always welcome. However, the effective management of precious resources made available for public health and for unmet medical needs is a very complex task. "Venture philanthropy" is a new management discipline aimed at optimizing the impact of such investments by taking concepts and techniques from venture capital finance and business management and applying them to achieve philanthropic goals.

There is often a long time gap between a breakthrough in our basic understanding of life processes to beneficial application of those newly gained insights. There is a need to be patient: It takes a long time to move from "bench to bedside," not only because of concerns about safety and efficacy, but also because of the inertia in a medical provider system that is based on "best practices." Those established best medical practices are not easily changed or replaced.

It is also important to invest in public health in order to make sure that healthcare costs are kept at affordable levels. Only with enough focus on preventive measures and encouragement of healthy lifestyles will there be enough room in government budgets for investments in new nanomedical research.

In this book, nanomedicine is defined quite broadly, including genetics, molecular biology, regenerative medicine, and computational approaches to modeling and simulation and data analytics.

There has already been a lot of "innovation crossover" between IT and biotechnology (BT), and this trend is expected to continue and further accelerate.

For the future of medicine to globally benefit mankind, we need a true alignment between public and private stakeholders committed to forming a chain of trust that serves and respects persons/patients. Such an alignment will enable nanomedicine to be a powerful force for societal impact.

Figure 9.1 illustrates the path for Nanomedicine from Science to Business and Impact. Impact is very important because it motivates the funding of research, the generation of new breakthrough ideas. Figure 9.1 should therefore be interpreted as a funnel with feedback: If the many ideas generated to the left are not producing examples

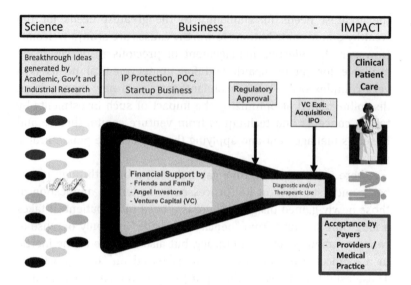

Figure 9.1 From Science to Business and Impact.

of real impact to the right, society will no longer set aside resources to generate those new ideas. It is therefore of utmost importance to manage all aspects of this process as effectively as possible, to avoid waste and duplication of effort, and to build trust among all stakeholders in the healthcare ecosystem.

References

1. European Science Foundation, European Medical Research Councils (EMRC), Forward Look Report, Consensus Conference, Le Bischenberg, France (2005).
2. http://www.nobelprize.org/nobel_prizes/medicine/laureates/1967/
3. http://www.nobelprize.org/alfred_nobel/
4. http://www.nobelprize.org/nobel_prizes/about/
5. Nobel Prize in Physics 1932: http://www.nobelprize.org/nobel_prizes/physics/laureates/1932/heisenberg-lecture.html
6. Arnold Sommerfeld was nominated a record 81 times but never received the Nobel Prize in physics. He was admired as a masterful teacher and mentor of a generation of great physicists, among them Nobel Laureates Werner Heisenberg, Wolfgang Pauli, Peter Debye, and Hans Bethe.
7. Nobel Prize in Medicine 1932: http://www.nobelprize.org/nobel_prizes/medicine/laureates/1931/index.html
8. Nobel Prize in Chemistry 1937: http://www.nobelprize.org/nobel_prizes/chemistry/laureates/1937/index.html
9. Nobel Prize in Medicine 1922: http://www.nobelprize.org/nobel_prizes/medicine/laureates/1922/index.html
10. Nadia Rosenthal, as quoted in her Foreword to "The Molecules of Life" by Russ Hodge, Facs On File, Infobase, NY (2009).
11. http://en.wikipedia.org/wiki/Rhodopsin
12. Nobel Prize in Chemistry 2012: http://www.nobelprize.org/nobel_prizes/chemistry/laureates/2012/press.html
13. http://www.nobelprize.org/nobel_prizes/medicine/laureates/1967/press.html
14. Bruce McManus, private communication. Adapted from Sykes RB, New Medicines: The Practice of Medicine and Public Policy, 2000, London, Nuffield Trust Publications Committee.

15. http://copyright.gov/help/faq/faq-duration.html#duration

16. http://www.uspto.gov/faq/trademarks.jsp

17. www.evpa.eu.com

18. William Swope, IBM Almaden Research, Private Communication.

19. https://www.michaeljfox.org/foundation/news-detail.php?staff-on-the-move-foundation-unique-role-in-venture-philanthropy

20. The electron volt (symbol eV) is a unit of energy, defined as the amount of energy associated with the movement of a single electron across an electric potential difference of one volt.

21. http://www.unh.edu/news/docs/2007AngelMarketAnalysis.pdf

22. This is not the place to discuss "high-energy physics."

23. http://www.nobelprize.org/nobel_prizes/physics/laureates/1933/schrodinger-bio.html

24. The Hamiltonian or "Hamiltonian Operator" is related to the total energy of the system described by the Schrödinger equation. See https://en.wikipedia.org/wiki/Hamiltonian_(quantum_mechanics)

25. As of 2013, the official number is 98. However, lifetimes of some elements are so short that it's more an academic debate whether to count more than 90 elements as "natural." For more details see http://en.wikipedia.org/wiki/Periodic_table

26. "The Elements", Jack Challoner, Carlton Books Ltd (2012), ISBN 978 1 78097 125 4.

27. Electronegativity is a useful concept first introduced by Linus Pauling to describe the tendency of an atom or a functional group to attract electrons towards itself. An atom's electronegativity is affected by both its atomic number and the distance between its outer ("valence") electrons and the charged nucleus. The most electronegative elements are the halogens (F, Cl, …) and least electronegative are the alkali metals (Na, K, …). See https://en.wikipedia.org/wiki/Electronegativity.

28. The only way to make all four bonds equivalent is to create a spatial sp3 hybridization with tetrahedral bond angles of $109°28'$.

29. Semiconductor devices used to amplify and switch electronic signals and electrical power.

30. "Lithium in the prevention of suicidal behavior and all-cause mortality in patients with mood disorders: a systematic review of randomized trials" by Cipriani A., Pretty, H., Hawton, K., and Geddes, J.R. (2005). *Am J Psychiat,* **162**(10), pp. 1805–1819.

31. Eukaryotes are organisms whose cells contain a nucleus and other organelles enclosed within membranes. Most eukaryotic cells also

contain other membrane-bound organelles such as mitochondria. All multicellular organisms are eukaryotes, including animals, plants and fungi.

32. To be discussed in detail in a subsequent chapter.
33. Body fat or just fat (loose connective tissue).
34. http://en.wikipedia.org/wiki/Glutamine
35. http://en.wikipedia.org/wiki/Arginine
36. http://en.wikipedia.org/wiki/Jons_Jacob_Berzelius
37. http://www.nobelprize.org/nobel_prizes/chemistry/laureates/1958/
38. http://www.nobelprize.org/nobel_prizes/chemistry/laureates/1954/
39. http://www.nobelprize.org/nobel_prizes/physics/laureates/1914/
40. http://www.nobelprize.org/nobel_prizes/physics/laureates/1915/wh-bragg-facts.html
41. http://www.nobelprize.org/nobel_prizes/chemistry/laureates/2009/
42. http://en.wikipedia.org/wiki/Protein_Crystallization
43. Worldwide Protein Data Bank: http://www.wwpdb.org/
44. http://www.nobelprize.org/nobel_prizes/chemistry/laureates/1962/
45. http://www.nobelprize.org/nobel_prizes/chemistry/laureates/1972/anfinsen-bio.html
46. http://www.nobelprize.org/nobel_prizes/physics/laureates/1944/rabi-facts.html
47. http://www.nobelprize.org/nobel_prizes/physics/laureates/1952/
48. http://scitation.aip.org/content/aip/magazine/physicstoday/article/56/3/10.1063/1.1570771
49. HarperCollins Publishers (2006).
50. Recipient of the Nobel Prize in Physiology or Medicine 2002.
51. Nobel Prize in Physiology or Medicine 1968, http://www.nobelprize.org/nobel_prizes/medicine/laureates/1968/
52. See Nobel lecture 1968: http://www.nobelprize.org/nobel_prizes/medicine/laureates/1968/nirenberg-lecture.pdf
53. From p. 143 in Matt Ridley's book (reference number i above).
54. In 1980, Fredrick Sanger shared one half of the Nobel Prize in chemistry with Walter Gilbert for "determination of the sequence

of the nucleotide blocks in DNA." http://www.nobelprize.org/nobel_prizes/chemistry/laureates/1980/presentation-speech.html#

55. http://www.biology.iupui.edu/biocourses/N100/2k3ch13dogma.html (Indiana University)

56. http://www.nobelprize.org/nobel_prizes/medicine/laureates/1974/presentation-speech.html

57. http://www.nobelprize.org/nobel_prizes/chemistry/laureates/2009/presentation-speech.html

58. http://www.nobelprize.org/nobel_prizes/medicine/laureates/1974/duve-facts.html

59. http://www.nobelprize.org/nobel_prizes/medicine/laureates/1906/golgi-facts.html

60. https://de.wikipedia.org/wiki/Biochemical_Pathways

61. http://www.roche.com/media/roche_stories/roche-stories-2013-07-22.htm

62. http://www.nobelprize.org/nobel_prizes/medicine/laureates/2013/press.html

63. "What is Life" by Erwin Schrödinger, Cambridge University Press (first published 1944).

64. http://www.nobelprize.org/nobel_prizes/physics/laureates/1918/

65. http://en.wikipedia.org/wiki/Valence_bond_theory

66. http://en.wikipedia.org/wiki/Luria%E2%80%93Delbr%C3%BCck_experiment

67. "Evaluating plague and smallpox as historical selective pressures for the *CCR5-Δ32* HIV-resistance allele" by Galvani AP and Slatkin M (2003). *Proc Natl Acad Sci USA*, **100**(25), pp. 15276–15279.

68. "On the origin of species by means of natural selection, or the preservation of favoured races in the struggle for life" by Charles Darwin (1859). John Murray, London; modern reprint Charles Darwin, Julian Huxley (2003). On The Origin of Species. Signet Classics. ISBN 0-451-52906-5.

69. "Identifying signatures of natural selection in Tibetan and Andean populations using dense genome scan data" by Bigham A, Bauchet M, Pinto D, Mao X, Akey JM, Mei R, Scherer SW, Julian CG, Wilson MJ, López Herráez D, Brutsaert T, Parra EJ, Moore LG, and Shriver MD (2010). *PLoS Genet*, **6**(9), e1001116.

70. "Adaptation and mal-adaptation to ambient hypoxia; Andean, Ethiopian and Himalayan patterns" by Xing G, Qualls C, Huicho L,

Rivera-Ch M, Stobdan T, Slessarev M, Prisman E, Ito S, Wu H, Norboo A, Dolma D, Kunzang M, Norboo T, Gamboa JL, Claydon VE, Fisher J, Zenebe G, Gebremedhin A, Hainsworth R, Verma A, and Appenzeller O (2008). *PLoS ONE*, **3**(6), e2342.

71. "Altitude adaptation in Tibetans caused by introgression of Denisovan-like DNA" by Huerta-Sánchez E, Jin X, Asan, Bianba Z, Peter BM, Vinckenbosch N, Liang Y, Yi X, He M, Somel M, Ni P, Wang B, Ou X, Huasang, Luosang J, Cuo ZX, Li K, Gao G, Yin Y, Wang W, Zhang X, Xu X, Yang H, Li Y, Wang J, Wang J, and Nielsen R (2014). *Nature*, **512**, pp. 194–197.

72. http://www.nobelprize.org/nobel_prizes/chemistry/laureates/1993/mullis-lecture.html

73. http://www.nobelprize.org/nobel_prizes/medicine/laureates/1959/

74. http://www.nobelprize.org/nobel_prizes/medicine/laureates/1978/

75. "The Gene Wars" by Robert Cook-Deegan (1994). W W Norton & Co., New York; London.

76. http://genomesymposium.ucsc.edu/bio-sinsheimer.html

77. https://en.wikipedia.org/wiki/Craig_Venter

78. http://www.genome.gov/11006943

79. Huanming Yang (BGI), private communication.

80. "Economic impact of the Human Genome Project" by S Tripp and M Grueber, Battelle Memorial Institute, May 2011.

81. "Adsorption analysis and chromatographic method. Application to the chemistry of chlorophyll" by Mikhail Tswett (1906). *Proc Ger Bot Soc*, **24**, pp. 384–393.

82. Fiers had spent 1960–62 at the California Institute of Technology to study viral DNA, and at the University of Wisconsin in Madison to work in the laboratory of Gobind Khorana (Nobel laureate 1968 for his work on nucleic acid synthesis in the study of the genetic code).

83. "DNA sequencing with chain-terminating inhibitors" by Sanger F, Nicklen S, and Coulson AR (1977). *Proc Natl Acad Sci USA*, **74**(12), pp. 5463–5467.

84. "A new method for sequencing DNA" by Maxam AM and Gilbert W (1977). *Proc Natl Acad Sci USA*, **74**(2), pp. 560–564.

85. http://www.nobelprize.org/nobel_prizes/chemistry/laureates/1980/press.html

86. http://www.nobelprize.org/nobel_prizes/chemistry/laureates/1980/gilbert-lecture.pdf

87. http://www.nobelprize.org/nobel_prizes/medicine/laureates/1959/kornberg-lecture.pdf, page 12

88. Fredrick Sanger died on November 19, 2013, at age 95. He won two Nobel prizes in chemistry: 1958 for protein (insulin) sequencing, 1980 for DNA sequencing.

89. Example and subsequent explanation adapted from Robert M. Cook-Deegan, ref. #71 above.

90. See his Nobel lecture, http://www.nobelprize.org/nobel_prizes/chemistry/laureates/1980/sanger-lecture.pdf

91. "The synthesis of oligonucleotides containing an aliphatic amino group at the 5' terminus: synthesis of fluorescent DNA primers for use in DNA sequence analysis" by Smith LM, Fung S, Hunkapiller MW, Hunkapiller TJ, and Hood LE (1985). *Nucleic Acids Res*, **13**(7), pp. 2399–2412.

92. http://www.pacificbiosciences.com/aboutus/executives/list.html

93. http://www.jcvi.org/cms/press/press-releases/full-text/article/j-craig-venter-science-foundation-announces-500000-technology-prize-for-advances-leading-to-the/

94. http://genomics.xprize.org/sites/genomics.xprize.org/files/docs/Competition_Fact_Sheet.pdf

95. "Real-time DNA sequencing using detection of pyrophosphate release" by Ronaghi M, Karamohamed S, Pettersson B, Uhlén M, and Nyrén P (1996). *Anal Biochem*, **242**(1), pp. 84–89.

96. http://en.wikipedia.org/wiki/454_Life_Sciences

97. http://www.celldex.com/about/locations.php

98. http://www.genomeweb.com/sequencing/roche-shutting-down-454-sequencing-business

99. http://en.wikipedia.org/wiki/Illumina_%28company%29

100. "Accurate whole human genome sequencing using reversible terminator chemistry" by DR Bentley, et al. (2008). *Nature*, **456**(7218), pp. 53–59.

101. "Accurate multiplex polony sequencing of an evolved bacterial genome" by Shendure J, Porreca GJ, Reppas NB, Lin X, McCutcheon JP, Rosenbaum AM, Wang MD, Zhang K, Mitra RD, and Church GM (2005). *Science*, **309**(5741), pp. 1728–1732.

102. http://www.appliedbiosystems.com/absite/us/en/home/applications-technologies/solid-next-generation-sequencing/next-generation-systems/solid-sequencing-chemistry.html

103. http://www.lifetechnologies.com/us/en/home/about-us/news-gallery/company-fact-sheet/company-history.html

104. http://www.completegenomics.com/about-us/

105. Picture copied from Complete Genomics web site.

106. http://www.businessweek.com/articles/2012-12-20/complete-genomics-chinese-bid-sparks-a-security-fight

107. http://www.genomics.cn/en/index

108. http://www.pacificbiosciences.com/aboutus/history/

109. "The $1,000 Genome: The Revolution in DNA Sequencing and the New Era of Personalized Medicine" by Kevin Davies (2010). ISBN-10: 1416569596.

110. http://www.xconomy.com/national/2014/03/03/pacbio-the-post-hype-sleeper-of-genomics/

111. http://www.agbt.org/about.html

112. http://investor.pacificbiosciences.com/releasedetail.cfm?ReleaseID=793199

113. http://dnae.co.uk/company/about/

114. "An integrated semiconductor device enabling non-optical genome sequencing" by JM Rothberg, et al. (2011). *Nature*, **475**, pp. 348–352.

115. http://www.genomeweb.com/sequencing/life-technologies-acquire-ion-torrent-725m

116. http://www.genomeweb.com/arrays/bionanomatrix-introduces-single-molecule-analyzer-researchers-ashg

117. Picture copied from Bionano web site.

118. Vinod Makhijani, private communication.

119. https://www.nanoporetech.com/

120. "Characterization of individual polynucleotide molecules using a membrane channel" by Kasianowicz JJ, Brandin E, Branton D, and Deamer DW (1996). *Proc Natl Acad Sci USA*, **93**, pp. 13770–13773.

121. "Designed protein pores as components for biosensors" by O Braha, et al. (1997). *Chem Biol*, **4**, pp. 497–505.

122. https://www.nanoporetech.com/technology/analytes-and-applications-dna-rna-proteins/dna-strand-sequencing

123. "Nanopore in metal-dielectric sandwich for DNA position control" by Polonsky S, Rossnagel S, and Stolovitzky G (2007). *Appl Phys Lett*, **91**, p. 153103.

124. http://www.roche.com/media/media_releases/med_dia_2010-07-01.htm

125. Picture made available by IBM Research.

126. http://www.roche.com/media/media_releases/med-cor-2013-04-23.htm

127. "Recognition tunneling" by S Lindsay, et al. (2010). *Nanotechnology*, **21**(26), p. 262001.

128. "Comparison of next-generation sequencing systems" by Liu L, Li Y, Li S, et al. (2012). *J Biomed Biotechn* (Hindawi Publishing Corporation) 1–11.

129. AGBT 2014 talk by David Jaffe, Broad Institute; http://www.broadinstitute.org/genome_bio/bios/bio-jaffe.html

130. "The real cost of sequencing: higher than you think" by Sboner A, Mu XJ, Greenbaum D, Auerbach RK, and Gerstein MB (2011). *Genome Biol*, **12**(8), p. 125.

131. https://github.com/samtools/hts-specs

132. http://www.ebi.ac.uk/ena/about/cram_usage

133. "The variant call format and VCFtools" by Danecek P, Auton A, Abecasis G, Albers CA, Banks E, DePristo MA, Handsaker RE, Lunter G, Marth GT, Sherry ST, McVean G, and Durbin R (2011). *Bioinformatics*, **27**(15), pp. 2156–2158.

134. https://web.archive.org/web/20120529215509/http://ohsr.od.nih.gov/guidelines/belmont.html

135. http://nuffieldbioethics.org/report/animal-research-2/animals-research/

136. http://nuffieldbioethics.org/about/

137. http://www.un.org/en/documents/udhr/index.shtml

138. http://www.un.org/en/documents/udhr/index.shtml

139. http://eur-lex.europa.eu/LexUriServ/LexUriServ.do?uri=OJ:L:2010:276:0033:0079:en:PDF

140. In 2005, Verity was acquired by AUTONOMY. Autonomy was then acquired by Hewlett-Packard (under short term CEO Leo Apotheker) for $10.2B in October 2011.

141. https://www.systemsbiology.org/hood-group

142. http://www.nytimes.com/2012/03/08/technology/cost-of-gene-sequencing-falls-raising-hopes-for-medical-advances.html?pagewanted=all

143. http://www.businessweek.com/articles/2012-12-20/complete-genomics-chinese-bid-sparks-a-security-fight

144. http://www.newsrx.com/library/newsletters/Biotech-Business-Week/2139633.html

145. http://www.genomics.cn/en/news/show_news?nid=99717

146. http://www.embo.org/news/press-releases/press-releases-2006/leading-scientists-elected-to-embo-ranks

147. http://twas.org/recipients-twas-awards-and-prizes

148. http://www.leopoldina.org/en/members/list-of-members/

149. http://www.nasonline.org/member-directory/?q=BGI&site=nas_members

150. "Partnership Perspective: Merging a US Company and a Chinese Company" by Clifford A. Reid (Complete Genomics)", presentation at ICG-9 Conference, Shenzhen (China), Sep 9–11, 2014.

151. http://www.genome.gov/27544383

152. http://www.ncbi.nlm.nih.gov/genome/

153. "An integrated map of genetic variation from 1,092 human genomes" by RA Gibbs, et al. (2012). *Nature*, **491**(7422), pp. 56–65.

154. Paracelsus, Selected Writings, ed. with an introduction by Jolande Jacobi, trans. Norbert Guterman, (New York: Pantheon, 1951), pp. 79–80.

155. http://www.gao.gov/new.items/d0749.pdf, page 27

156. IBM Institute for Business Value analysis, based on data compiled by "Pharmaceutical Executive" (May 2012, pp. 24–34) and PhRMA Industry Profile 2012. (PhRMA = Pharmaceutical Research and Manufacturers of America).

157. See David Filmore, Modern Drug Discovery (MDD) Vol. 7, Issue 11, pp. 24–28 (Nov 2004).

158. http://www.nobelprize.org/nobel_prizes/chemistry/laureates/2012/

159. "The druggable genome" by Hopkins AL and Groom CR (2002). *Nat Rev Drug Discov*, **1**(9), pp. 727–730.

160. From a presentation (Feb 2002) given at IBM Research by Bennett M. Shapiro, M.D., Executive VP, Merck & Co., Inc.

161. http://www.nobelprize.org/nobel_prizes/medicine/laureates/2007/

162. http://www.nobelprize.org/nobel_prizes/chemistry/laureates/1984/press.html

163. "The rule of five" by CA Lipinski, et al. (1997). *Adv Drug Deliv Rev*, **23**, 3–25.

164. http://pubchem.ncbi.nlm.nih.gov/summary/summary.cgi?cid=3672

165. http://www.fda.gov/AboutFDA/WhatWeDo/History/default.htm

166. http://www.umc-products.com/DynPage.aspx?id=73590&mn1= 1107&mn2=1132

167. http://www.fda.gov/AboutFDA/Transparency/Basics/ucm194516. htm

168. http://www.nobelprize.org/nobel_prizes/medicine/laureates/1908/ press.html

169. http://www.nobelprize.org/nobel_prizes/medicine/laureates/1984/

170. http://en.wikipedia.org/wiki/Biomarker

171. http://www.fda.gov/cder/genomics/PGX_biomarkers.pdf

172. http://www.fda.gov/oc/initiatives/criticalpath/

173. "The FDA critical path initiative and its influence on new drug development" by Woodcock J and Woosley R (2008). *Annu Rev Med*, **59**, pp. 1–12.

174. "Stratified medicine: strategic and economic implications of comining drugs and clinical biomarkers" by Mark. R. Trusheim, Ernst R. Berndt and Frank L. Douglas (2007). *Nature Reviews Drug Discovery* **6**, pp. 287–293.

175. http://en.wikipedia.org/wiki/Omeprazole

176. http://en.wikipedia.org/wiki/Nexium

177. http://en.wikipedia.org/wiki/Prozac

178. http://en.wikipedia.org/wiki/Paxil

179. http://en.wikipedia.org/wiki/Zoloft

180. http://en.wikipedia.org/wiki/Trastuzumab

181. http://en.wikipedia.org/wiki/Imatinib

182. http://www.gene.com/media/product-information/herceptin-development-timeline

183. "Discovery of the Philadelphia chromosome: a personal perspective" by PC Nowell (2007). *J Clin Invest*, **117**(8), pp. 2033–2035.

184. http://en.wikipedia.org/wiki/Tau_protein

185. http://csdd.tufts.edu/reports/impact_reports

186. Biomarkers In Drug Development, John Wiley & Sons, Hoboken, New Jersey (2010).

187. "Knocked out" existing gene by replacing it or disrupting it with an artificial piece of DNA. The loss of gene activity often causes changes in a mouse's phenotype, which includes appearance, behavior and other observable physical and biochemical characteristics. Knockout mice are important animal models for studying the role of genes which have been sequenced but whose functions have not been determined.

188. "The Acquisition of Aventis by Sanofi", Case Study 306-238-1, University of St. Gallen, Institute of Management (2006).

189. http://www.ifb.unisg.ch/~/media/Internet/Content/Dateien/ InstituteUndCenters/IfB/Lehre/Lehrbuecher%20und%20Fallstudien/ Case%20Studies/Aventis_I_306-238-1S_Inspection%20Copy_C_en. ashx

190. http://www.abbvie.com/

191. "From vision to decision: pharma 2020"; www.pwc.com/pharma2020; lead authors Steve Arlington, Joe Palo, Nick Davis.

192. http://businesstoday.intoday.in/story/sun-pharma-ranbaxy-deal-indian-pharma-industry-cipla-dr-reddys-under-watch/1/207241. html

193. http://seekingalpha.com/article/1280811-china-biotech-in-review-simcere-receives-503-million-privatization-offer

194. Letter from the CEO, Daniel Armbrust, Sematech. http://www. sematech.org/corporate/ceo_letter.htm

195. http://www.imi.europa.eu/

196. http://www.efpia.eu/

197. http://www.nih.gov/science/amp/index.htm

198. http://www.nobelprize.org/nobel_prizes/physics/laureates/1986/ presentation-speech.html

199. "Beiträge zur Theorie des Mikroskops und der mikroskopischen Wahrnehmung" by Ernst Abbe (1873). *Archiv für Mikroskopische Anatomie*, **9**(1), pp. 413–468.

200. "Theory of the scanning tunneling microscope" by Tersoff J and Hamann DR (1985). *Phys Rev B*, **31**(1985), pp. 805–813.

201. http://www.nobelprize.org/nobel_prizes/physics/laureates/1919/

202. "Weyl's theory applied to the Stark effect in the hydrogen atom" by Hehenberger M, McIntosh HV, and Brandas E (1974). *Phys Rev A*, **10**, pp. 1494–1506.

203. 1 Å $= 10^{-10}$ m $= 0.1$ nm. The unit is named after the Swedish physicist Anders Jonas Ångström, one of the founders of the science of spectroscopy and appointed keeper of the Uppsala Astronomical Observatory.

204. "Press Release: The 1986 Nobel Prize in Physics," http://www. nobelprize.org/nobel_prizes/physics/laureates/1986/press.html

205. Gordon E. Moore, *Electronics*, April 19, 1965, pp. 114–117.

206. "Design of ion-planted MOSFETs with very small physical dimensions" by R Dennard, et al. (1974). *IEEE J Solid State Circuit,* **SC-9**(5), pp. 256–268.

207. http://www.ieee.org/about/awards/bios/moh_recipients.html

208. http://www.nobelprize.org/nobel_prizes/chemistry/laureates/2013/press.html

209. "Text-based knowledge discovery: search and mining of life sciences documents" by Mack R and Hehenberger M (2002). *Drug Discov Today,* **7**(11), pp. S89–S98.

210. http://www.bloomberg.com/news/2011-02-16/don-t-read-this-unless-you-want-to-know-if-ibm-won-jeopardy-.html

211. "Final Jeopardy: Man vs. Machine and the Quest to Know Everything" by Stephen Baker. (2011) Boston: Houghton Mifflin Harcourt.

212. "Natural language question-answering systems: 1969" by RF Simmons (1970). *Commun ACM,* **13**(1), pp. 15–30.

213. *IBM Journal of Research & Development,* **56**(3–4), (2012).

214. http://www.research.ibm.com/cognitive-computing/index.shtml#fbid=pBmAByM5xsF

215. "Implicit text linkages between Medline records: using Arrowsmith as an aid to scientific discovery" by Swanson DR and Smalheiser NR (1999). *Libr Trends,* **48**(1), pp. 48–59.

216. "Trouble at the Lab," Economist.com, October 19–25, 2013, pp. 26–30.

217. "Why most published research findings are false" by John P A Ioannidis (2005). *PLoS Med,* **2**(8), e124.

218. "Drug development: raise standards for preclinical cancer research" by Begley CG and Ellis LM (2012). *Nature,* **483**(7391), pp. 531–533.

219. "Believe it or not: how much can we rely on published data on potential drug targets?" by Prinz F, Schlange T, and Asadullah K (2011). *Nat Rev Drug Discov,* **10**(9), p. 712.

220. Stanford University has created METRICS, the MEta-Research Innovation Center at Stanford.

221. Dialogue for Reverse Engineering Assessments and Methods. http://www.the-dream-project.org/

222. "Wisdom of crowds for robust gene network inference" by G Stolovitzky, et al. (2012). *Nat Meth,* **9**(8), pp. 796–804.

223. http://depts.washington.edu/bakerpg/drupal/

224. http://homes.cs.washington.edu/~zoran/

225. "Biologists enlist online gamers" by Katherine Bourzac (2008). *MIT Technology Review*, Biochemistry News, May 8, 2008.

226. "Online video game plugs players into remote-controlled biochemistry lab" by J Bohannon (2014). *Science*, **343**(6170), p. 475.

227. "Recent lessons learned from prevention and recent-onset type 1 diabetes immunotherapy trials" by Staeva TP, Chatenoud L, Insel R, and Atkinson MA (2013). *Diabetes*, **62**(1), pp. 9–17.

228. http://jdrfconsortium.jaeb.org/ViewPage.aspx?PageName=Home

229. "A pancreas in a box" by D Clery (2014). *Science*, **343**(6167), pp. 133–135.

230. http://www.medtronicdiabetes.com/treatment-and-products?utm_source=google&utm_medium=cpc&utm_campaign=RHW-Branded&utm_term=medtronic&utm_content=Medtronic|mkwid|smd7pQxkf_dc|pcrid|39204180121&gclid=CK_VttC9q7wCFY1QOgodkHoATw

231. "Cascade- and Nonskid-chain-like syntheses of molecular cavity topologies" by Buhleirier E, Wehner W, and Vögtle F (1978). *Synthesis*, **1978**(2), pp. 155–158.

232. "Click chemistry: diverse chemical function from a few good reactions" by Kolb HC, Finn MG, and Sharpless KB (2001). *Angew Chem Int Ed*, **40**(11), pp. 2004–2021.

233. "Self-assembly of a nanoscale DNA box with a controllable lid" by Andersen ES, Dong M, Nielsen MM, Jahn K, Subramani R, Mamdouh W, Golas MM, Sander B, Stark H, Oliveira CL, Pedersen JS, Birkedal V, Besenbacher F, Gothelf KV, and Kjems J (2009). *Nature*, **459**(7243), pp. 73–76.

234. "Nanonization strategies for poorly water-soluble drugs" by Chen H, Khemtong C, Yang X, Xhang X, and Gao J, (2011). *Drug Discov Today*, **16**(7–8), pp. 354–360.

235. https://www.pharmadeals.net/journal/pharmadealsreview/fulltext/pdf/1471

236. http://www.alkermes.com/Research-and-Development/Technology-Platforms

237. "Nanocrystal technology, drug delivery and clinical applications" by Junghanns JU and Müller RH (2008). *Int J Nanomed*, **3**(3), pp. 295–310.

238. "Mucoadhesive thiochitosan coated liposomes for oral administration of drugs" by Ruth Prassl, presented at CLINAM 7/2014.

239. "Nanotherapeutics in the EU: an overview on current state and future directions" by Hafner A, Lovrić J, Lakoš GP, and Pepić I (2014). *Int J Nanomed*, **9**, pp. 1005–1023.

240. "Basics of magnetic nanoparticles for their application in the field of magnetic fluid hyperthermia" by Mody VV, Singh A, and Wesley B (2013). *Eur J Nanomed*, **5**(1), pp. 11–21.

241. http://www.nobelprize.org/nobel_prizes/chemistry/laureates/1996/

242. "Buckysomes: new nanocarriers for anticancer drugs" by Danila D, Golunski E, Partha R, McManus M, Little T, and Conyers J (2013). *J Pharmaceutics*, **2013**(1–5), DOI: http://dx.doi.org/10.1155/2013/390425.

243. "Carbon nanotubes in biology and medicine: in vitro and in vivo detection, imaging and drug delivery" by Liu Z, Tabakman S, Welsher K, and Dai H (2009). *Nano Res*, **2**, pp. 85–120.

244. http://www.fda.gov/downloads/Drugs/GuidanceCompliance RegulatoryInformation/Guidances/ucm070570.pdf

245. http://www.ema.europa.eu/docs/en_GB/document_library/Scientific_guideline/2013/03/WC500140351.pdf

246. http://www.clinam.org/

247. "Analysis on the current status of targeted drug delivery to tumors" by Kwon IK, Lee SC, Han B, and Park K (2012). *J Control Release*, **164**, pp. 108–114.

248. "Drug targeting to tumors: principles, pitfalls, and (pre-)clinical progress" (2012) by Lammers T, Kiessling F, Hennink W, and Storm G (2012). *J Control Release*, **161**, pp. 174–187.

249. http://www.nobelprize.org/nobel_prizes/physics/laureates/1901/rontgen-facts.html

250. http://www.nobelprize.org/nobel_prizes/chemistry/laureates/1943/hevesy-facts.html

251. "Folded Stern-Gerlach experiment as a means for detecting nuclear magnetic resonance in individual nuclei" by JA Sidles (1992). *Phys Rev Lett*, **68**, 1124–1127.

252. "Force detection of nuclear magnetic resonance" by Rugar D, Zuger O, Hoen S, Yannoni CS, Vieth HM, and Kendrick RD (1994). *Science,* **264**, pp. 1560–1563.

253. http://news.harvard.edu/gazette/story/2014/04/mri-on-a-molecular-scale/

254. http://www.marketwatch.com/story/bruker-announces-the-worlds-first-preclinical-magnetic-particle-imaging-mpi-system-2013-09-19

255. http://www.definiens.com/clinical/tissue-diagnostics.html

256. http://www.zeiss.com/corporate/en_de/history/founders.html

257. http://www.nobelprize.org/nobel_prizes/chemistry/laureates/2014/

258. http://www.zeiss.com/microscopy/en_de/solutions/bioscience-tasks-applications/live-cell-imaging.html

259. http://www.leica-microsystems.com/applications/life-science/fluorescence/

260. http://www.microscopyu.com/articles/fluorescence/

261. http://www.olympusmicro.com/primer/techniques/fluorescence/fluorhome.html

262. http://www.jeol.co.jp/en/corporate/outline/

263. http://www.meijitechno.com/fluorescence_applications.htm

264. http://www.roche.com/about/business/diagnostics.htm

265. http://www.qdvision.com/color-iq

266. "Noninvasive imaging of quantum dots in mice" by Ballou B, Lagerholm BC, Ernst LA, Bruchez MP, and Waggoner AS (2004). *Bioconjug Chem*, **15**(1), pp. 79–86.

267. "An integrated semiconductor device enabling non-optical genome sequencing" by Jonathan M Rothberg, et al. (2011). *Nature, **475**, pp. 348–352.

268. http://nanophotonics.ece.cornell.edu/

269. "Mid-IR Photonics" by William Green, Optical Fiber Communication Conference, San Francisco, California, March 9-13, 2014. http://www.opticsinfobase.org/abstract.cfm?URI=OFC-2014-Th4I.4&origin=search

270. "A case of cancer in which cells similar to those in the tumours were seen in the blood after death" by TR Ashworth (1869). *Aust Med J*, **14**, pp. 146–147.

271. "A new device for rapid isolation by size and characterization of rare circulating tumor cells" by I Desitter, et al. (2011). *Anticancer Res*, **31**(2), pp. 427–442.

272. http://report.nih.gov/NIHfactsheets/ViewFactSheet.aspx?csid=62&key=R#R

273. http://www.nobelprize.org/nobel_prizes/medicine/laureates/2012/

274. "Induction of pluripotent stem cells from mouse embryonic and adult fibroblast cultures by defined factors" by Takahashi K and Yamanaka S (2006). *Cell*, **126**(4), pp. 663–676.

275. http://en.wikipedia.org/wiki/Stem_cell

276. "Embryonic stem cell lines derived from human blastocysts" by Thomson JA, Itskovitz-Eldor J, Shapiro SS, Waknitz MA, Swiergiel JJ, Marshall VS, and Jones JM (1998). *Science,* **282**(5391), pp. 1145–1147.

277. http://www.nobelprize.org/nobel_prizes/medicine/laureates/2007/

278. http://hsci.harvard.edu/aging-and-gdf11-what-we-know

279. http://www.upmc.com/Services/regenerative-medicine/research/cell-therapy/Pages/urinary-incontinence.aspx

280. http://www.wakehealth.edu/WFIRM/

281. http://www.the-scientist.com/?articles.view/articleNo/41052/title/The-Bionic-Eye/#eye2

282. http://bionicvision.org.au/eye;   http://www.bionicsinstitute.org/Pages/default.aspx

283. http://www.retina-implant.de/default.aspx

284. http://www.pixium-vision.com/fr/

285. http://www.rle.mit.edu/media/pr152/29_PR152.pdf

286. "Testing of semichronically implanted retinal prosthesis by supra-choroidal-transretinal stimulation in patients with retinitis pigmentosa" by T Fujikado, et al. (2011). *Invest Ophthalmol Vis Sci,* **52**, pp. 4726–4733.

287. "Gene Therapy for Human Genetic Disease?" by T. Friedmann and R. Roblin (1972). *Science* **175**(4025): 949–955.

288. "The discovery of zinc fingers and their applications in gene regulation and genome manipulation" by A Klug (2010). *Annu Rev Biochem,* **79**, pp. 213–231.

289. "Gene Therapy's Second Act" by Ricki Lewis, Scientific American (March 2014), pp. 52–57.

290. http://www.cira.kyoto-u.ac.jp/e/about/director.html

291. http://alliancerm.org/member-profiles

292. http://www.mesoblast.com/about-us/company-overview

293. http://www.nobelprize.org/nobel_prizes/chemistry/laureates/1967/

294. http://www.medi-post.com/am_1.asp

295. http://www.genengnews.com/gen-news-highlights/janssen-inks-up-to-337-5m-cell-therapy-collaboration-with-capricor/81249320/

296. http://www.nobelprize.org/nobel_prizes/medicine/laureates/1958/lederberg-facts.html

297. http://www.ucsf.edu/news/2014/08/116526/do-gut-bacteria-rule-our-minds

298. http://commonfund.nih.gov/hmp/index

299. http://www.human-microbiome.org/

300. http://www.earthmicrobiome.org/

301. "Your microbes, your health" (December 2013). *Science*, **342**, pp. 1440–1441; www.sciencemag.org

302. https://www.humanbrainproject.eu/discover/the-project/overview; jsessionid=1ssq42yfd44qq1qshcq6rwz4pr

303. http://www.braininitiative.nih.gov/index.htm

304. http://www.neuroscienceblueprint.nih.gov/connectome/

305. "Speech production: Wernicke, Broca and beyond" by Blank SC, Scott SK, Murphy K, Warburton E, and Wise RJ (2002). *Brain*, **125**(Pt 8), pp. 1829–1838.

306. http://www.nobelprize.org/nobel_prizes/medicine/laureates/1997/

307. http://www.nobelprize.org/nobel_prizes/medicine/laureates/2014/

308. http://www.wingsforlife.com/en/research

309. http://www.who.int/mediacentre/factsheets/fs297/en/

310. http://globocan.iarc.fr/Pages/fact_sheets_cancer.aspx

311. http://cancer.sanger.ac.uk/cancergenome/projects/cosmic/

312. http://cancergenome.nih.gov/

313. https://icgc.org/icgc

314. "18FDG-PET predicts response to imatinimb mesylate (Gleevec) in patients with advanced gastrointestinal stromal tumors" by AD Van den Abbeele, et al. (2002). *Proc Am Soc Clin Oncol*, **21**(1), 403a. The figure was presented by Richard A. Frank and GE Healthcare, at IBM's Imaging Biomarker Summit in Philadelphia, June 2006.

315. "Cancer immunotherapy" by Jennifer Couzin-Frankel (2013). *Science*, **342**, pp. 1432–1433. See also "Antibody therapeutics in cancer" by Sliwkowski MX and Mellmann I, (Genentech) (2013). *Science*, **341**, pp. 1192–1198.

316. "Human Ig superfamily CTLA-4 gene: chromosomal localization and identity of protein sequence between murine and human CTLA-4 cytoplasmic domains" by Dariavach P, Mattéi MG, Golstein P, and Lefranc MP (December 1988). *Eur J Immunol*, **18**(12), pp. 1901–1905.

317. "Enhancement of antitumor immunity by CTLA-4 blockade" by Leach DR, Krummel MF, and Allison JP (1996). *Science*, **271**, pp. 1734–1736.

318. http://www.yervoy.com/patient.aspx

319. http://www.who.int/mediacentre/factsheets/fs317/en/

320. "Cardiovascular therapy through nanotechnology: how far are we still from bedside?" by Cicha I, Garlichs CD, and Alexiou C (2014). *Eur J Nanomed*, **6**(2), pp. 63–87.

321. "National, regional, and global trends in fasting plasma glucose and diabetes prevalence since 1980: systematic analysis of health examination surveys and epidemiological studies with 370 country-years and 2.7 million participants" by Danaei G, Finucane MM, Lu Y, Singh GM, Cowan MJ, Paciorek CJ, et al. (2011). *Lancet*, **378**(9785), pp. 31–40.

322. http://www.who.int/mediacentre/factsheets/fs312/en/

323. http://medicalxpress.com/news/2014-09-diabetes-faster-insulin-producing-cells.html

324. "Generation of functional human pancreatic β cells in vitro" by Pagliuca FW, Millman JR, Gürtler M, Segel M, Van Dervort A, Ryu JH, Peterson QP, Greiner D, and Melton DA (2014). *Cell*, **159**(2), pp. 428–439.

325. "The intestinal microbiome in type 1 diabetes" by Dunne JL, Triplett EW, Gevers D, Xavier R, Insel R, Danska J, and Atkinson MA (2014). *Clin Exp Immunol*, **177**(1), pp. 30–37.

326. http://www.smartglobalhealth.org/issues/entry/infectious-diseases

327. http://www.gatesfoundation.org/What-We-Do/Global-Development/Vaccine-Delivery

328. http://www.gavi.org/

329. http://www.nobelprize.org/nobel_prizes/medicine/laureates/1945/fleming-facts.html

330. http://www.nobelprize.org/nobel_prizes/medicine/laureates/1939/domagk-facts.html

331. http://www.nobelprize.org/nobel_prizes/medicine/laureates/1945/florey-facts.html

332. http://www.nobelprize.org/nobel_prizes/medicine/laureates/1945/chain-facts.html

333. http://www.nobelprize.org/nobel_prizes/chemistry/laureates/1964/hodgkin-facts.html

334. "Extensively drug-resistant tuberculosis" by P LoBue (2009). *Curr Opin Infect Dis*, **22**(2), pp. 167–173.

335. "Biodegradable nanostructures with selective lysis of microbial membranes" by Nederberg F, Zhang Y, Tan JP, Xu K, Wang H, Yang C, Gao S, Guo XD, Fukushima K, Li L, Hedrick JL, and Yang YY (2011). *Nat Chem*, **3**(5), pp. 409–414.

336. http://www.nobelprize.org/nobel_prizes/medicine/laureates/1975/baltimore-facts.html

337. http://www.who.int/mediacentre/factsheets/fs103/en/

338. http://www.who.int/medicines/publications/essentialmedicines/18th_EML.pdf

339. "Free energy simulations reveal a double mutant avian H5N1 virus hemagglutinin with altered receptor binding specificity" by Das P, Li J, Royyuru AK, and Zhou R (2009). *J Comput Chem*, **30**(11), pp. 1654–1663.

340. "Mitigation strategies for pandemic influenza in the United States" by Germann TC, Kadau K, Longini IM Jr, and Macken CA (2006). *Proc Natl Acad Sci USA*, 103(15), pp. 5935–5940.

341. "Kidneys on demand" by A Griffin (2007). *Brit Med J*, **334**(7592), pp. 502–505.

342. http://www.nlm.nih.gov/medlineplus/druginfo/meds/a601117.html

343. http://health.usnews.com/best-hospitals/rankings

344. http://grad-schools.usnews.rankingsandreviews.com/best-graduate-schools/top-medical-schools/research-rankings

345. www.sciencemag.org, Vol. 344, pp. 24–25, 4 April 2014

346. www.sciencemag.org/special/breastcancer, 28 March 2014

347. http://www.cccblog.org/2013/12/02/darpa-announces-two-programs-as-part-of-white-house-brain-initiative/

348. http://ec.europa.eu/programmes/horizon2020/en/area/health

349. http://www.mrc.ac.uk/About/Factsfigures/index.htm

350. http://www.mpg.de/institutes

351. http://www.uicc.org/membership/institut-gustave-roussy

352. http://ki.se/en/about/ki-in-brief

353. http://www.ncc.go.jp/en/about/chronology.html

354. http://www.riken.jp/en/about/

355. http://www.research.a-star.edu.sg/

356. http://nrpgm.sinica.edu.tw/en/content.php?cat=agtc

357. http://www.csiro.au/Organisation-Structure/Divisions/Animal-Food-and-Health-Sciences.aspx

358. https://www.vlsci.org.au/

359. http://www.csir.res.in/home.asp

360. http://www.genomics.cn/en/index

361. http://www.fiercehealthpayer.com/story/kaisers-integrated-healthcare-model-could-benefit-other-insurers/2013-02-20

362. http://www.hl7.org/about/index.cfm?ref=nav

363. http://www.cdisc.org/mission-and-principles

364. http://www.fda.gov/

365. http://www.ema.europa.eu/ema/index.jsp?curl=pages/home/Home_Page.jsp&mid=

366. http://www.nytimes.com/2014/03/16/science/billionaires-with-big-ideas-are-privatizing-american-science.html?_r=0

367. Jennifer Couzin-Frankel, "Chasing the Money", Science Vol. 344, pp. 25–26; 4 April 2014.

368. http://www.hhmi.org/about/history

369. http://www.gatesfoundation.org/Who-We-Are

370. http://www.gatesfoundation.org/Who-We-Are/General-Information/Foundation-Factsheet

371. http://en.wikipedia.org/wiki/The_Giving_Pledge

372. http://www.wellcome.ac.uk/About-us/index.htm

373. http://en.wikipedia.org/wiki/Frederick_Sanger

374. http://www.jax.org/milestones/nobels.html

375. http://www.salk.edu/about/history.html

376. http://www.nobelprize.org/nobel_prizes/medicine/laureates/1962/crick-facts.html

377. http://www.nobelprize.org/nobel_prizes/medicine/laureates/1969/luria-facts.html

378. http://www.nobelprize.org/nobel_prizes/medicine/laureates/1965/monod-facts.html

379. http://wi.mit.edu/about

380. http://www.broadinstitute.org/

381. http://www.broadinstitute.org/history-leadership/board-directors/bios/eric-lander

382. http://wyss.harvard.edu/viewpage/71/leadership

383. http://www.alleninstitute.org/about_us/background.html

384. http://www.kavlifoundation.org/about-foundation

385. http://www.kavlifoundation.org/institutes

386. http://www.kavlifoundation.org/eric-r-kandel

387. http://mcgovern.mit.edu/about-the-institute/facts-at-a-glance

388. http://www.idg.com/www/home.nsf/docs/remembering_pat_mcgovern

389. http://www.stanleyresearch.org/dnn/Overview/tabid/36/Default.aspx

390. http://www.broadinstitute.org/psych/stanley

391. http://www.wingsforlife.com/en/about-us/#history

392. http://en.wikipedia.org/wiki/Dietrich_Mateschitz

393. http://www.efa-net.eu/member-news/news-from-austria/105-red-bull-founder-gives-biggest-donation-in-austrian-history

394. https://www.michaeljfox.org/foundation/michael-story.html

395. https://www.michaeljfox.org/foundation/publication-detail.html?id=341&category=7

396. http://www.alz.org/about_us_about_us_.asp

397. http://banneralz.org/why-bai/about-bai.aspx

398. http://banneralz.org/news-plus-media/banner-alzheimer%27s-institute-news.aspx

399. http://ki.mit.edu/

400. http://www.nobelprize.org/nobel_prizes/medicine/laureates/1969/

401. http://en.wikipedia.org/wiki/Jon_Huntsman,_Sr.

402. http://www.stjude.org/stjude/v/index.jsp?vgnextoid=58034c2a71fca210VgnVCM1000001e0215acRCRD

403. http://www.cancer.org/aboutus/whoweare/our-history

404. http://www.pcf.org/site/c.leJRIROrEpH/b.5699537/k.BEF4/Home.htm

405. http://www.curemelanoma.org/about-mra/mra-overview/

406. http://www.fastercures.org/about/what-we-do/

407. http://www.ohsu.edu/xd/about/news_events/news/2013/09-21-nike-co-founder-issues-b.cfm

408. http://www.ohsu.edu/xd/education/schools/school-of-medicine/about/druker091609.cfm

409. http://www.cancer.org/index

410. http://ww5.komen.org/AboutUs/FinancialInformation.html

411. http://en.wikipedia.org/wiki/Woody_Johnson

412. http://www.diabetes.org/about-us/?loc=util-header_aboutus

413. http://main.diabetes.org/dorg/PDFs/Financial/2012-american-diabetes-association-annual-report.pdf

414. http://www.cff.org/AboutCF/

415. http://en.wikipedia.org/wiki/Francis_Collins

416. http://www.whitehouse.gov/the-press-office/2013/04/02/fact-sheet-brain-initiative

417. http://www.whitehouse.gov/share/brain-initiative

418. http://fmurray.scripts.mit.edu/docs/Murray_IPE2012_Science Philanthropy_13102.proof.pdf

419. "An Economic Engine: NIH Research, Employment, and the Future of the Medical Innovation Sector" by Everett Ehrlich, United for Medical Research (Spring 2011). Updated 2012: http://www. unitedformedicalresearch.com/advocacy_reports/nihs-role-in-sustaining-the-u-s-economy/

420. http://www.who.int/nmh/global_monitoring_framework/en/

421. http://www.reuters.com/article/2014/09/11/us-southkorea-smoking-idUSKBN0H60XT20140911

422. "Health in all policies-the Finnish initiative: background, principles, and current issues" by Puska P and Ståhl T (2010). *Annu Rev Public Health*, **31**, pp. 315–328.

423. http://www.healthdata.org/

424. WHO Statistical data: http://apps.who.int/ghodata/?theme=country

425. "Pocket world in figures," *The Economist* (2013 Edition).

426. "Consumer-driven health care: lessons from Switzerland" by Herzlinger RE and Parsa-Parsi R (2004). *JAMA*, **292**(10), pp. 1213–1220.

427. www.ey.com/progressions, 2012, http://www.ey.com/GL/en/Industries/Life-Sciences/Bringing-convenience-and-efficiency-to-health-care

428. http://www.cdc.gov/

429. http://aspe.hhs.gov/poverty/12poverty.shtml

430. "Public Law 111–148." 111th United States Congress. Washington, D.C.: United States Government Printing Office. March 23, 2010.

431. https://www.govtrack.us/congress/bills/111/hr3590/text

432. "Obamacare: Experimental Medicine." *The Economist*, Sep 20–26, 2014, pp. 25–27.

433. "A Health Care Success Story" by Bob Kocher and Farzad Mostashari; NY Times OP-ED, 24 Sep 2014, p. A31.

Index

$1K genome *see* $1,000 human genome
$1000 human genome 113
3D bioprinting 260, 271
3D visual system 292
23andMe 146
454 *see* 454 Life Sciences
454 Life Sciences 115
1000 Genomes Project 155

A*STAR *see* Agency for Science, Technology and Research
Aastrom Biosciences 279
abatacept *see* Orencia
AbbVie 181
ABI *see* Applied Biosystems
Abraxane 230
Academia Sinica 338
academic medical research centers 335
Accelerating Medicines Partnership 203
accountable care organizations 367
acetaminophen 170
ACO *see* accountable care organizations
ACT *see* Advanced Cell Technology
Actavis 198
acute lymphoblastic leukemia 266, 305
acute myocardial infarction 270
AD *see* Alzheimer's disease

ADA *see* American Diabetes Association
adenine 46
adenosine-5′-triphosphate 26, 90, 162, 215
ADME 171
adult stem cells 255
Advanced Cell Technology 273
Advances in Genome Biology and Technology 122
Advil *see* ibuprofen
Affordable Care Act 366
AGBT *see* Advances in Genome Biology and Technology
Agency for Science, Technology and Research 338
AGTC *see* Applied Genetic Technologies Corporation
alanine 57
Albert Sabin 330
alcohol 40–41
Aleve *see* Naproxen
Alexion 198
ALL *see* acute lymphoblastic leukemia
Allen Institute for Brain Science 347, 354
Alliance for Regenerative Medicine 268
allografts 330
alpha-helix 65–66
ALSAC *see* St. Jude Children's Research Hospital
Altor Bioscience Corporation 283

Alzheimer's Association 203, 348
Alzheimer's disease 104, 188, 239,
    291, 297–298, 348
American Cancer Society 351
American Diabetes Association
    203, 352
Amgen 179, 198, 233, 280, 283
AMI *see* acute myocardial
    infarction
amino acids 55
AMP *see* Accelerating Medicines
    Partnership
AMRC *see* academic medical
    research centers
Andes 98–99
angel investors 14, 374
angiography 236
animal models 164–166, 197, 347
antagonistic antibodies 309
antibiotic resistance 318–320
antibiotics 188, 196, 286,
    317–319, 321–322, 330
antidepressant 184
antimicrobial polymer
    nanoparticles 316, 321
antiretroviral drugs 328
APOE *see* apolipoprotein E
apolipoprotein E 299
Applied Biosystems 111
Applied Genetic Technologies
    Corporation 273
archaea 60, 103, 287
archenteron 253
arginine 58, 61
Argos Therapeutics 275
ARM *see* Alliance for Regenerative
    Medicine
Artificial Pancreas Project
    214–217, 313, 332
asparagine 58, 60, 73
aspartic acid 57–59
aspirin 169–170, 175
Astra-Zeneca 194
Athersys 277–278

atoms 17
ATP *see* adenosine-5'-triphosphate
Au *see* gold
autografts 330
AUTONOMY 136
Avastin 181, 356
Avicenna 158
axons 263, 292, 294, 304

bacteria 103
Baltimore classification 323
BAM *see* binary sequence
    alignment/map
Banner Alzheimer's Institute *see*
    Alzheimer's Association
Battelle Memorial Institute 150
Bayer 194, 198, 233, 272, 318
BBB *see* blood–brain barrier
BCR-ABL 186
Beijing Genome Institute 339
Beike Biotechnology 267
belatacept 309
BENEFICENCE 137
benzene 23, 34–35
Berkshire Hathaway 342
Berzelius 63
beta sheet 65, 67
Betalogics 314
BGI 99, 121, 145–150, 250, 339
BGI *see* Beijing Genome Institute
big data 131, 197
Bill & Melinda Gates Foundation
    342, 358
binary sequence alignment/map
    133
bioavailability 171
biobanks 140
bioethics 135
Biogen 180, 198, 204, 355
bioinformatics 152, 167
biologics 176
biomarker 183
biomaterials 217

biomedical imaging 234–235, 243
biomedical research budget 336
Bionano Genomics 123
bionic 262–263, 332, 371
bionic eye 262–263, 332, 371
Bionic Vision Australia 263
biosimilars 182
BioTime 273
blastocyst *see* blastula
blastula 252–254
blockbuster business model 193, 201
blood–brain barrier 171, 298
BMGF *see* Bill and Melinda Gates Foundation
BMI *see* body mass index
BMP-11 *see* bone morphogenetic protein 11
BMS *see* Bristol-Myers Squibb
body mass index 304, 362
bone marrow transplantation 270
bone morphogenetic protein 11 258
BOWTIE 134
Bragg 68–69
brain 290
brain *see* Broca's area
brain *see* Wernicke's (speech) area
BRAIN 291, 354
BRCA1 141
BRCA2 *see* BRCA1
breast cancer research 336, 351
Bristol-Myers Squibb 204, 230
Broad Institute 345, 348
Broca's area 295
bubonic plague 98
Buckyballs 232
BVA *see* Bionic Vision Australia

C *see* carbon
Ca *see* calcium
calcium 24, 29–30, 94, 163, 235, 275

California Institute for Regenerative Medicine 273
cancer 304
cancer *see* metastases
cancer genome 156, 305–307
Cancer Genome Anatomy Project 156
Cancer Genome Atlas 156, 306
The Cancer Genome Atlas 306
Cancer Genome Project 156, 306–307
Capricor 278–279
capsid 323
carbohydrates 39
carbon 25
carbon nanotube 232
carboxyl group 41–42, 63–64
cardiovascular disease 310
CCD *see* charged-coupled device
CDC *see* Center for Disease Control
CDI *see* Cellular Dynamics International
CDISC 340
Celebrex 171
Celera 106, 141
Celgene 198, 268, 270, 283
celiac disease 286–287, 309
cell 86
cell cycle 55, 90, 306
cell growth 55, 86, 90, 305
Celladon 275
cellular differentiation 55, 91, 153, 163
Cellular Dynamics International 277
Center for iPS Cell Research and Application 266
Centers for Disease Control 341, 362
central nervous system 94, 172, 215, 290–291, 293, 295, 297, 299, 301, 303
cerebral cortex 292–295, 297
CETUS 100–101

CFF *see* Cystic Fibrosis Foundation
CGM *see* continuous glucose monitoring
chain termination 110–111
charged-coupled device 115
charitable foundations 342
chemical synthesis 167, 177, 346
cheminformatics 167
chemotherapy 179, 181, 186, 227–228, 230, 282, 308
CHF *see* congestive heart failure
Chiesi Farmaceutici 229, 276
chirality 56
chlorine 23–24, 32, 330
cholesterol 43–44, 163, 188, 195
chromosome 81, 85, 89, 97, 186–187, 265
chronic kidney disease 229
cigarette smoking 363
CiRA, Kyoto *see* Center for iPS Cell Research and Application
circuit simulation 210
circulating tumor cells 248
CIRM *see* California Institute for Regenerative Medicine
CKD *see* chronic kidney disease
CLINAM 234
clinical research organizations 339
clinical trials 174
cloning 103
CMOS *see* complementary metal–oxide–semiconductor
CNS *see* central nervous system
CNT *see* carbon nanotube
codons 57–62, 81–82
cognitive computing 197, 212–213
cognitive dysfunction 95, 142, 188, 239, 300, 371
colloid chemistry 220
combinatorial chemistry 167, 169–170, 192

Commonwealth Scientific and Industrial Research Organisation 338
complementary metal–oxide–semiconductor 245
Complete Genomics 119
computed tomography 74, 235, 240, 307
conduction band 22, 36–37, 244
cones 5, 262–263
congenital disorder 251
congestive heart failure 270
conscience 292
continuous glucose monitoring 216, 313
copper 2, 24, 31–32, 235, 280
cortex *see* left hemisphere
cortex *see* right hemisphere
COSMIC 306
Council of Scientific and Industrial Research 339
covalent 23
coverage 131
COX-1 169–170
COX-2 *see* COX-1
Craig Venter 106, 113
CRAM 133
critical path initiative 183, 191
CRO *see* clinical research organizations
cross-licensing 200
crowdsourcing 213–214
CSIR *see* Council of Scientific and Industrial Research
CSIRO *see* Commonwealth Scientific and Industrial Research Organisation
CT *see* computed tomography
CTC *see* circulating tumor cells
CTLA-4 *see* cytotoxic T-lymphocyte antigen 4
Cu *see* copper
CVD *see* cardiovascular disease
Cyborg 371

cysteine 57–58, 60, 66
Cystic Fibrosis Foundation 352
cytopathology 241
cytoplasm 87–90, 93, 323, 328
Cytori Therapeutics 279
cytosine 47–49, 53, 78, 154
cytotoxic T-lymphocyte antigen
    4 308

DARPA *see* Defense Advanced
    Research Projects Agency
Darwin 96, 98
data analysis 130–131
David Deamer 125
DBS *see* deep brain stimulation
DCE-MRI 238
deep brain stimulation 300
deer tick 315
Defense Advanced Research
    Projects Agency 336
Definiens 241
dementia 298
dendrimers 217
dendritic encapsulation 218
Dengue virus 322
Denisovan 100
deoxynucleotides 111
deoxyribonucleotides 51
deoxyribose 50–51
Department of Defense 336
Department of Energy 336
depression 142, 286, 298–300,
    337
diabetes 312
diagnostic tests in vitro 243
diagnostic tests in vivo 243
dideoxynucleotides 111
diploid 84–85, 95–96
DKFZ *see* German Cancer Research
    Center
DNA, RNA 46–47, 90, 136, 323
DNA Electronics 123, 144
DNA microarray 153, 246

DNA polymerase 101
DNA sequencing 106
DNA transistor 128–129
docking 93, 168
DOD *see* Department of Defense
DOE *see* Department of Energy
Dolly 104–105
dominant 83
Dorothy Crowfoot Hodgkin
    318–319
double helix 78, 82, 102, 109
Doug Melton 314
downstream analysis 116,
    131–132, 134
Doxil 224, 226–229, 234
Doxorubicin 227, 229
DREAM project 214
drug carriers 222
drug delivery 93, 193, 197, 205,
    214, 217, 219, 221–227, 229,
    231–234, 281, 285, 311–312
drug discovery 160
drug target 160, 167
druggable genes 164

Earth Microbiome Project 289
EBI *see* European Bioinformatics
    Institute
Ebola virus 324–325
Ebola virus disease 324
The Economist 214, 359, 366–367
ectoderm 253–255
Edward Jenner, Jonas Salk 330
EEG *see* electroencephalography
EFPIA 202–203
EKG *see* electrocardiography
electrocardiography 241
electroencephalography 235
electron microscope 172, 206
elements *see* atoms
Eli Lilly 179
Ellison Medical Foundation 346

ELSI *see* ethical, legal, and social
  issues
EMA *see* European Medicines
  Agency
embryonic stem cells 104, 165,
  252, 256
EMP *see* Earth Microbiome Project
EMT *see* epithelial-mesenchymal
  transition
ENCODE Project 156
endoderm 253–255, 281
Endoplasmic reticulum 87
endoscope 302
Enterococcus faecalis 320
enveloped viruses 324, 327
enzymes 161
EPAS1 gene 99–100
epigenetics 49, 154
epitaxy 245
epithelial 254, 326
epithelial–mesenchymal
  transition 254
EPO *see* European Patent
  Organization
Epogen 179
Eric Lander 149, 345
Ernst Ruska 206
erythropoietin *see* Epogen
Escherichia coli 101, 178, 318
esophagus 260
esterases 161–162
esters 42, 46, 162
EteRNA project 214
ethical, legal, and social issues
  108, 148
Ethiopia 98–99, 359
EU *see* European Union
eukaryote 90
European Bioinformatics Institute
  134
European Medicines Agency 233,
  340
European Patent Organization 141
European Union 337

EVD *see* Ebola virus disease
Evotec 272, 314
exons 81–82

Faster Cures 350
FASTQ 133
fatal familial insomnia 295
fats 42, 46, 358
fatty acid 42, 163
FDA *see* Food & Drug
  Administration
Fe *see* iron
Fibrocell Science 276
fludeoxyglucose 235
fluorescence microscope 323
fluoroscope 302
fMRI *see* functional MRI
Food and Drug Administration
  160, 217, 340
Francis Collins 106, 141, 149, 352
Francis Crick 77, 79, 344
Frank L. Douglas 184
Frederick Sanger 65, 343
functional MRI 237

G-protein–coupled receptors 160,
  161
G *see* guanine
gadolinium 238
Galen of Pergamon 158
gamete 85, 251
Gamida Cell 280
GAO *see* US Government
  Accountability Office
gastrulation 253–255
GAVI *see* Global Alliance for
  Vaccines and Immunization
Gd *see* gadolinium
GDF11 *see* growth differentiation
  factor 11
GDP *see* gross domestic product
GE *see* General Electric

gene expression 91, 153–154, 156, 271, 276
gene therapy 251
Genentech 179, 181–182, 185–186, 194, 198
General Electric 243
genes 83
genetic code 77
genetics 77
Genexol-PM 230
Genia 127
genomic medicine 197, 285, 334–335, 338, 344
genomics 152
genotype 95, 100, 103, 186, 246
genotyping 100
George M. Church 118
Gerd Binnig 206
German Cancer Research Center 337
Gilead 198, 229, 233, 327
Gilead Sciences 327
Giving Pledge 342
GlaxoSmithKline, Novartis, AstraZeneca, Merck, and Pfizer 321
Gleevec *see* imatinib
Global Alliance for Vaccines and Immunization 317
glucagon 70, 214–215, 217, 281, 313
glucose 39
glutamic acid 58, 62
glutamine 58, 61
Glybera 276
glycerol 41–43, 46
glycine 58–59, 61
glycoproteins 72, 93, 323, 326
gold 24, 31–32, 143, 157
Golgi 88, 93
Gordon E. Moore 209
GPCR *see* G protein-coupled receptors
gross domestic product 359

growth differentiation factor 11 258
GSK 194, 198, 272
guanine 46–48, 53, 78

H *see* hydrogen
H1N1 326
H_2O 23–24, 38–39
H5N1 *see* H1N1
HAI *see* hospital-acquired infection
Hansjorg Wyss 346
haploid 84–85, 91
HapMap Project 155
Harvard Stem Cell Institute 258, 314, 355
HCP *see* Human Connectome Project
HDL 44
HDSA *see* Huntington's Disease Society of America
Health Level Seven International 340
healthcare ecosystem 333
healthcare expenses 360, 368–369
heart transplants 261
Heinrich Rohrer 206
Helicos 121
Heliscope 121
Helmholtz Association 337
hemagglutinin 326
hemoglobin 70
hepatitis B 178, 304, 323
Herceptin 181, 185–187, 356
heterozygous 95–96
HGH *see* human growth hormone
HGP *see* Human Genome Project
HHMI *see* Howard Hughes Medical Institute
hidden links 212–213
high altitude 98
high-throughput screening 168
hippocampus 296–297

Histidine 58–59
HIV *see* human immunodeficiency
    virus
HIV-AIDS 188
HIV resistance 98
HL7 *see* Health Level Seven
    International
HMP *see* Human Microbiome
    Project
homozygous 95–96
hormones 292
hospital-acquired infection 320
Howard Hughes Medical
    Institute 272, 342
HTS *see* high throughput screening
Huanming Yang 147
Human Brain Project 291
human cloning 106
Human Connectome Project 291
human genome 81
Human Genome Project 106
human growth hormone 178, 270
human immunodeficiency virus
    98, 177, 188, 271, 315, 342
human insulin 103, 178–179
Human Microbiome Project 156,
    288
Humira 181, 356
humoral immunity 316
Huntington's Disease Society of
    America 352
Huntsman Cancer Institute 349
hybridomas 180
hydrogen 17–18
hyperglycemia 312, 314
hypoglycemia 217, 312
hypoxia *see* oxygen starvation

I *see* iodine
IBM 128
IBM, IBN *see* antimicrobial
    polymer nanoparticles
IBM Research 129, 207, 212, 237

IBM ZRL 206
ibuprofen 169–170
ICGC *see* International Cancer
    Genome Consortium
IDEC 180–181
IEEE 209
IGR *see* Institut Gustave Roussy
Illumina 116
imatinib 185
IMI *see* Innovative Medicines
    Initiative
immune system 71, 97, 179–180,
    185, 188, 223, 275, 282, 287,
    309, 314, 316, 322–324, 332
immunoglobulin 71–72, 308
immunomodulators 266
immunosuppressant drugs 261,
    331
immunotherapy 266
impact 16, 373
IND *see* Investigational Drug
    Application
induced pluripotent stem 256,
    258, 277
infant mortality 359–362
infectious diseases 315
influenza 326
influenza virus 326
innovation 6
Innovative Medicines Initiative
    202
Institut Gustave Roussy 338
Institute of Oncology 338
insulators 22, 37
insulin 69
intellectual property 7
International Cancer Genome
    Consortium 307
introns 81–82
inversion 196
investigational new drug (IND)
    application 173
iodine 24, 32, 161
ion channel 29, 160, 162

ion-sensitive field-emission
transistors 123
Ion Torrent 123
ionic 23
ionization 36
IP *see* cross-licensing
ipilimumab 309
IPO 11, 13–16, 374
iPS *see* induced pluripotent stem
iron 30
iron oxide nanoparticles 231, 238
ISFET *see* ion-sensitive
field-emission transistors
isoleucine 57–59

Jackson Memorial Laboratory 343
James Allison 309
James Watson 77, 106, 143, 149
Janssen 272, 278–279, 281
JCR Pharmaceuticals 270
JDRF *see* Juvenile Diabetes
Research Foundation
JEOL 243
JHU *see* Johns Hopkins University
Jian Wang 147, 149
J&J 181, 198, 229, 250, 269
J&J *see* Betalogics
J&J *see* Janssen
Johns Hopkins University 4, 134,
257, 342, 354
Jonathan Rothberg 115, 123, 125
Juno Therapeutics 282
JUSTICE 137
Juvenile Diabetes Research
Foundation 216, 352

K *see* potassium
Karolinska Institute 4–5, 338
Kavli Institutes 347, 353
Kavli Prizes 347
KI *see* Karolinska Institute
kidney transplants 261

kinase 161, 186, 304
Knight Cancer Institute at OHSU
350
knockout mouse 165–166, 192
KNOME 146
Koch Institute 349

lab-on-a-chip 246
language 295
language *see* speech
Larry Ellison 346, 353
LDL 44
lead identification 167–168
LED *see* light emitting diode
left hemisphere 293, 295
Leica 243
Leroy Hood 111
Leucine 58–60
Li *see* lithium
Life Sciences industry 201
Life Technologies 119, 123, 134,
143–144
ligase *see* ligation
ligation 53, 104, 118–120
light-emitting diode 245
lipid 42
Lipinski 169
Lipitor 195
Lipodox 227
lipofection 224
lipophilic 44, 172, 221
liposomal nanodrugs 229
liposomes 89, 223–228, 233–234
lithium 28
liver transplants 261
lobes 34, 293–294
LOC *see* lab-on-a-chip
long-term potentiation 58, 297
Lonza 270
Louis Pasteur 33, 330
LTP *see* long-term potentiation
luciferase 115–116
lupus 203, 309

Luye 199
Lyme disease 315, 318
lysine 57–60
lysosome 88

mab *see* monoclonal antibody
machine learning 212, 241
macular degeneration 182, 229, 270
magic bullets 179
magnesium 24, 29, 235
magnetic fluid hyperthermia 231
magnetic nanoparticles 231–232, 238
magnetic resonance force microscopy 237
magnetic resonance imaging 74, 235, 237, 295, 341
magnetoencephalography 241
malaria 98, 315, 342, 358–359
mammography 236
Marburg virus 323
Mario Capecchi, Martin Evans, and Oliver Smithies 258
Mateschitz *see* Red Bull
Max von Laue 68
Max-Planck Institutes 337
Mayo Clinic 146
McGovern Institute for Brain Research 347
MDR *see* multidrug resistant
measles 316–317, 329
Medarex *see* ipilimumab
Medical Research Council 109, 142, 337, 343
Medipost 272–273
MEG *see* magnetoencephalography
Meiji 243
meiosis 85–86, 91
membrane 86
membrane transport protein 164
Memorial Sloan Kettering Cancer Center 186, 282

memory 296
Mendel 83–85, 372
mesenchymal 254, 270, 272
mesoblast 270, 272
mesoderm 253–255
metabolic pathways 92, 287
metabolites 61, 89, 92, 154, 172, 288
metabolomics 154
metabonomics 154
metagenomics 154
metals 20, 22
metastasis 239, 246
methanogens 287
methicillin-resistant *S. aureus* 320
methionine 49, 57–58, 60
methylation 48–49, 110, 154–155
MFH *see* magnetic fluid hyperthermia
Mg *see* magnesium
MHLW *see* Ministry for Health, Labor and Welfare
micelle 220–222, 230
Michael Bloomberg 342
Michael J. Fox Foundation 348
Microbial Genome Project 156
microbiome 154, 156, 285–289, 314
microfluidics 116, 123, 130, 246–247
Mike Hunkapiller 111, 122
Mike Milken *see* Faster Cures
Milan *see* Institute of Oncology
minimally invasive surgery 302, 341
Ministry for Health, Labor and Welfare 340
MIS *see* minimally invasive surgery
MIT *see* Koch Institute
mitochondria 55, 89–90
mitochondrial genome 81
mitochondrion 55
mitosis 91, 251–252

modeling and simulation 197, 209–210, 373
molecular cloning 103, 177–178
molecules 17
monoclonal antibody 308–309, 356
Moore's law *see* Gordon E. Moore
morphogenesis 86, 91–92, 251
morula 252–253
MOSFET 209
mosquito 315
motor control 4, 293
mountain sickness 98–99
MPI *see* Max-Planck-Institutes
MPI for neuroscience 337
MRA *see* Myeloma Research Alliance
MRC *see* Medical Research Council
MRC *see* U.K. Medical Research Council
MRFM *see* magnetic resonance force microscopy
MRI *see* magnetic resonance imaging
MRSA *see* methicillin-resistant Staphylococcus aureus
MS *see* multiple sclerosis
MSKCC *see* Memorial Sloan Kettering Cancer Center
Mullis 100–101
multidrug resistant 319
multiple sclerosis 182, 272, 286, 298
multipotent 253, 255, 273
muscle repair 259
mutation 49, 96–98, 264, 266, 306
Mycobacterium tuberculosis 317
Myeloma Research Alliance 350
Mylan 196, 198
Myocet 227–229
Myriad Genetics 141

Na *see* sodium

nano 1–2, 205, 225, 264
nanochannel 123, 129–130
nanocrystals 225–226, 233, 244–245
nanodiagnostics 243, 245, 247, 249, 285
nanomedicine 1, 3
nanometer *see* nano
nanoparticles 2, 217
nanopore 124–130, 132, 145
nanoscopy 242
nanotoxicity 232
Naproxen 169
National Breast Cancer Coalition 336
National Cancer Center 338
National Center for Biotechnology Information 152
National Human Genome Research Institute 106, 306
National Institutes of Health 106, 166, 251, 288, 336, 372
National Library of Medicine 152
National Science Foundation 337
NBCC *see* National Breast Cancer Coalition
NCBI *see* National Center for Biotechnology Information
NCC *see* National Cancer Center
NCD *see* non-communicable diseases
Neanderthal Man 143
NeoStem 274–275
neuraminidase 326–327
neuraminidase inhibitors 327
neurotransmitter 94, 162
new molecular entity 167
Nexium *see* Prilosec
next-generation sequencing 113
NGS *see* next generation sequencing
NHGRI *see* National Human Genome Research Institute
NHP *see* non-human primates

NIDEK 264
NIFTY test 149
NIH *see* National Institutes of
Health
NIH funding 342, 355–356
Nikon 243
nitrogen 26
NLM *see* National Library of
Medicine
NME *see* new molecular ENTITY
NMR *see* nuclear magnetic
resonance
Nobel *see* Nobel Prize
Nobel Prize 3
noncommunicable disease 357
nonhuman primates 142
nonsteroidal anti-inflammatory
drugs 169
Novartis 194–195, 283
NSAID *see* nonsteroidal
anti-inflammatory drugs
NSF *see* National Science
Foundation
nuclear hormone receptors 163
nuclear magnetic resonance 35,
154, 236
nuclear medicine 31, 235, 240
nucleobases 46, 48, 78, 80
nucleosides 46, 50, 54, 78
nucleotides 50
Nuffield Council on Bioethics 142

O *see* oxygen
Obamacare *see* Affordable Care Act
obesity 363
obsessive compulsive disorder
300
OCD *see* obsessive compulsive
disorder
olfactory epithelium 290
oligopotent 253
Olympus 243, 280
omics 151–153, 155, 308

oncogenes 305
ONPRC *see* Oregon National
Primate Research Center
opposable thumb 292
orbitals 19–20, 22–23, 33–36
Oregon National Primate Research
Center 105
Orencia 309
organogenesis 253
Organovo 271
orphan diseases 189, 196
oseltamivir 327
Osiris Therapeutics 272
ovum 85
Oxford Nanopore 124–127, 132
oxygen 27
oxygen starvation 297

PacBio *see* Pacific Biosystems
Pacific Biosystems 121
paclitaxel 230, 232
palindromic 102
pancreas 214
pandemic 326, 328–329
Paracelsus 157–159, 348
parenteral 182, 229
Parkinson's disease 242, 298–299,
348
patents *see* intellectual property
pathogenic bacteria 317–319
Paul Ehrlich 179
Pauling 65
payers of health insurance 339
PCF *see* Prostate Cancer
Foundation
PCR *see* polymerase chain reaction
PD *see* Michael J. Fox Foundation
PD *see* Parkinson's Disease
PD *see* pharmacodynamics
PDB *see* Protein Data Bank
PEG *see* polyethylene glycol
penicillin 318–320
peptide 63–66, 69–70, 162, 168

peramivir 327
periodic table 19–20, 22–24, 240
personal genomics 118, 146
personalized medicine 181,
    183–185, 187–191
PET *see* positron emission
    tomography
Pfizer 194
pharmacodynamics 172
pharmacokinetics 172, 222, 233
pharmacovigilance 176, 202
phenotype 84, 95–96, 164–165,
    254
phenylalanine 57–59
Philadelphia chromosome
    186–187
Philips 238, 243, 250
phosphatases 161–162
phosphodiester bond 51–52, 111
phospholipids 44–45, 89, 94,
    223–224, 323
phosphorus 24, 26–27, 229, 235
photolithography 119–120, 245
pi-bonds 34
piezoelectric 207–208
pig kidneys 261
Pixium Vision 264
PK *see* pharmacokinetics
placenta 252, 255, 270
plague 98, 329
pluripotent 253, 255–258, 273,
    277
Pluristem Therapeutics 274
polio 179, 316–317, 330, 342, 344
Polony sequencing 118
polyethylene glycol 224
polymer drugs 228
polymerase chain reaction 100,
    178
polypeptide 64–67, 72, 88
positron emission tomography
    235, 307, 341
potassium 24, 28–29, 32, 163, 235
poverty line 363–366

PPIs 184
PPPs *see* public-private
    partnerships
preclinical 160
prenatal testing 139
Prilosec 184, 187, 195
prions 296
Prograf *see* Tacrolimus
prokaryote 86, 89, 287
Proline 58, 60
Prontosil 318
proof of concept 129, 276
Prostate Cancer Foundation 350
proteases 161
protein 63
protein chip *see* protein microarray
Protein Data Bank 69, 134
protein microarray 246
protein receptors 91, 93
proteomics 149, 153, 193
proton *see* hydrogen
protoplasm 86
providers 334
Pseudomonas 317, 321
public health 329
public–private partnerships 166,
    372
purines 46–47, 51, 54, 61, 78
pyrimidines 47–48, 51, 53–54, 61,
    78
pyrosequencing 114–117, 119,
    123, 144, 154
pyrrolisine 60

QSAR *see* quantitative structure
    activity relations
quantitative structure–activity
    relationships 169
quantum 18
quantum chemistry 19, 33, 36, 40,
    168, 210
quantum dot 244–245

radiopharmaceutical 235
radiotracers 235
Ragon Institute 351
Ranbaxy 199
read length 124, 131–132
recessive 83–84, 96, 265, 282
recognition tunneling 130
recombinant DNA 103, 177
Red Bull 348
regenerative medicine 251
rejuvenation 258
Remicade 181, 356
Renagel 229–230
renal transplantation 309
Renvela *see* Renagel
replication, transcription, and
   translation 82
restriction enzymes 102–103
retina 4–5, 182, 262–264, 290,
   313
Retina Implant AG 264
retinal prostheses 263
retinitis pigmentosa 263–264, 273
retroviruses 327–328
rheumatoid arthritis 181, 203,
   277, 286, 290, 309
rhodopsin 5, 161
ribonuclease 71–72
ribose 49–51
ribosome 69, 82–83, 90, 318
right hemisphere 293
RIKEN 107, 113, 338
Rituxan 180–181, 356
RNA virus 323
Robert Dennard 209
Robert Koch 180, 330
Roche 92, 116, 119, 122–124,
   128–129, 132, 143–145, 169,
   194, 198, 229, 233, 243, 250,
   272
Roche Diagnostics 123–124, 144
rods 5, 262–263
Roentgen 235
Rosalind Franklin 77

Rule-of-Five *see* Lipinski

S *see* sulfur
saccharides 39
SALK Institute 344, 354
Sangamo BioSciences 271
Sanger 65, 82, 107, 109–115, 121,
   131–132, 143, 154, 306, 343
Sanger Centre 343
Sanger sequencing 110–115,
   131–132, 143, 154
SANOFI 193–194
SBH *see* sequencing-by-
   hybridization
scanning probe microscopy 208
scanning tunneling microscope
   206, 208
schizophrenia 142, 286, 291, 298,
   347
Schrödinger 18–19, 33, 35, 96,
   245
Schrödinger *see* quantum
SCI *see* spinal cord injury
SCID *see* severe combined
   immunodeficiency
science philantropy 354–355
SCNT *see* somatic-cell nuclear
   transfer
selenocysteine 56, 58, 62
SEMATECH 199–200
semiconductor industry 119,
   199–201, 209, 245
semiconductors 3, 22, 37, 199,
   244
sepsis 61, 159, 277, 320
sequencing-by-hybridization 145
Serine 58, 61
serotonin 62, 161, 184–185
severe combined
   immunodeficiency 257, 265
Sherpas 100
Shinya *see* Yamanaka
Shire 196, 198, 271

Shotgun sequencing 131
Si *see* silicon
sialic acid 326
sickle-cell disease 62, 98
Siemens 239, 243, 250
sigma *see* orbitals
sigma-bonds 34
in silico 143, 168
silicon 28
silicon photonics 247
SIMCERE 199
single-molecule sequencing 121–123
single payer systems 362
single-photon emission computed tomography 240
smallpox 98, 316, 323, 329–330
snip 100
SNP *see* snip
social computing 213
societal impact 373
sodium 29
Solexa 116–117, 126, 143
SOLiD 118–119, 143–144
solubilization 221–222
somatic cell nuclear transfer 104
spatial navigation 297
spatiotemporal reasoning 293
SPECT *see* single-photon emission computed tomography
speech 5, 292, 295, 300
sperm 84–85
spinal cord injury 271, 300
SPM *see* scanning probe microscopy
SSRIs 184
St. Jude Children's Research Hospital 350
Stanley Medical Research Institute 347
Staphylococcus aureus 125, 318
statins 188–189
stealth liposomes 224
stem cell 104, 251

STEMCELL Technologies 279
StemCells Inc 278
steroid 43
STM *see* scanning tunneling microscope
strand sequencing 127
stratified medicine 184, 188
Streptococcus 317–318, 320
Streptococcus pneumoniae 320
structural proteins 163
sulfur 22, 24, 28, 60, 235
Sun Pharma 199, 227
superbugs 319–320
superresolution microscopy 242
surfactant 220–221
Susan G. Komen Foundation 351
Svante Pääbo 143
synapses 95, 162, 185, 242, 262, 292
Synthes 346

T *see* thymine
T1D *see* type 1 diabetes
Tacrolimus 332
Takeda 198, 204, 229, 233, 355
Tamiflu *see* oseltamivir
Taq polymerase 102
target validation 164, 167
Taxol *see* paclitaxel
Tc *see* Technetium
TCGA *see* The Cancer Genome Atlas
Technetium 31, 240
Teva 199, 233, 270, 272, 280, 283
text analytics 211–213
thalamus 294–296
Thermo Fisher 119, 123, 144, 250
threonine 57–58, 61–62, 73
thymine 46–49, 53–54, 78, 80, 82, 90
Tibet 98
TiGenix 277
tissue engineering 260
Toshiba 243, 250

totipotent 253
transcriptomics 149, 153
transgenic animals 165
translational medicine *see*
    nanomedicine
transplantation 330
transplantation medicine 261, 331
triglycerides 42–43, 46
trophoblast 252
tryptophan 57–58, 62, 184
Tufts Center 191
tumor suppressor genes 305
tunneling 129–130, 206–208, 237
Tylenol *see* acetaminophen
type 1 diabetes 216, 290, 352
tyrosine 58–59, 62–63, 186
tyrothricin 318

U *see* uracil
uHTS 168, 193
U.K. Medical Research Council 142
ultrasound 235
umbilical cord 255, 272–273, 280
UMC *see* Uppsala Monitoring
    Centre
unipotent 253
UniQure 276
Uppsala Monitoring Centre 176
uracil 46–49, 53, 80, 82, 90
US *see* ultrasound
US Government Accountability
    Office 159
Utah *see* Huntsman Cancer
    Institute

vaccination 316–317, 341
validated biomarkers 191
valine 57–58, 62
vancomycin-resistant
    Enterococcus 320
variant call format 134
VCF *see* Variant Call Format

VELVET 133–134
venture capital 12, 14, 282–283,
    373–374
venture philanthropy 342
vesicle 89, 92–95, 223
vesicle traffic 92–93
Viacyte 281, 314
Vioxx 171
viral envelope 323–324
viral glycoproteins 323
viral infections 229, 283, 304, 317,
    322
virus 322
vision 5, 241, 262, 293
visual perception 3, 5, 263–264
in vitro 173
Voyager Therapeutics 282
VRE *see* vancomycin-resistant
    Enterococcus

Wake Forest Institute for
    Regenerative Medicine 261
Walter Gilbert 109
Warren Buffett 342
water *see* H$_2$O
Watson–Jeopardy 212
Wellcome Trust 107, 142, 306,
    343, 353
Wernicke's (speech) area 295
West Nile virus 323
WFIRM *see* Wake Forest Institute
    for Regenerative Medicine
WGS *see* whole genome sequencing
Whitehead Institute 107, 345
WHO *see* World Health
    Organization
whole genome sequencing 132
Wings for Life 303, 348
World Health Organization 139,
    176, 310, 327, 357
Wyss Institute 346, 353

X chromosome 85
X-ray crystallography 68–70,
    73, 77
X-Rays *see* Roentgen

Y chromosome 81, 85
Yamanaka 256–258, 266,
    281

yeast cells 93, 179
yellow fever virus 323
Yervoy 309

zanamivir 327
Zeiss 242–243
zinc finger nucleases 265
zygote 85, 91, 251

T - #0028 - 101024 - C32 - 229/152/24 [26] - CB - 9789814613767 - Gloss Lamination